Springer Series in

CHEMICAL PHYSICS 74

Springer Series in

CHEMICAL PHYSICS

The purpose of this series is to provide comprehensive up-to-date monographs in both well established disciplines and emerging research areas within the broad fields of chemical physics and physical chemistry. The books deal with both fundamental science and applications, and may have either a theoretical or an experimental emphasis. They are aimed primarily at researchers and graduate students in chemical physics and related fields.

75 **Basic Principles in Applied Catalysis**
By M. Baerns

76 **The Chemical Bond**
A Fundamental Quantum-Mechanical Picture
By T. Shida

77 **Heterogeneous Kinetics**
Theory of Ziegler-Natta-Kaminsky Polymerization
By T. Keii

78 **Nuclear Fusion Research**
Understanding Plasma-Surface Interactions
Editors: R.E.H. Clark and D.H. Reiter

79 **Ultrafast Phenomena XIV**
Editors: T. Kobayashi, T. Okada, T. Kobayashi, K.A. Nelson, S. De Silvestri

80 **X-Ray Diffraction by Macromolecules**
By N. Kasai and M. Kakudo

81 **Advanced Time-Correlated Single Photon Counting Techniques**
By W. Becker

82 **Transport Coefficients of Fluids**
By B.C. Eu

83 **Quantum Dynamics of Complex Molecular Systems**
Editors: D.A. Micha and I. Burghardt

84 **Progress in Ultrafast Intense Laser Science I**
Editors: K. Yamanouchi, S.L. Chin, P. Agostini, and G. Ferrante

85 **Quantum Dynamics Intense Laser Science II**
Editors: K. Yamanouchi, S.L. Chin, P. Agostini, and G. Ferrante

86 **Free Energy Calculations**
Theory and Applications in Chemistry and Biology
Editors: Ch. Chipot and A. Pohorille

87 **Analysis and Control of Ultrafast Photoinduced Reactions**
Editors: O. Kühn and L. Wöste

88 **Ultrafast Phenomena XV**
Editors: P. Corkum, D. Jonas, D. Miller, and A.M. Weiner

89 **Progress in Ultrafast Intense Laser Science III**
Editors: K. Yamanouchi, S.L. Chin, P. Agostini, and F. Ferrante

90 **Thermodynamics and Fluctuations Far from Equilibrium**
By J. Ross

91 **Progress in Ultrafast Intense Laser Science IV**
Editors: K. Yamanouchi, A. Becker, R. Li, and S.L. Chin

Rudolf Holze

Surface and Interface Analysis

An Electrochemists Toolbox

With 207 Figures

Springer

Professor Dr. Rudolf Holze
Technische Universität Chemnitz
Institut für Chemie
09107 Chemnitz, Germany
E-Mail: rudolf.holze@chemie.tu-chemnitz.de

Series Editors:
Professor A.W. Castleman, Jr.
Department of Chemistry, The Pennsylvania State University
152 Davey Laboratory, University Park, PA 16802, USA

Professor J.P. Toennies
Max-Planck-Institut für Strömungsforschung
Bunsenstrasse 10, 37073 Göttingen, Germany

Professor K. Yamanouchi
University of Tokyo, Department of Chemistry
Hongo 7-3-1, 113-0033 Tokyo, Japan

Professor W. Zinth
Universität München, Institut für Medizinische Optik
Öttingerstr. 67, 80538 München, Germany

Springer Series in Chemical Physics ISSN 0172-6218

ISBN 978-3-540-00859-0 e-ISBN 978-3-540-49829-2

Library of Congress Control Number: 2008930218

Typesetting: Data prepared by VTEX using a Springer TEX macro package
Cover design: eStudio Calamar Steinen

SPIN: 10877831 57/3180/VTEX
Printed on acid-free paper

9 8 7 6 5 4 3 2 1

springer.com

Preface

As undertaken by electrochemists, spectroscopic, topographical and numerous other non-electrochemical methods extensively covered in this book have driven the investigation of structure and dynamics at phase boundaries between condensed matter influenced by the presence and action of both an electric field and charged particles (like ions in solution or electrons in a metal) towards significant advances in recent years. They are based in particular upon an intense use of spectroscopic and surface sensitive methods adapted to the particular needs of *in situ* investigations of these electrochemical interfaces. Consequently the area is called (not completely exactly) spectroelectrochemistry. More recently, scanning probes capable of mapping the interface and providing a more or less topographic image have become available.

Numerous reviews covering single methods or families of related methods have appeared in journals, monographs and volumes of series. In addition, books containing collected review papers have been published. The articles were written by specialists and experts in the respective methods and describe fundamental and applied aspects, including examples of successful applications. Unfortunately this approach cannot provide the full picture because coherence is lacking between methods and different but related properties and other aspects of a given system investigated with various methods. In addition, the papers are generally of a somewhat different level; sometimes they are filled with a flood of details or extensive repetitions of fundamental information already provided in standard textbooks. The present book attempts to close this gap by providing the generalist's view. It offers a broad collection of spectroscopic and surface sensitive techniques currently employed in investigations of electrochemical systems. Relationships between different methods pertaining both to the principles of the methods and the properties of the investigated systems are highlighted. In many cases examples illustrating the power and the potential of a combined use of several spectroelectrochemical techniques are discussed in detail.

As presumably nobody can be an expert in all discussed methods and techniques, this book will not cover all methods with the same intensity and expertise. Nevertheless, all reported methods are described in sufficient detail, enabling the reader to access the current literature, to evaluate the methods and to choose methods of potential use for his given problem. This is the main purpose of the book: to serve as a guide to the successful application of spectroelectrochemistry and surface analytical methods in electrochemistry—just like a toolbox.

This book is based on numerous original papers and reviews. Because of the immense number of original papers and reviews, I have only provided references to those that I have assumed to be of particular importance as an introduction or that offer essential details not covered in this text. I wish to express my sincere apologies to all authors who have published important results and are not quoted explicitly within this book. Their contributions are nevertheless highly esteemed. Personal communications and experience with several methods in the laboratory of our research group and numerous contributions from coworkers and colleagues added to the base of this book. Some of them provided papers and preprints, examined experimental details, supplied original data or checked parts of the manuscript. Help and contributions from V. Brandl, M. Bron, C.H. Hamann, J. Lippe, A. Malinauskas, M. Probst, S. Schomaker, B. Speiser and B. Westerhoff are gratefully acknowledged. W. Vielstich introduced me to electrochemistry, E.B. Yeager provided inspiring access to new areas of research and to new methods during my postdoctoral stay at his laboratory, G. Comsa initiated a continuing interest in surface science and C.H. Hamann was an inspiring discussion partner. Without their stimulating support this book would have remained unwritten. Close cooperation and helpful assistance in planning and preparing this book by the staff of Springer-Verlag (Heidelberg), in particular by Claus Ascheron, Angela Lahee and Adelheid Duhm, are gratefully appreciated. My wife's tolerance for many hours spent in front of computer terminals and scientific papers was the essential prerequisite. The book would have been impossible without this.

Chemnitz, September 2008 R. Holze

The best books are those which make the reader supplement them.
(Voltaire, Philosophical Dictionary, Foreword)

It is vain to do with more what can be done with less (Entia non sunt multiplicanda praeter necessitatem).
(William of Occam)

Contents

List of Acronyms

Commonly used acronyms in surface analysis, surface spectroscopy and spectro-electrochemistry

aSNOM	Apertureless scanning near field optical microscopy
AAS	Atomic absorption spectroscopy
AC-SECM	Alternating current scanning electrochemical microscopy
ADXD	Angular dispersive X-ray diffraction
AEAPS	Auger electron appearance potential spectroscopy
AES	Auger electron spectroscopy
AES-SAM	Auger electron spectroscopy scanning Auger microscopy
AFM	Atomic force microscopy
AGFM	Alternating gradient field magnetometry
AIRS	Absorption infrared spectroscopy, also: abnormal infrared effect
AIS	Atom inelastic scattering
ALE	Atomic layer epitaxy
APD	Azimuthal photoelectron diffraction
APS	Appearance potential spectroscopy
ARAES	Angle-resolved Auger electron spectroscopy
ARPEFS	Angle-resolved photoemission extended fine structure
ARUPS	Angle-resolved ultraviolet photoelectron spectroscopy
ATR	Attenuated total reflection (spectroscopy), see also: FTIR
AXAFS	Atomic X-ray absorption fine structure spectroscopy
BAW	Bulk acoustic wave
BB-SFG	Broad band sum frequency generation
CD	Circular dichroism
CDAD	Circular dichroism photoelectron angular distribution
CDP	Current density probe
CELS	Characteristic energy loss spectroscopy
CER	Contact electric resistance
CFM	Chemical force microscopy
CL	Cathodoluminescence, Chemiluminescence
CLSM	Confocal laser scanning microscopy
CTR	Crystal truncation rod
CV	Cyclic voltammetry
DAPS	Disappearance potential spectroscopy

DCTS	Differential conductance tunneling spectroscopy
DEMS	Differential electrochemical mass spectroscopy
DESERS	Deenhanced surface enhanced Raman spectroscopy
DFT	Density functional theory
DPDS	Differential photothermal deflection spectroscopy
DRIFT	Diffuse reflection infrared Fourier transform spectroscopy
DRIFTS	same as DRIFT
DS	Desorption spectroscopy
DXAFS	Dispersive X-ray absorption fine structure (spectroscopy, measurement)
DTS	Distance tunneling spectroscopy
EBMA	Electron beam micro analysis, see EMMA
EBSD	Electron backscatter diffraction
ECESR	Electrochemical ESR spectroscopy
ECL	Electrochemiluminescence
ECTDMS	Electrochemical thermal desorption mass spectrometry
EC-STM	Electrochemical scanning tunneling microscopy
EDX	Energy dispersive X-ray (analysis)
EDXD	Energy dispersive X-ray diffraction
EELS	Electron energy loss spectroscopy
EER	Electrolyte electroreflectance
EFM	Electric force microscopy
EFTEM	Energy filtered transmission electron microscopy
EIS	Electron impact spectroscopy (also: electrochemical impedance spectroscopy)[1]
ECALE	Electrochemical atomic layer epitaxy
ELEED	Elastic low energy electron diffraction
ELL	Ellipsometry
ELS	Energy loss spectroscopy
EMIRS	Electrode potential modulated infrared spectroscopy
EMLC	Electrochemically modulated liquid chromatography
EM(MA)	Electron microprobe (mass analysis)
EMP	Electron micro probe, see EMMA
EMSI	Ellipsomicroscopy
EQCB	Electrochemical quartz crystal balance
EQCM	Electrochemical quartz crystal microbalance
EQMB	Electrochemical quartz (crystal) microbalance
ERS	Electroreflectance spectroscopy
ESCA	Electron spectroscopy for chemical analysis
ESD	Electron stimulated desorption
ESDIAD	Electron stimulated desorption ion angular distribution
ESI	Electrospray ionization

[1] The use of spectroscopy in connection with impedance measurements is not reasonable, thus it is frequently discouraged.

ESI-FTMS	Electrospray ionization Fourier transform mass spectrometry
ESID	Electron stimulated ion desorption
ESI-MS	Electrospray ionization mass spectroscopy
ESR	Electron spin resonance (spectroscopy)
EXAFS	Extended X-ray absorption fine structure (analysis), see also: NEXAFS, SEXAFS
EXELFS	Extended electron loss fine structure spectroscopy
FAB	Fast atom bombardment
FAB-MS	Fast atom bombardment mass spectrometry
FCS	Fluorescence correlation spectroscopy
FD	Flash desorption
FDS	Flash desorption spectroscopy
FES	Field emission spectroscopy
FFC	Filiform corrosion
FIM	Field ion microscopy
FIS	Field ion spectroscopy
FMIR	Frustrated multiple internal reflectance (spectroscopy)
FSCV	Fast scan cyclic voltammetry
FSCV–SECM	Fast scan cyclic voltammetry scanning electrochemical microscope
FT-EMIRS	Fourier transform electrode potential modulated infrared spectroscopy
FTIR[2]	Fourier transform infrared (spectroscopy), see also: FTIRRAS, MIRFTIRS, PMFTIRRAS, SNIFTIRS, SPAIRS
FTIR	Frustrated total internal reflection[3]
FT-IR	Fourier transform infrared (spectroscopy)
FTIR-DRS	Fourier transform infrared diffuse reflection spectroscopy
FTIRRAS	Fourier transform infrared reflection absorption spectroscopy
FT-SERS	Fourier transform surface enhanced Raman spectroscopy
GAIRS	Grazing angle absorption infrared spectroscopy
GC	Glassy carbon
GISAXS	Grazing incidence small angle X-ray scattering
μGISAXS	Microbeam grazing incidence small angle X-ray scattering
GIXAFS	Grazing incidence X-ray absorption fine structure spectroscopy
GI-XAS	Grazing incidence X-ray absorption spectroscopy
GIX(R)D	Grazing incidence X-ray diffraction
HAADF	High angle angular dark field (imaging in TEM)
HEED	High energy electron diffraction
HEIS	High energy ion scattering
HIID	Heavy ion induced desorption
HIKE	High kinetic energy photoemission spectroscopy
HIXE	Proton induced X-ray emission spectroscopy

[2] Sometimes also written FT-IR, see there.

[3] This term is hardly used anymore, see instead ATR.

HOPG	Highly ordered pyrolytic graphite
HOPG	Highly oriented pyrolytic graphite
HPXPS	High pressure X-ray photoelectron spectroscopy
HREELS	High resolution electron energy loss spectroscopy
HRSKP	Height-regulated scanning Kelvin probe (also: HR-SKP)
ICP-AES	Inductively coupled plasma atomic emission spectrometry
ICP-AESE	Inductively coupled plasma atomic emission spectroelectrochemistry
ICP-MS	Inductively coupled plasma mass spectrometry
IES	Inelastic electron tunneling spectroscopy
IETS	same as IES
ILEED	Inelastic low energy electron diffraction
ILIT	Indirect laser-induced temperature jump (technique)
IMMA	Ion microprobe mass analysis
IMXA	Ion microprobe X-ray analysis
IMPS	Intensity modulated photocurrent spectroscopy
IMVS	Intensity modulated photovoltage spectroscopy
INS	Ion neutralisation spectroscopy
INSEX	*In situ* X-ray reflection/transmission diffraction
IPS	Inverse photoelectron spectroscopy
IR	Infrared spectroscopy, see also: EMIRS, IRRAS, LPSIRS
IRD	Information rich detection
IRE	Infrared emission spectroscopy, internal reflection element
IRPDS	Infrared photothermal deflection spectroscopy
IRRAS	Infrared reflection absorption spectroscopy, see also: PMFTIRRAS
IRS	Internal reflectance spectroscopy
IS	Ionisation spectroscopy
ISS	Ion surface scattering (spectroscopy)
IES	Inelastic tunneling spectroscopy
ITIES	Interface between two immiscible electrolyte solutions
ITO	Indium (doped) tin oxide
KFM	Kelvin force microscopy
LAMMA	Laser microprobe mass analysis
LD	Laser desorption, see also: linear dichroism
LEED	Low energy electron diffraction
LEEM	Low energy electron microscopy
LEIS	Low energy ion scattering
LEIS	Localized (scanning) electrochemical impedance spectroscopy
LFM	Lateral force microscopy
LIBS	Laser induced breakdown spectroscopy
LID	Laser induced fluorescence
LIMS	Laser ionisation mass spectroscopy
LIPS	Laser induced plasma spectroscopy

LPSIRS	Linear potential scan infrared spectroscopy
LRS	Laser Raman spectroscopy
LSM	Layered synthetic microstrucutures
LSPR	Localized surface plasmon resonance
MBRS	Molecular beam relaxation spectroscopy
MBS	Mössbauer spectroscopy
MBSS	Molecular beam surface scattering
MCD	Magnetic circular dichroism
MCT	Mercury cadmium telluride (detector)
MEF	Metal enhanced fluorescence
MEIS	Medium energy ion scattering
MER(S)	Multiple external reflectance (spectroscopy)
MFM	Magnetic force microscopy
MFON	Metal film over nanospheres
MFTIRS	Microscope Fourier transform infrared spectroscopy
MIR(S)	Multiple internal reflectance (spectroscopy)
MIRFTIRS	Multiple internal reflection Fourier transform infrared spectroscopy
MMC	Metal matrix composite
MOKE	Magneto-optical Kerr effect
MORE	Micro-optical ring electrode
MR	Micro Raman (spectroscopy)
MSHG	Magnetization induced second harmonic generation
MSS	Molecule-surface scattering
NEXAFS	Near edge X-ray absorption fine structure spectroscopy
NIR	Near infrared (spectroscopy)
NIR-SERS	Near infrared surface enhanced Raman spectroscopy
NIS	Neutron inelastic scattering
NR(S)	Normal Raman (spectroscopy)
NPD	Neutron powder diffraction
NSOM	Near field scanning optical microscope
OLEMS	Online electrochemical mass spectrometry
ORC	Oxidation-reduction electrode potential cycling
OPG	Ordinary pyrolytic graphite
OSEE	Optically stimulated exoelectron spectroscopy
OTE	Optically transparent electrode
PAM	Photoacoustic method
PAS	Photoacoustic spectroscopy
PASCA	Positron annihilation spectroscopy for chemical analysis
PAX	Photoelectron spectroscopy with adsorbed xenon
PBD	Probe beam deflection
PCS	Photocurrent spectroscopy
PD	Photodesorption
PDATRFTIRS	Potential dependent attenuated total reflectance Fourier transform infrared spectroscopy

PDBS	Photothermal beam deflection spectroscopy
PDFTIRRAS	Potential difference Fourier transform infrared reflection absorption spectroscopy
PDFTIRES	Potential difference Fourier transform infrared emission spectroscopy
PDIRS	Potential difference infrared spectroscopy
PDMS	Plasma desorption mass spectrometry
PDS	Photothermal deflection spectroscopy
PECM	Photoelectrochemical microscope
PEEM	Photoelectron emission microscopy
PEIS	Photoelectrochemical impedance spectroscopy[4]
PEM	Photoelastic modulator
PEM	Photoelectrochemical measurement
PES	Photoelectron spectroscopy, see also: ESCA, UPS, XPS
PFM-AFM	Pulsed-force mode atomic force microscopy
PG	Pyrolytic graphite
PhD	Photoelectron diffraction
PHEEM	Photoemission electron microscopy
PI	Penning ionization
PIG	Paraffin-impregnated graphite
PIGME	Particle induced gamma ion emission
PIXE	Particle induced X-ray emission spectroscopy
PMFTIRRAS	Polarisation modulated Fourier transform infrared reflection absorption spectroscopy
PMFS	Potential modulated fluorescence spectroscopy
PMR	Potential modulated reflectance
PMRS	Potential modulated reflectance spectroscopy
PPES	Photopyroelectric (photothermal) spectroscopy
PSD	Photon-stimulated desorption
PSI	Phase-shift interferometry
PSP	Plasmon surface polaritons
PSTM	Photon scanning-tunneling microscope
PTS	Photothermal spectroscopy
PVS	Photovoltage spectroscopy
QEXAFS	Quick-scanning extended X-ray absorption fine structure (analysis)
RAIRS	Reflection absorption infrared spectroscopy
RAS	Raman spectroscopy [also NR(S)], see also: DESERS,SERRS, SERS, SRS, SRRS, SUERS
RAS	Reflection anisotropy spectroscopy, see also: RDS
RBS	Rutherford backscattering
RDE-EIS	Rotating disc electrode electrochemical impedance spectroscopy

[4] See footnote on p. xii.

RDS	Reflectance difference spectroscopy, see also: RAS
RC-SECM	Redox competition mode scanning electrochemical microscopy
ReflEXAFS	EXAFS at a grazing angle below the critical angle of total reflection
REMPI	Resonance enhanced multiphoton ionization
RHEED	Reflected high energy electron diffraction
RIBS	Rutherford ion backscattering spectroscopy
RRS	Resonance Raman spectroscopy
RTIL	Room temperature ionic liquid
SAES	Scanning Auger electron spectroscopy
SAM	Scanning Auger microprobe
SAM	Self-assembled monolayer
SANS	Small angle neutron scattering
SAPG	Stress-annealed pyrolytic graphite
SAXS	Small angle X-ray scattering
SC	Surface conductivity (measurement)
SDD	Surface differential X-ray diffraction
SDEMS	Scanning differential electrochemical mass spectroscopy
SECM	Scanning electrochemical microscope
SEF	Surface enhanced fluorescence
SEI	Solid electrolyte interface
SEIDAS	Surface enhanced infrared difference absorption spectroscopy
SEIRAS	Surface enhanced infrared absorption spectroscopy
SEIRRAS	Surface enhanced infrared reflection absorption spectroscopy
SEM	Scanning electron microscope
SERS	Surface enhanced Raman spectroscopy
SEHRS	Surface enhanced hyper-Raman spectroscopy scattering
SERHRS	Surface enhanced resonance hyper-Raman spectroscopy
SERRS	Surface enhanced resonance Raman spectroscopy
SES	Surface enhanced spectroscopy
SESHG	Surface enhanced second harmonic generation
SEVS	Surface enhanced vibrational spectroscopy
SEXAFS	Surface extended X-ray absorption fine structure (analysis) spectroscopy
SFG	Sum frequency generation
SFIRS	Synchrotron far infrared spectroscopy
SFM	Scanning force microscopy, see also: AFM
SG/TC	Substrate generation/tip collection (mode)
SHEED	Surface high energy electron diffraction
SHG	Second harmonic generation
SI	Surface ionisation
SICM	Scanning ion conductance microscopy
SIET	Scanning ion-sensitive electrode technique
SIMS	Secondary ion mass spectroscopy

SKP	Scanning Kelvin probe
SKPFM	Scanning Kelvin probe force microscopy
SMD	Single molecule detection
SMOKE	Surface magneto-optical Kerr effect
SMS	Single molecule spectroscopy
sm-SERS	Single molecule surface enhanced Raman spectroscopy
SMS-SERS	Single molecule surface enhanced Raman spectroscopy
SNMS	Secondary (sputtered) neutral (ion) mass spectroscopy
SNIFTIRS	Subtractively normalized interfacial Fourier transform infrared spectroscopy
SNIM	Scanning near field infrared microscopy
SNOM	Scanning near field optical microscopy
SORS	Spatially offset Raman spectroscopy
SP	Surface plasmon
SPAIR(S)	Single potential alteration infrared (spectroscopy)
SPFELS	Surface plasmon field enhanced light scattering
SPLS	Surface plasmon enhanced light scattering (see also SPFELS)
SPM	Scanning probe microscopy, see also: STM, EFM, MFM, AFM, SEM
SPPL	Surface plasmon enhanced photoluminescence
SPR	Surface plasmon resonance
SPS	Surface photovoltage spectroscopy
SQUID	Superconducting quantum interference device
SRET	Scanning reference electrode technique
SR-GIX(R)D	Synchrotron radiation grazing incidence X-ray diffraction
SRRS	Surface resonance Raman spectroscopy
SRS	Surface Raman spectroscopy
SRS	Specular reflectance spectroscopy
STEM	Scanning transmission electron microscopy
STM	Scanning tunneling microscopy
STS	Scanning tunneling spectroscopy
SUERS	Surface unenhanced Raman spectroscopy
SVET	Scanning vibrating electrode technique
SW-FTIRS	Square wave Fourier transform infrared reflection spectroscopy
SWNT	Single-wall carbon nanotube
SXAPS	Soft X-ray appearance potential spectroscopy
SXES	Soft X-ray emission spectroscopy
SXS	Surface X-ray scattering
TDMS	Thermodesorption mass spectroscopy
TDS	same as TDMS
TEAS	Thermal energy atom scattering
TEF	Tip enhanced fluorescence
TERS	Tip enhanced Raman scattering
THG	Third harmonic generation

TLC	Thin layer cell
TPD	Temperature-programmed desorption
TPRS	Temperature-programmed reaction spectroscopy
TR-SFG	Time-resolved sum frequency generation
TRBD	Total reflection Bragg diffraction
UME	Ultramicroelectrode
UPS	Ultraviolet photoelectron spectroscopy
UV-Vis	Spectroscopy with UV and visible light
VASE	Variable angle spectroscopic ellipsometry
VSF	Visible sum frequency
VSFG	Vibrational sum frequency generation
VTS	Voltage tunneling spectroscopy
WAXS	Wide-angle X-ray scattering
WDX	Wavelength dispersive X-ray (analysis)
XANES	X-ray absorption near edge structure
XES	X-ray emission spectroscopy
XPD	X-ray photoelectron diffraction
XPS	X-ray photoelectron spectroscopy
XRD	X-ray diffraction
XRR	X-ray reflection
XSW	X-ray standing wave

α	Accommodation coefficient measurement
σ	Adsorption isotherm measurement
θ	Degree of coverage
$\varnothing$	Work function
$\varnothing_{PE}$	Work function determination by photoemission measurement
$\varnothing_{TE}$	Work function determination by thermoemission measurement
$\varnothing_{FE}$	Work function determination by field emission measurement
$\varnothing_{FERP}$	Work function determination by field emission retarding potential measurement
$\varnothing_{DTERP}$	Work function determination by diode thermoemission retarding potential measurement
$\varnothing_{K}$	Work function determination by Kelvin method

List of Symbols

Common symbols and abbreviations used in electrochemistry[1]

A	Area, constant, optical absorption
a	Activity, absorption coefficient
a_i	Debye length
CV	Cyclic voltammogram
C	Capacitance / µF
C_D	Double layer capacity / µF
C_{diff}	Differential double layer capacity
C_{int}	Integral double layer capacity
c	Concentration / **M**
c_s	Surface concentration
c_0	Bulk concentration / **M**
D	Diffusion coefficient / $cm^2\ s^{-1}$
d	Distance
E	Electrode potential / V
$\mathbf{E}$	Electrical field strength
E_A	Activation energy / $kJ\ mol^{-1}$
E_F	Fermi energy
E_H	Electrode potential vs. standard hydrogen electrode, see also: E_{NHE}
$E_{Hg_2SO_4}$	$= E_{MSE}$
E_{MSE}	electrode potential vs. mercurous sulfate electrode, $c_{sulfate} = 0.1$ **M**
E_m	Electrode potential of measurement, also: E_{meas}
E_{NCE}	Electrode potential vs. normal calomel electrode
E_{NHE}	Electrode potential vs. normal (i.e. standard) hydrogen electrode[2]
E_{pzc}	Electrode potential of zero charge
E_{ref}	Reference electrode potential, electrode potential of reference measurement, see also: E_r
E_r	Reference electrode potential, electrode potential of reference measurement, also: E_{ref}

---

[1] For further details of electrochemical symbols, see R. Parsons, *Pure Appl. Chem.* **37** (1974) 499.

[2] The term "normal electrode" should be avoided because normal may be misunderstood as a designation of a certain concentration.

E_{SCE}	Electrode potential vs. saturated calomel electrode
E_{SHE}	Electrode potential vs. standard hydrogen electrode
E_0	Electrode potential at rest (i.e. $I = 0\ \mathrm{mA\,cm^{-2}}$)
E_{00}	Standard electrode potential
ΔE	Difference of electrode potentials, e.g. peak potentials
e_0	Elementary charge
F	Faraday constant
I	Ionic strength
I	Current / A
I_{ct}	Charge transfer current / A
j_{ct}	Charge transfer current density / A $\mathrm{cm^{-2}}$
k	Extinction coefficient
M	Molarity / $\mathrm{mol \cdot l^{-1}}$
M	Molar mass
m	Molality
N_L	Loschmidt number
n	Number of mols, electrode reaction valency, refractive index
R	Resistance, universal gas constant, reflected intensity of light
R_{CT}	Charge transfer resistance
Rf	Roughness factor
R_{sol}	Electrolyte solution resistance
RHE	Relative hydrogen electrode
T	Absolute temperature
u	Ionic mobility
upd	Underpotential deposit
V	Volume
v	$\mathrm{d}E/\mathrm{d}t$, scan rate in cyclic voltammetry

Part I

Introduction and Overview

1 Introduction

Electrochemistry is an extremely interdisciplinary area of science closely related to chemistry, physics, materials science, biology, surface science and a host of other fields. It is basically devoted to investigations of structures and dynamics as being present at interfaces between phases containing different types of mobile charged particles. Thus the process of electrochemical production of chlorine proceeding at the surface of a metal electrode in contact with a halide-containing solution in chlorine-alkali electrolysis is a typical case of electrochemistry at work. In the electrode the mobile charges are electrons, which are withdrawn from the chloride anions that are the mobile charged species in the solution phase. The interface established between molten iron and the fairly complicated mixture of molten metal oxides floating on top of the molten metal is another example. The stimulation of nerves and the propagation of information along nerves is another, totally different case. One more different case is the delamination of an organic coating applied to a metal surface for corrosion protection that occurs because of processes occurring at the metal-coating interface in the presence of a thin solution film on top of the coating containing small concentrations of electrolyte salts. In all of the examples just mentioned charged particles and large electric fields at interfaces play a pivotal role in establishing the structure at the interface and in controlling the various processes occurring at the interface. These examples share an additional feature that may possibly complicate their investigation: All interfaces are "buried" between two condensed phases (liquid or solid). Electrochemical investigations have focused in recent decades on elucidating their static and dynamic features exactly. The breadth of the few listed examples illustrates the need for intense research. Beyond these aspects of interfacial electrochemistry, the processes and structures within a condensed phase containing ions far away from the interface with their particular properties are the other main topics of electrochemistry. They are related to ionic conduction, thermodynamics of perfect (ideal) and real phases and other topics that are of only peripheral interest within the scope of this book. Initial investigations of electrochemical interfaces were limited to the measurements of electrical and chemical quantities, e.g. current, voltage, potential, charge or concentration. Kinetics could be elucidated by performing these measurements as a function of time, concentration and temperature. Unfortunately the obtained results provided in most cases only a very rough and macroscopic picture. A complete understanding, in particular at the microscopic level, was impossible or only highly speculative. With the increasing importance and breadth of the application of electrochemistry in industry, technology and daily life, this has become a considerable obstacle. Optimization of a given system or process and the understanding of the failure of a product with electrochemical features is possible only based on a deeper understanding of the electrochemical interface.

The advent of a broad variety of methods in vacuum physics, surface and materials science and analytical chemistry has provided a rich zoo of methods that have been converted for use in experimental electrochemistry. In contrast to the experimental methods briefly touched upon before, these methods are summarized as "non-traditional methods".

The overwhelming abundance of available methods is a tremendous possibility and a considerable seduction. Reasonable use of available methods and selection of the most suitable ones is possible only based on at least a minimum knowledge of the available methods, their strengths and weaknesses and their known pitfalls. So far the researcher has to consult either the original literature, including numerous more or less strongly focused reviews, or quite a few books that provide overviews of some subfamilies of the large number of available methods. The present book attempts to provide a broad overview delving into the depth of a certain method only as far as necessary to provide an initial understanding. It attempts to include all known methods of investigation applied to electrochemical interfaces that have arrived at a stage where a useful application has become visible to a broader audience.

In the following chapters and sections this will be done in a way that will hopefully be useful for the researcher looking for a method for his particular problem and for the graduate student looking for possible methods to treat the task of his masters thesis. In the second chapter, the structure and dynamics of electrochemical interfaces will be reviewed briefly. The term "interphase" will be introduced, stressing the fact that the topmost layers or regions of both phases that are in contact at the interface are different from layers and regions within the bulk of the phases. The third chapter pays a closer look at the possibilities of classical electrochemical methods and stresses the limitations beyond which traditional electrochemical methods provide only the basis for speculative interpretation of experimental data.

In Chap. 4, spectroscopic and surface analytical methods that form the bulk of this book are treated in the form of a general overview. This chapter will be of interest for a reader with an already established background in traditional electrochemistry who is looking for a general introduction. Chapter 5 provides the complete picture of spectroscopic methods. Chapters 6 and 7 treat X-ray and surface topographic methods. A brief introduction explains the way methods are assigned to various families. Descriptions of the available methods as known to the author are provided together with instructive examples of their applications. The structure of a description always follows the same scheme. The fundamentals are presented briefly. The electrochemical system properties that can be investigated with the method are indicated. A description of the experimental setup is provided. Some examples serve as illustration. References are provided especially for methods and experimental details that cannot be treated exhaustively within the limited scope of this book. No attempt is made to provide a complete listing of all reported applications of a given technique.

A generous list of acronyms is provided, giving the reader access to fast explanations of the myriad of examples of the modern letter soup omnipresent in scientific papers. The subject index will serve as a fast access lane especially for those readers searching for information about a particular method.

2 Structure and Dynamics of Electrochemical Phase Boundaries

In our environment surfaces, phase boundaries and interfaces are omnipresent. A living cell, a crystal of salt, a living organism, a cutting tool or a sheet of metal are separated from the surroundings by surfaces—and thus interfaces are established. At a first glance this may be accepted as a trivial and obvious fact of life without much importance. The properties and activities of a biological cell are presumably controlled by its structure and the processes occurring in its interior; density and the chemical and physical properties of a crystal depend on the chemical identity of its constituents and their arrangement in the crystal. The same may be said of the other examples. A closer look at this understanding leaves a number of open questions, because many processes of central importance are occurring at the interface. By just looking at the bulk properties of the adjacent systems these properties may be understood improperly or, sometimes, not at all.

This quickly becomes evident when we take another look at our examples: The exchange of all materials (nutrients, water, waste, etc.) of a cell with the environment occurs via the membrane enclosing the cell, wherein a liquid/liquid interface is established. The permeability of the membrane is of vital importance for a proper functioning of these transport processes. It depends upon the structure of the membrane itself and the structure and composition of the interface between the membrane and the surrounding atmosphere. The selective permeability for toxins of the cells and membranes constituting an animal's skin may be a reason for the massive die-off among amphibians. The practical applicability and the many mechanical properties of a cutting tool certainly depend on the bulk properties of the used raw material. But the durability of the cutting edge, its hardness and stability even under demanding operating conditions, depend mostly upon the chemical composition and structure of its surface, which basically forms a solid/gas interface. Surface hardening or treatment by, for example, ion implantation or galvanic deposition of additional material, may considerably change and improve the bulk material's initial properties. The corrosion stability of the metal sheet depends of course upon the chemical identity or, in the case of an alloy, the composition of the metal. But corrosion takes place only on the outermost layer of the sheet, and consequently the surface properties once more control the process occurring at this solid/liquid interface.

These apparently unrelated examples have one feature in common: Structure or dynamics or even both at the interface are strongly influenced by electronic charges that are present as ions in the liquid around the membrane or the corroding metal. Sometimes the participation of charged particles in chemical processes as necessary in the surface treatment of the cutting tool is essential. The investigation of the structure and dynamics of interfaces influenced by electric charges or by charged particles is the task of electrochemists. Traditional methods and their limits will be reviewed briefly in the following chapter.

3 Scope and Limitations of Classical Electrochemical Methods

The experimental tools of electrochemists were, until a few years ago, mainly rather simple measurements of electrical, physical and chemical quantities. Using a broad variety of experimental methods today called "classical electrochemical methods", they were able to provide models of electrified interfaces with respect to both structure and dynamics. Unfortunately their results were in many cases of a very macroscopic nature, any interpretations of the model with respect to the microscopic structure and mechanistic aspects of the dynamics and reaction were only more or less reasonable derivations. This gap, which caused many misunderstandings of puzzling features in electrochemical processes and interfaces, has started to close. The use of an enormous variety of spectroscopic and surface analytical tools in investigations of these interfaces has considerably broadened our knowledge. In many cases microscopic models based on the results of these studies with "non-traditional electrochemical methods" have enabled us to understand many hitherto strange phenomena in a convincing way.

The "traditional methods" generally are based upon measuring the potential of a working electrode and the current necessary to establish this potential. In many cases the registration of these data is done as a function of time, concentration or some other additional experimental variable. The relationship between current and potential, in particular the deviation of the actual electrode potential established during the flow of a current from the rest potential at nil current, can be explained by assuming a variety of hindrances impeding the flow of current or increasing the electrode potential actually necessary to obtain the desired current. This difference of potentials is called overpotential and the various contributing hindrances are named according to the cause of the hindrance: A charge transfer overpotential is caused by a sluggish charge transfer needing additional activation energy in order to proceed at appreciable rates. The slow mass transport caused by limited rates of diffusion, convection or migration generates a concentration overpotential (sometimes it is more precisely called "diffusion overpotential"). The name implies that an insufficient concentration of a reactant needed in the electrochemical reaction is the cause. Of course a small concentration of the reactant in the bulk of the solution enhances this effect. Slow removal of reaction products will also result in diffusion overpotentials. Chemical reactions preceding or following the charge transfer as heterogeneous reactions on the electrode surface or as homogeneous reactions in the solution phase produce reaction overpotentials. As both transport and reaction tend to impede the overall rate of the electrochemical reaction by limiting either the supply of reactants or the removal of products, they are sometimes summarized as concentration overpotentials. Finally, adsorption or desorption of reactants and crystallization in the case of metal deposition or dissolution may be slow, resulting in adsorption or crystallization overpotentials. Numerous electrochemical methods have been used to separate these various overpotentials in order to get a handle for further improvements of the reaction rate for practical applications or to get a better understanding of the reaction kinetics. Of course, it was always tempting to deduce mechanistic models describing the reaction on a molecular-microscopic level. Unfortunately, such deductions must be mostly tentative because they generally are not based on direct molecular or atomic evidence of the actual identity of the species proposed as participants in the electrochemical reaction sequence.

This lack of evidence has stimulated intense efforts to modify known methods in order to comply with the specific requirements of measurements in the presence of an electrolyte solution or other specifics of electrochemical investigations.

At any given interface between two phases the properties of both phases close to the interface and, in particular, those of the topmost layers are different from those in the bulk. In order to separate this special portion of a system from both bulk phases the term interphase has been coined for this quasi-phase in between the bulk phases. This term considerably expands the two-dimensional view of the phase boundary as a simple interface between two completely homogenous phases. The particular properties of these interphases are of pivotal importance for their behavior in many areas of science and technology. In applied sciences an improvement of these properties is possible only with knowledge of these properties that is as broad and deep as possible. In electrochemistry the interphase properties are further complicated by the involvement of charged particles and extremely high electric fields. A broader overview of the electrochemical interface will identify further adjacent domains:

- The electrolyte solution
- Diffusion and/or reaction layers
- The electrochemical double layer comprised of the diffuse (Gouy-Chapman) and the inner (Helmholtz) layer
- Space charge layers and surface states (with semiconducting electrodes)
- Adsorbed species on the electrode surface
- Molecular or polymer films
- The bulk of the electrode material

Properties of interphases relevant for an understanding of structures and dynamics therein can be grouped into atomic (microscopic) and macroscopic ones. Nevertheless the close relationships between both types of properties have allowed us to infer conclusions with respect to atomic models from macroscopic information. Investigations of these properties have gained tremendously in recent years from the application of a broad variety of spectroscopic methods. Certainly many properties, in particular macroscopic ones, can be studied by measuring surface properties (like conductivity, hardness, etc.) with non-spectroscopic methods. In most cases the results will be rather general and a complete understanding of relationships between the surface properties of interest, any structural features of the surface and the interphase on an atomic level will not be accessible. This will seriously obstruct further improvements of the knowledge of these properties.

4 Spectroscopy and Surface Analysis at Interfaces Between Condensed Phases

The need for further methods to study electrochemical interfaces and interphases has been indicated in the preceding chapter. Spectroscopic methods in particular were suggested as suitable tools [1–3]. The major advantage was considered to be their capability to provide information on a microscopic level by probing the subject of interest, resulting in signals that contain the desired information specifically from the place of interest. In this chapter a first overview is given of available methods and possible approaches for a typical selection of experimental tasks, which are taken predominantly from electrochemical investigations. All methods will be dealt with in detail in the sections of the following chapters; thus the immediately following text is addressed preferably at the reader looking for a general introduction.

The broad variety of methods suitable for spectroscopy at surfaces and other kinds of surface analysis has been touched upon already in the preceding chapter. The term "surface" referred mostly to solids being exposed to a vacuum or even ultra high vacuum environment. Few methods can be used under ambient conditions and at atmospheric pressure or even in the presence of condensed phases as encountered in electrochemical systems because of the strong interactions between several probes and signals with particles in condensed phases or in the gas phase at ambient temperature and pressure. The same line of argument applies to the interface between immiscible liquids.

The application of spectroscopic methods to surface studies always involves a probe used to stimulate or perturb the interphase in a well-defined way. This causes a signal to be emitted from the interphase. In many cases the signal is simply the modulated probe. Special care has to be exercised in order to obtain information exclusively from those parts of the interface as close as possible to the interface. Many techniques are essentially surface sensitive (i.e. selective). In some cases methods or sample systems have to be modified in order to achieve this surface sensitivity.

The broad range of energies (or wavelengths, frequency or any other equivalent unit) as depicted in Fig. 4.1 provides ample opportunities to develop methods.

Various probes available for surface studies are depicted in Fig. 4.2. Electromagnetic radiation ($h \cdot \nu$), neutral atom beams (i^0), ion beams ($i^{\pm}$), magnetic (H) or electric (E) fields and thermal excitation (W) can be used as probes.

Because of the numerous different types of interactions and resulting signals, the possible combinations of probes and signals are myriad. A simple matrix indicating some of these combinations and the resulting methods is provided in Table 4.1.

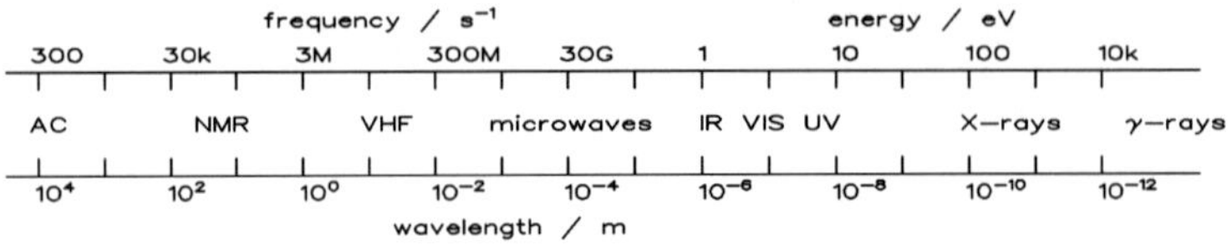

Fig. 4.1. The electromagnetic spectrum (approximate values and numbers only) with typical applications

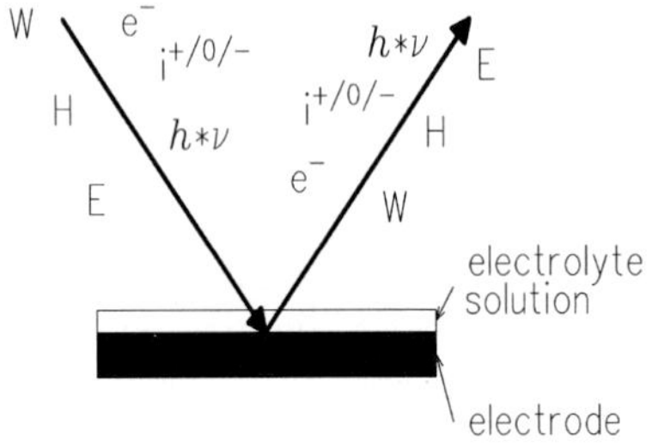

Fig. 4.2. Probes and signals in electrochemical interfacial investigations; for explanation of symbols, see text

The explanation of the acronyms is listed above (p. xi). No distinction is made here between methods applicable *in situ* or *ex situ*. Suggestions for names and abbreviations provided in the literature have been taken into account [4]. Once the effect of the interaction of the probe with the surface or the intensity of the measured signal is investigated as a function of either the energy of the probe or the signal, the method becomes a spectroscopy.

A very rich family of methods is based on electron or particle beams ranging in kinetic energy from a few electron volts (eV) up to several thousand eV (keV). As spectroscopic studies of electrochemical systems should be done preferably *in situ*, i.e. in the presence of the electrolyte solution (or molten electrolyte in some cases), electrons, ions and neutral particle beams cannot be used at these conditions; they can only be used *ex situ* after an appropriate sample transfer. The same applies to short wavelength UV-radiation [5]. Even the remaining combinations are manifold. Despite the inherent drawbacks of *ex situ* methods some *ex situ* techniques are used because of the valuable answers they can provide and/or because the necessary transfer is very well understood and does not result in artifacts. The spectroscopic methods for electrochemical applications developed from methods of various origins by adapting them to electrochemical requirements, in particular for their use under *in situ* conditions, are commonly summarized as spectroelectrochemical methods. Various authors have reviewed the combination of electrochemistry and spectroscopic methods, e.g. [6, 7]. Obviously the applicability of any method depends on the compatibility of the employed probes and signals with the components of the electrochemical cell. The advent of new sources for the generation of probes and new means for analysing and detecting signals from the interface/interphase has frequently changed the possibilities of methods considerably. This applies in partic-

Table 4.1. Probes and signals in surface analysis

Probe	Signal					
	e^-	$h \cdot \nu$	i^0	$i^\pm$	W	$E; H$
e^-	AES-SAM	APS	ESD	ESD		$\varnothing_{\mathrm{FERP}}$
	EELS	SXAPS	ESDIAD	ESDIAD		$\varnothing_{\mathrm{DTERP}}$
	ELS	DAPS				
	EXELFS	EM/MA				
	AEAPS					
	LEED					
	RHEED					
	SEM, STM					
	IETS					
$h \cdot \nu$	UPS	IR, ELL	PD	PD		
	XPS	UV-Vis	LD	LIMS		
	ESCA	RAS				
	$\varnothing_{\mathrm{PE}}$	MBS				
	ARUPS	SHG				
	PCS	ESR				
	EXAFS	INSEX				
	SEXAF	EXAFS				
	XANES	SEXAFS				
		XANES				
i^0		CL	MBSS	SI		
			MBRS			
			TEAS			
			FAB			
			α, σ			
$i^\pm$	INS	IMXA		ISS		
				SIMS		
				HIID		
				RBS		
				LEIS, HEIS		
W	$\varnothing_{\mathrm{TE}}$		FD	SI		
			DS			
			TDS			
$E; H$	FES			FIS		$\varnothing_{\mathrm{K}}$
	$\varnothing_{\mathrm{FE}}$			FIM		

ular to the use of synchrotrons as sources of electromagnetic radiation. Synchrotrons are circular electron beam accelerators. Electrons travelling on a circular orbit emit radiation, the energy distribution of which depends on the energy of the travelling electrons. By adjusting the electron energy, electromagnetic radiation with a broad range of frequencies extending from the far infrared into the range of hard X-rays can be generated; in a typical setup the energy can be continuously varied from 10 eV to 2700 eV. In addition, this radiation shows a defined time structure; it is completely polarized [8–13].

A brief overview of available methods can be obtained easily by looking at the various interface and interphase properties and grouping the methods with respect to their ability to provide information relevant to a particular property. In electrochemical systems where an electrode is in contact with an electrolyte and has possibly been modified by an adsorbed species or a surface layer, these properties are:

macroscopic properties

- Optical absorptivity (color)
- Optical reflectivity
- Electrical conductivity
- Crystallographic structure of the interphase
- Concentration of adsorbed species, see also: microscopic properties

microscopic properties

- Chemical identity of atoms and molecules in the interface, state of oxidation, coordination with further ligands, distance from interacting atoms in the electrode surface
- Type, strength and orientation of interaction between these particles and the electrode and their environment

These various properties are collected together with a selection of methods suitable for their investigation in Fig. 4.3.

The explanation of the acronyms is provided above (see p. xi). Only methods applicable under *ex situ* conditions are emphasized in the figure (italics). As already indicated, a sample transfer from the electrochemical cell into a ultrahigh vacuum (UHV) analysis system accompanied by drying of the sample and exposure to the atmosphere is necessary and any conceivable influence of this step may result in artifacts. This is most impressively demonstrated in studies of corrosion layers on iron electrodes. *Ex situ* methods have repeatedly yielded erroneous results; for example, because of dehydration of the corrosion products [14].

Macroscopic Properties. In many cases macroscopic properties of molecular and thick adsorbate layers or electrode coatings are closely related to their chemical composition and thickness. They are also related to several application-related parameters, e.g. color, conductivity or state of passivity.

The optical absorption in the UV-Vis region of the electromagnetic spectrum can be measured as a function of electrode potential and wavelength in external reflection or in transmission. In the latter case the use of optically transparent electrodes (OTE) made of glass coated with indium-doped tin oxide (ITO) or sputtered with gold or platinum is required. In order to minimize absorption, especially of colored electrolyte solutions, thin layer cells with short optical pathlengths are preferred. Nevertheless, in many cases simple cells made of standard cuvets, working electrodes cut from coated glass, metal wires serving as counter electrodes and small

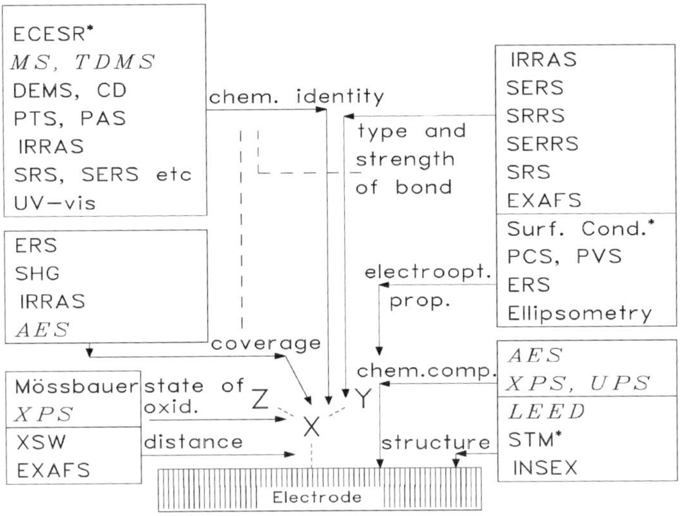

Fig. 4.3. Spectroelectrochemical methods applied to an electrochemical interface composed of an electrode with an adsorbate ZXY; for explanation of acronyms, etc., see text

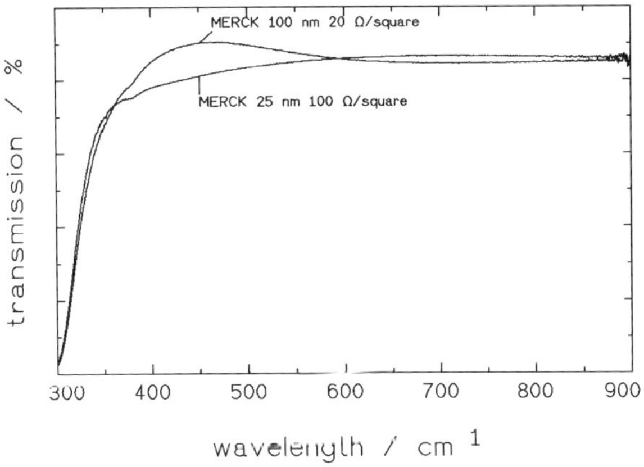

Fig. 4.4. Transmission spectra of glass sheets with different ITO-coatings as indicated

reference electrodes mounted without protruding into the optical beam can be manufactured easily [15, 16]. Experiments can be performed in standard double beam spectrometers. A complete cell is placed in the reference beam with a working electrode and without the electrochemically active species present in the other cell. Any absorption of the electrolyte solution or the OTE will be canceled out by the nulling procedure of the spectrometer. Typical UV-Vis spectra of ITO coated electrodes are shown in Fig. 4.4.

Electronic transitions in thin films of conducting polymers deposited on the OTE have been investigated with UV-Vis spectroscopy. As an example, in Fig. 4.5, UV-

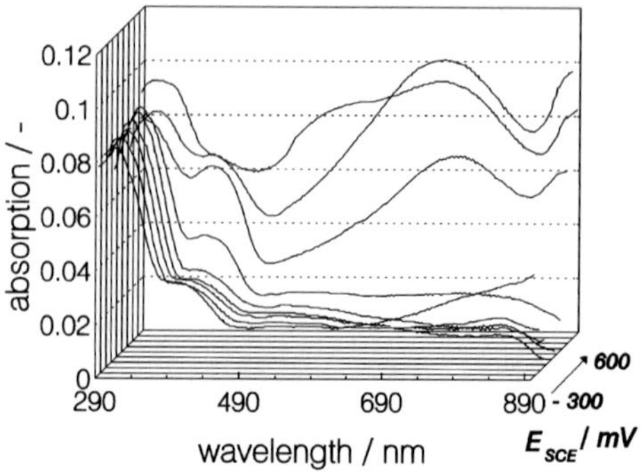

Fig. 4.5. UV-Vis spectra of a conducting polymer layer of poly(2-propylaniline) deposited on a gold sputtered glass slide electrode; electrode potentials as indicated

Vis absorption spectra of a film of poly(2-propylaniline) on an ITO electrode as a function of electrode potential (i.e. degree of doping) are shown.

Various electronic transitions caused by excitation of electrons from π into π^* states and from binding into antibinding polaron[1] and bipolaron[2] states that change as a function of electrode potential can be seen.

In the case of dissolved reaction intermediates or products, the use of a transmission arrangement and OTE results in the detection of these species in the electrolyte solution. This method is, of course, not surface sensitive; it has been used frequently for the investigation of organic electrochemical reactions involving various species of distinctly different UV-Vis-absorption [17, 18]. In the case of optically active dissymmetric organic species, circular dichroism (CD) may be observed. The stereoselectivity of heterogeneous electron transfer processes can be studied using a spectrometer with linear polarized light [19].

In the case of interphases on top of opaque or thick electrodes, external reflection of the probe light beam arriving at the electrode surface under investigation after passage through the adjoining electrolyte film is possible. Because no reference as necessary in the conventional two-beam spectrometers is available, some other parameter has to be modulated in order to get a spectrum that can be put into relationship with defined states of the interphase. Generally, the electrode potential is modulated. Consequently, the method is called electroreflectance spectroscopy (ERS). As the interaction of species in the interphase with the light depends upon the plane of polarization of the incoming light spectra obtained with p-polarized and s-polarized light are different in many cases, differences between correspond-

[1] In solid state physics, "polarons" are electrostatically induced local lattice distortions caused by an electron in a ionic crystal. In conducting polymers radical cations (lone electrons associated with positively charged holes) have a similar effect.

[2] Recombination of two polarons (similar to recombination of two electrons resulting in Cooper pairs in superconductors) leads to spinless bipolarons with a formal charge of two. Conversion between polaron and bipolaron states as a function of electrode potential is associated with changes in the spectra.

ing spectra obtained therewith can be used in the interpretation. Molecular adsorbates [20], their orientations [21] and electronic transitions that depend on the crystallographic orientation of the electrode have been studied [22]. Measurements with internal reflectance have been reported infrequently [23]. The electrode is deposited on a transparent crystal and the probe light is guided into the crystal under an angle resulting in attenuated total reflectance (ATR) at the electrode-coated surface. Light absorption can be attributed to properties of species on top of the electrode. Unfortunately, self-absorption of the crystal complicates the spectra.

Besides different absorption of s- and p-polarized light by adsorbates or surface layers on electrodes, the phase shift between both is influenced during the passage. Simultaneous measurements of both are done with ellipsometry. Recent experimental developments of automatic ellipsometers have greatly simplified experimental procedures; nevertheless, the interpretation of results is difficult. Passive layers [24–33], underpotential metal deposits [34–36], organic adsorbates [37] and surface modifying films [38] have been investigated. The range of accessible wavelengths has been extended from the UV-Vis into the infrared region [39].

The overall reflectance of an electrode surface measured without respect to the wavelength of the incoming light is of small experimental importance. The use of light sources of high intensity in narrow energy intervals may induce nonlinear effects at the surface. The generation of light at the second (SHG) or the third harmonic frequency (THG) is a possible result. In the presence of molecular adsorbates, sum frequency generation (SFG) with light frequency that combines the frequency of the incoming light and molecular vibrations can be detected. The former techniques have been used in studies of electrode coverage with simple adsorbates and of surface crystallographic structure [40, 41]; the latter is suitable for investigations of organic adsorbates [42, 43].

In the previous discussion, only reflection and absorption of light have been considered. In many cases the interaction of light with species in the interphase may cause further processes by exciting electrons into upper states and forming electron–hole pairs. Photovoltages measured directly (photovoltage spectroscopy PVS) or photocurrents detected by applying a potential difference across the interphase (photocurrent spectroscopy PCS) as a function of exciting wavelength can be used to investigate optoelectronic properties of semiconductors [44].

Although the electronic conductivity of an interphase that is present on an electrode can be related to various optoelectronic properties that are also measurable with spectroscopic techniques, the direct measurement of surface conductivities is not a spectroelectrochemical method. It is nevertheless a surface sensitive method that provides results closely related to those of other methods discussed in this book. Data on the electrosorption of alcohols on gold electrodes [45] or the electrode potential dependent conductivity of intrinsically conducting polymers [46] have been obtained with *in situ* surface conductivity measurements. Figure 4.4 shows the electrical resistance of a poly(2-propylaniline) film measured *in situ* under experimental conditions suppressing any influence of solution phase conduction. The influence of the electrode potential upon the resistance is obvious. A detailed discussion of this relationship has been presented elsewhere [7].

The crystallographic structure of interphases can be investigated with various methods. *In situ*, the application of X-ray diffraction INSEX is possible. Because of the depth of penetration of X-ray beams both in the transmission and the external reflection arrangement, the sample has to be made very thin in order to minimize unwanted contributions from the bulk of the electrode. Crystalline products of corrosion processes [47, 48], surface films [49], surface reconstruction [50] and catalyst systems [51] have been investigated.

Dynamic diffraction of an X-ray beam at lattice planes close to the surface of a perfect single crystal can result in standing X-ray waves (XSW). Measurements of the geometry of the standing wave or of photoelectrons released by the XSW can yield information on the position of atoms relative to the diffracting lattice plane in the interphase [52, 53].

The surface morphology of both crystalline and amorphous electrode surfaces or interphase layers can be studied *in situ* with scanning tunneling microscopy STM. The method is based upon the very pronounced dependency of the tunneling current between a sharp tip and a conducting surface. Scanning a surface with such a tip at a constant mechanical (macroscopic) distance morphological features of microscopically (atomically) different distance from the tip will result in dramatic changes of the tunnel current. This in turn can be used to map the atomic morphology of the surface. Numerous studies pertaining to processes involving metal deposition, dissolution, corrosion and reconstruction have been reported [54–64]. The investigation of organic adsorbates has been hampered so far by considerable difficulties during interpretation of the obtained tunneling current vs. surface topography data. The conceivable use of the STM arrangement in tunneling spectroscopy [65] may provide additional information on vibrational properties of organic adsorbates and thus ease the interpretation of STM data. The first experiments with intrinsically conducting polymers have been described [66].

The morphology, including features of adsorbates, can be determined with a small tip that scans the electrode surface mechanically using a very low contact pressure. This method is termed atomic force microscopy (AFM) [67] because van der Waals force interactions are probed. The first applications in metal deposition studies have been reported [68–71].

Ex situ macroscopic properties of adsorbates and interphases not sensitive towards a transfer of the sample from the electrochemical cell into an analysis system (in most cases a UHV chamber) can be studied with additional techniques. The crystallographic data can be derived from low energy electron diffraction (LEED) [72]; this is possible only if the sample has a minimum degree of ordering.

Surface concentrations of adsorbed or deposited species can be derived from a number of methods discussed previously. In the case of absorption spectroscopies, the extinction of the probing light corresponds more or less directly to the degree of coverage and an independent way of calibrating is generally necessary. The intensity of SHG signals corresponds to the coverage. Further details have been reviewed elsewhere [73].

Microscopic Properties. In most electrochemical experiments involving complex adsorption and/or reaction steps, the identification of adsorbates, intermediates and reaction products and their relationship to the environment are of central importance. As has been pointed out in the previous chapter some information about the chemical identity of species involved in the reaction can be gleaned indirectly from macroscopic properties of the interphase. Unambiguous identification is nevertheless possible almost exclusively only with additional spectroscopic techniques.

As shown in Fig. 4.3, a broad array of methods is available for this task. Direct identification of volatile molecular species present on or near an electrode is possible with differential electrochemical mass spectrometry (DEMS) [74–81]. For this method the electrode has to be porous and gas permeable. It is mounted on the inlet port of a mass spectrometer. Because of the pressure gradient between the vacuum and the electrochemical cell with the porous electrode operated at ambient pressure, volatile species can be sucked into the mass spectrometer and analysed therein.

Sometimes reactive intermediates—particularly, of electroorganic reactions—have unpaired electrons and consequently show the typical properties of a radical. Although the free spin of the electron is generally quenched upon strong adsorptive interaction with a metallic surface and consequently not detectable, sometimes the electrochemical reaction sequence allows detection of these radicals by electrochemical electron spin resonance spectroscopy (ECESR). This is possible when the radical desorbs and is present in the electrolyte phase in a concentration sufficient for detection by electron spin resonance spectroscopy or when the radical is separated from the metal by a suitable spacer (i.e. incorporated in an insulating interphase on the surface). A variety of radical intermediates generated electrochemically has been studied *in situ* with various different electrochemical cells [82–84]. Radical cations formed during electrooxidation of electrochemically active polymers can also be studied with ECESR [7, 82]. As an example, ECESR spectra of polyindoline are displayed as a function of the electrode potential in Fig. 4.6.

With an increase of the electrode potential towards anodic values, the polymer is oxidized (p-doped). The unpaired electron causes a radical–cation-like behavior. The high degree of delocalization results in the single line spectrum. A higher degree of doping corresponds to a larger concentration of free spins and, accordingly, the twice-integrated ECESR signal shows a well-defined dependency upon the electrode potential.

These radical cations are termed polarons based on the physical description of similar states in solid state physics. Their relevance to the electronic conduction mechanism in these films has been discussed elsewhere [86].

The chemical identity of molecular adsorbates on electrode surfaces can be derived from their vibrational behavior. *In situ* infrared (IR) and Raman spectroscopy are possible. Because of the strong IR absorption of most electrolyte solvents, modulated techniques are necessary. Potential and polarization modulation have been employed [53] and the methods are named correspondingly. Modulation is not nec-

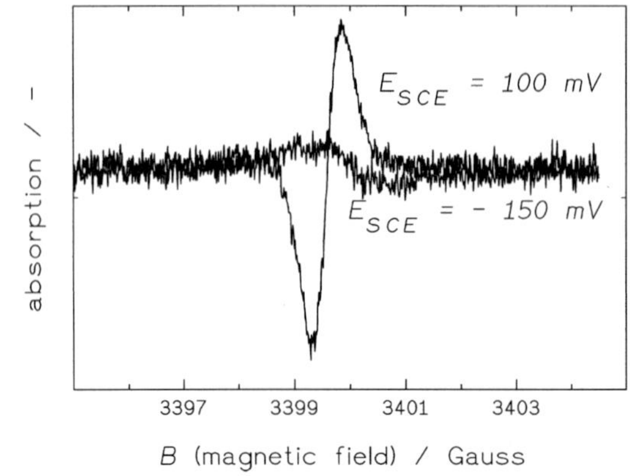

Fig. 4.6. ECESR spectra of a film of polyindoline in an aqueous solution of 1 N H_2SO_4 at $E_{SCE} = -100$ and -150 mV [85]

essary with Raman spectroscopy in the case of aqueous electrolyte solutions because water is a very poor scatterer. The low efficiency of the Raman process makes simple surface Raman spectroscopy (SRS) rather insensitive. Giant enhancements of the scattered signal intensity have been observed after various surface treatments, particularly of coinage metal surfaces (Cu, Ag and Au). The resulting microstructure of the sample surface and additional surface-specific enhancement effects result in a 10^6-fold increase in the intensity of the scattered signal. The effect and numerous applications have been reviewed extensively [87]. Besides identification of the adsorbate, these methods also provide information about the orientation of the adsorbate and the interaction with the environment. A combined use of both methods sometimes provides complementary information that yields a more complete model of the adsorbate structure. This can be demonstrated with selected data of a study of the adsorption of 2-buten-1,4-diol on a polycrystalline gold electrode. Previous electrosorption studies with classical electrochemical methods have indicated a rather strong physisorption of the alcohol from a neutral aqueous solution [88]. With respect to the orientation of the molecule on the surface and the influence of the electric field in the double layer upon the intramolecular binding, no information could be derived from these electrochemical measurements. Various vibrational modes were detected using both vibrational spectroscopies (see Figs. 4.7 and 4.8).

Interpretation of the spectra based on a comparison with vibrational spectra of the alcohol in pure form and dissolved in the electrolyte solution resulted in a proposed adsorbate structure with the π-bond system of the unsaturated alcohol interacting strongly with the electrode-side on (vibrational mode of the C=C bond around 1598 cm^{-1}) and with the C–OH bond (vibrational mode of the C–OH bond around 1030 cm^{-1}) at a tilted or perpendicular orientation with respect to the electrode surface towards the electrolyte solution. This information is helpful for understanding the mechanism of the electrooxidation of this alcohol [89].

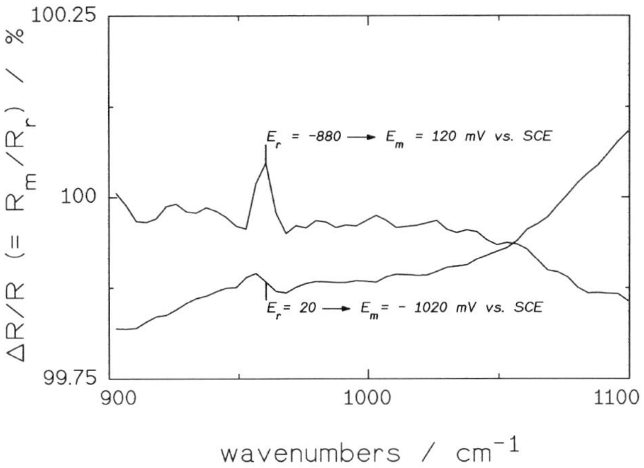

Fig. 4.7. Potential dependent IR spectra (PDIR) of a gold electrode in an aqueous solution of 2 m**M** 2-buten-1,4-diol and 0.1 **M** $KClO_4$ in water, electrode potentials as indicated, 256 or 512 scans at each electrode potential, resolution 8 cm^{-1}

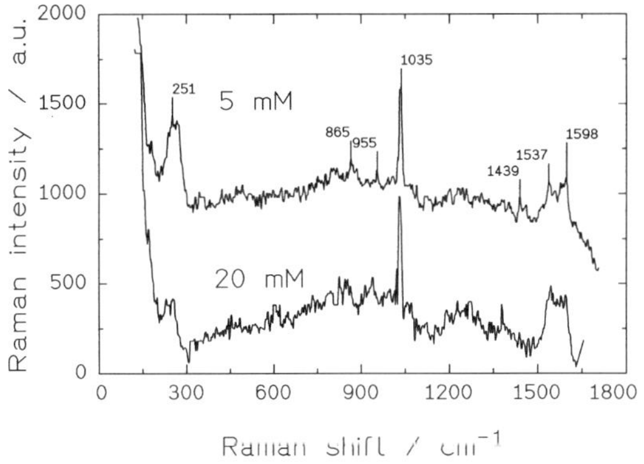

Fig. 4.8. SER spectra of a polycrystalline gold electrode in a solution of 0.1 **M** $KClO_4$ in water, concentration of 2-buten-1,4-diol indicated in the figure, $E_{SCE} = 0$ V (*upper trace*), 0.245 V (*lower trace*), resolution 7 cm^{-1}

Besides these techniques that are applicable to solid electrodes with a smooth or, in the case of IR spectroscopy, carefully polished surface, technologically important rough or porous samples are sometimes of interest. Vibrational spectroscopy at the surfaces of rough or porous samples is possible with photoacoustic and photothermal spectroscopy (PAS, PTS) [90–94].

After identification of an adsorbate, an intermediate or a reaction product by one of the methods discussed above or based on further information about the electrochemical process, additional information may be required.

Of course vibrational spectroscopies generally yield data suitable not only for identification of the adsorbate, but also indicative of intramolecular changes affected by the interaction with the electrode surface or the interactions within the interphase. These data may even contain information pertaining to the type and strength of interaction with the electrode surface.

In the case of metallic adsorbates (metal deposits, underpotentially deposited upd-layers, catalytically active metal deposits), the type of coordination to surface sites (one-, two- or three-fold) and the distance to these sites may be of interest. *Vice versa* the same type of data may be of importance in the case of adsorbed ions on metal electrodes or about the atomic environment of a given atom/ion in an interphase. Analysis of the fine structure of X-ray absorption (EXAFS, XANES) close to the X-ray absorption edge of the species (atom) of interest will yield this data provided the sample can be prepared in a very thin layer in order to exclude unwanted bulk interference. Otherwise the experiment can be done in reflection (SEXAFS). Information about the distance between the atom of interest and its first and sometimes even second shell of surrounding species can be derived from the spectra [95]. Availability of a suitable light source, generally a synchrotron (for details see p. 15), is an experimental prerequisite. The method has been applied in studies of passive and corrosion layers on various metals [96–102] and of molecular and ionic adsorbates on single crystal surfaces [103].

Very similar information can be obtained with the experimentally less demanding Mössbauer spectroscopy. The resonant absorption of gamma radiation by certain atomic nuclei (^{57}Fe, ^{119}Sn and several other isotopes of technical interest, e.g. ^{57}Co, can also be studied in the emission mode) yields information about the electromagnetic environment of this atom. By comparison with standard samples the chemical environment of a given atom, the specific type of chemical compound wherein it is incorporated and further morphological information can be derived from the Mössbauer spectra [104–113].

Local electron densities characteristic of a chemical environment can be studied with *in situ* positron annihilation spectroscopy (PASCA) [114].

Ex situ identification and investigation of adsorbed species is possible provided the interaction between adsorbate and electrode is strong enough to keep the interface unchanged even after moving the electrode into an ultrahigh vacuum chamber. Elemental identification is possible with various electron spectroscopies. Auger electron spectroscopy (AES) is particularly popular and can be used for quantitative measurements (degree of coverage), too [115–121]. The state of oxidation can be studied with photoelectron spectroscopy ESCA [122] or XPS [123].

Various mass spectroscopies are applicable *ex situ* in order to obtain molecular information. Secondary ion mass spectroscopy (SIMS) can be used [124, 125]. Thermal desorption mass spectroscopy is a viable alternative as a less intrusive and more surface sensitive tool [126, 127]. So far, the latter method has been applied exclusively to adsorbed hydrogen and carbon monoxide formed in an electrochemical reaction of organic CHO-compounds.

A general feature of practically all spectroscopic and surface sensitive methods described in the following chapters should be kept in mind: Compared to many electrochemical methods, in many cases these techniques provide information on a quite different time scale whereas fast electrochemical methods, instationary ones in particular, will yield data pertaining to processes occurring in microseconds and beyond. Most techniques described in the following chapters operate more slowly and will give information on a second or even slower time scale. Only in a few cases do some spectroscopies acquire data fast enough to provide time resolved information on a second or even millisecond time scale. This difference should always be kept in mind when comparing results obtained with these nontraditional methods with those of traditional electrochemical methods. Somewhat related to this general statement is the fact that many nontraditional methods yield particularly those data and results that can be obtained most easily with a given method. A simple example helps to explain this: When several intermediates are present during an electrochemical reaction measurement with infrared spectroscopy, we may identify preferable species with a large IR absorption cross section that are not necessarily the most important intermediates at all. Quite the contrary, these species may be formed at the dead end of a side reaction pathway. With another method, e.g. with a mass spectroscopy, those species in particular that can be ionized and detected most easily will be found. Such species may be different from those identified with infrared spectroscopy. Already this simple example illustrates the need for a careful selection and combination of experimental methods and for a careful interpretation.

The following chapters are devoted to single methods or to closely related families of methods. A brief overview, including some joint features of a given method, will precede the description of the individual techniques. The principle of organization of methods and techniques is a common feature. These are either probes shared by several techniques or particular surface/interphase properties studied with the methods. An introduction to spectroelectrochemistry in molecular inorganic chemistry is available [128].

Further Reading

A.W. Adamson, *Physical Chemistry of Surfaces*, 5th edn. (Wiley, New York, 1997)

K. Kolasinski, *Surface Science* (VCH/Wiley, Weinheim, 2002)

V.G. Bordo, H.-G. Rubahn, *Optics and Spectroscopy at Surfaces and Interfaces* (Wiley/VCH, Weinheim, 2005)

J.C. Rivière, S. Myhra (eds.), *Handbook of Surface and Interface Analysis* (Dekker, New York, 1998)

G. Ertl, J. Küppers, *Low Energy Electron Diffraction and Surface Chemistry*, 2nd edn. (Verlag Chemie, Weinheim, 1985)

R. Caudano, J.-M. Gilles, A.A. Lucas (eds.), *Vibrations at Surfaces* (Plenum, New York, 1982)

T.N. Rhodin, G. Ertl (eds.), *The Nature of the Surface Chemical Bond* (North-Holland, Amsterdam, 1979)

R.S. Mikhail, E. Robens, *Microstructure and Thermal Analysis of Surfaces* (Wiley, Chichester, 1983)

R.B. Hall, A.B. Ellis (eds.), *Chemistry and Structures at Interfaces* (Verlag Chemie, Deerfield Beach, 1986)

J.O'M. Bockris, A. González-Martin, in *Spectroscopic and Diffraction Techniques in Interfacial Electrochemistry*, ed. by C. Gutiérrez, C. Melendres. NATO ASI Series C, vol. 320 (Kluwer Academic, Dordrecht, 1990), p. 1
F. MacRitchie, *Chemistry at Interfaces* (Academic Press, San Diego, 1990)
R. Hoffmann, *Solids and Surfaces* (VCH, Weinheim, 1988)
J.T. Yates Jr., (ed.), *Vibrational Spectroscopy of Molecules on Surfaces* (Plenum, New York, 1987)
J.C. Vickerman (ed.), *Surface Analysis – The Principal Techniques* (Wiley, Chichester, 1997)
A.W. Czanderna, D.M. Hercules (eds.), *Progress in Techniques and Instrumentation: Methods of Surface Analysis* (Plenum, New York, 1991)
H.-H. Perkampus, *Encyclopedia of Spectroscopy* (VCH, Weinheim, 1995)
T. Owen, *Fundamentals of UV-Visible Spectroscopy* (Hewlett&Packard, 1996)
D. Brune, R. Hellborg, H.J. Whitlow, O. Hunderi (eds.), *Surface Characterization* (Wiley/VCH, Weinheim, 1997)
G.E. McGuire, J. Fuchs, P. Han, J.G. Kushmerick, P.S. Weiss, S.J. Simko, R.J. Nemanich, D.R. Chopra, Anal. Chem. **71**, 373R (1999)
K. Seibold, P. Albers, GIT Fachz. Lab. **1989** 637; 706
R.Parsons, Chem. Rev. **90**, 813 (1990)

References

1. H. Gerischer, Ber. Bunsenges. Phys. Chem. **92**, 1436 (1988)
2. J.O'M. Bockris, M. Gamboa-Aldeco, in *Spectroscopic and Diffraction Techniques in Interfacial Electrochemistry*, ed. by C. Gutiérrez, C. Melendres. NATO ASI Series C, vol. 320 (Kluwer Academic, Dordrecht, 1990), p. 55
3. D.M. Kolb, Surf. Sci. **500**, 722 (2002)
4. I. Mills, T. Cvitas, K. Homann, N. Kelley, K. Kuchitsu, *Quantities, Units and Symbols in Physical Chemistry*, 2nd edn. (Blackwell Scientific, Oxford, 1993)
5. D.E. Eastman, F.J. Himpsel, Phys. Today **5**, 64 (1981)
6. K. Rajeshwar, R.O. Lezna, N.R. de Tacconi, Anal. Chem. **64**, 429A (1991)
7. H. Gerischer, Angew. Chem. **100**, 63 (1988)
8. E.E. Koch (ed.), *Handbook of Synchrotron Radiation*, vol. 1 (North-Holland, Amsterdam, 1983)
9. E.E. Koch (ed.), *Handbook of Synchrotron Radiation*, vol. 2 (North-Holland, Amsterdam, 1983)
10. G.V. Marr (ed.), *Handbook of Synchrotron Radiation*, vol. 3 (North-Holland, Amsterdam, 1987)
11. C. Kunz, *Synchrotron Radiation – Techniques and Applications* (Springer, Berlin, 1979)
12. H. Winick, S. Doniach (eds.), *Synchrotron Radiation Research* (Plenum, New York, 1980)
13. G. Margaritondo, *Introduction to Synchrotron Radiation* (Oxford University Press, Oxford, 1988)
14. J.A. Bardwell, G.I. Sproule, M.J. Graham, J. Electrochem. Soc. **140**, 50 (1993)
15. J. Lippe, PhD Dissertation. Universität Oldenburg, 1991
16. M.B.G. Pilkington, B.A. Coles, R.G. Compton, Anal. Chem. **61**, 1787 (1989)
17. A.J. Bard, L.R. Faulkner, *Electrochemical Methods* (Wiley, New York, 1980), or *Electrochemical Methods*, 2nd edn. (Wiley, New York, 2000)

18. B.-S. Kim, S.-M. Park, J. Electrochem. Soc. **140**, 115 (1993)
19. J. Daub, J. Salbeck, I. Aurbach, Angew. Chem. **100**, 278 (1988)
20. J.A. Caram, C. Gutierrez, J. Electroanal. Chem. **323**, 213 (1992)
21. P.H. Schmidt, W. Plieth, J. Electroanal. Chem. **201**, 163 (1986)
22. D.M. Kolb, in *Spectroelectrochemistry*, ed. by R.J. Gale (Plenum, New York, 1988), p. 87
23. W.N. Hansen, Adv. Electrochem. Electrochem. Eng. **9**, 1 (1973)
24. B.D. Cahan, R.F. Spanier, Surf. Sci. **16**, 166 (1969)
25. B.D. Cahan, Surf. Sci. **56**, 354 (1976)
26. C.T. Chen, B.D. Cahan, J. Electrochem. Soc. **129**, 17 (1982)
27. R. Parsons, W.H.M. Visscher, J. Electroanal. Chem. **36**, 329 (1972)
28. J.L. Ord, F.C. Ho, J. Electrochem. Soc. **118**, 46 (1971)
29. J. Horkans, B.D. Cahan, E. Yeager, Surf. Sci. **46**, 1 (1974)
30. J.M.M. Droog, P.T. Alderliesten, G.A. Bootsma, J. Electroanal. Chem. **99**, 173 (1979)
31. S. Gottesfeld, S. Srinivasan, J. Electroanal. Chem. **86**, 89 (1978)
32. T. Ohtsuka, K.E. Heusler, J. Electroanal. Chem. **100**, 319 (1979)
33. J.L. Ord, J. Electrochem. Soc. **129**, 335 (1982), 767
34. R. Adzic, E. Yeager, B.D. Cahan, J. Electrochem. Soc. **121**, 474 (1974)
35. R. Adzic, B.D. Cahan, E. Yeager, J. Chem. Phys. **58**, 1780 (1973)
36. J. Horkans, B.D. Cahan, E. Yeager, J. Electrochem. Soc. **122**, 1585 (1975)
37. K. Kunimatsu, R.H. Parsons, J. Electroanal. Chem. **100**, 335 (1979)
38. F. Chao, M. Costa, E. Museux, E. Levart, Elektrokhimiya **29**, 57 (1993)
39. R.C. Kainthla, J.O'M. Bockris, J. Electrochem. Soc. **135**, 143C (1988)
40. R.M. Corn, in *Adsorption of Molecules at Metal Electrodes*, ed. by J. Lipkowski, P.N. Ross (VCH, New York, 1992), p. 347, and references therein
41. G.L. Richmond, in *Electroanalytical Chemistry*, vol. 17, ed. by A.J. Bard (Dekker, New York, 1991), p. 87
42. Y.R. Shen, J. Electrochem. Soc. **135**, 142C (1988)
43. Y.R. Shen, in *Chemistry and Structures at Interfaces*, ed. by R.B. Hall, A.B. Ellis (Verlag Chemie, Deerfield Beach, 1986)
44. M.W. Peterson, B.A. Parkinson, J. Electrochem. Soc. **135**, 1424 (1988)
45. R.I. Tucceri, D. Posadas, Electrochim. Acta **32**, 27 (1987)
46. R. Holze, J. Lippe, Synth. Met. **38**, 99 (1990)
47. J. McBreen, W.E. O'Grady, D.E. Sayers, C.Y. Yang, ECS-Meeting, Philadelphia, USA, 10–15 May 1987, Ext. Abstr. #523
48. M. Fleischmann, P. Graves, I. Hill, J. Robinson, A. Oliver, ECS-Meeting, Cincinnati, USA, 6–11 May 1984, Ext. Abstr. #427
49. T. Ikashoji, T. Iwasaki, Inorg. Chem. **27**, 1123 (1988)
50. M. Fleischmann, B.W. Mao, J. Electroanal. Chem. **247**, 311 (1988)
51. J. McBreen, Electrochemical society meeting, St. Louis, USA, May 17–22, 1992, Ext. Abstr. #35 and references therein
52. G. Materlik, J. Zegenhagen, W. Uelhoff, Phys. Rev. **B32**, 5502 (1985)
53. J.H. White, in *Electrochemical Interfaces*, ed. by H.D. Abruna (VCH, New York, 1991), p. 133
54. M.H. Dean, A. Pinkowski, U. Stimming, J. Electrochem. Soc. **135**, 143C (1988)
55. C.L. Li, L.D. Bell, C.P. Bankston, W.J. Kaiser, ECS-Meeting, Honolulu, 18–23.10.1987, Ext. Abstr. #1472
56. O. Lev, F.-R. Fan, A.J. Bard, J. Electrochem. Soc. **135**, 783 (1988)
57. F.F. Fan, A.J. Bard, Anal. Chem. **60**, 751 (1988)

58. D.H. Craston, C.W. Lin, A.J. Bard, J. Electrochem. Soc. **135**, 785 (1988)
59. L.D. McCormick, M.H. Dean, U. Stimming, T. Thundat, L. Nagahara, S.M. Lindsay, J. Electrochem. Soc. **135**, 144C (1988)
60. R.S. Robinson, J. Electrochem. Soc. **135**, 143C (1988)
61. L.-L. Ni, L.D. Bell, C.P. Bankston, W.J. Kaiser, J. Electrochem. Soc. **135**, 144C (1988)
62. R. Christoph, H. Rohrer, H. Siegenthaler, H. Wiese, Hauptversammlung der Bunsengesellschaft, Passau, 12–14.5.1988
63. P. Lustenberger, H. Rohrer, R. Christoph, H. Siegenthaler, J. Electroanal. Chem. **243**, 225 (1988)
64. K.M.H. Dean, A. Pinkowski, U. Stimming, J. Electrochem. Soc. **135**, 143C (1988)
65. for an introduction see: P.K. Hansma, in *Vibrational Spectroscopy of Molecules on Surfaces*, ed. by J.T. Yates, T.E. Madey (Plenum, New York, 1987), p. 135
66. R. Yang, W.H. Smyrl, D.F. Evans, J. Phys. Chem. **96**, 1428 (1992)
67. A.A. Gewirth, K.J. Hanson, Interface **2**, 37 (1993)
68. R. Mohr Rynders, R.C. Alkire, Electrochemical Society Meeting, St. Louis, USA, May 17–22, 1992, Ext. Abstr. #475
69. C.-C. Chen, S.M. Vesecky, A.A. Gewirth, J. Am. Chem. Soc. **114**, 451 (1992)
70. B.J. Cruickshank, A.A. Gewirth, R. Mohr Rynders, R.C. Alkire, J. Electrochem. Soc. **139**, 2829 (1992)
71. S. Manne, P.K. Hansma, J. Massie, V.B. Elings, A.A. Gewirth, Science **251**, 183 (1991)
72. D.M. Kolb, G. Lehmpfuhl, M.S. Zei, in *Spectroscopic and Diffraction Techniques in Interfacial Electrochemistry*, ed. by C. Gutiérrez, C. Melendres. NATO ASI Series C, vol. 320 (Kluwer Academic, Dordrecht, 1990), p. 361
73. R. Holze, J. Schneider, C.H. Hamann, Ber. Bunsenges. Phys. Chem. **92**, 1319 (1988)
74. O. Wolter, J. Heitbaum, Ber. Bunsen. Phys. Chem. **88**, 2, 6 (1984)
75. B. Bittins-Cattaneo, E. Cattaneo, P. Königshoven, W. Vielstich, in *Electroanalytical Chemistry*, vol. 17, ed. by A.J. Bard (Dekker, New York, 1991), p. 181
76. S. Wasmus, E. Cattaneo, W. Vielstich, Electrochim. Acta **35**, 771 (1990)
77. D. Tegtmeier, A. Heindrichs, J. Heitbaum, Ber. Bunsenges. Phys. Chem. **93**, 201 (1989)
78. K. Nishimura, R. Ohnishi, K. Kunimatsu, M. Enyo, J. Electroanal. Chem. **258**, 219 (1989)
79. H. Nakajima, H. Kita, K. Kunimatsu, A. Aramata, J. Electroanal. Chem. **201**, 175 (1986)
80. Y. Hirata, K. Suga, M. Fujihira, Chem. Lett. **1990**, 1155
81. M. Fujihira, T. Noguchi, J. Electroanal. Chem. **347**, 457 (1993)
82. R. Holze, Habilitationsschrift. Universität Oldenburg, 1988
83. R.G. Compton, A.M. Waller, in *Spectroelectrochemistry*, ed. by R.J. Gale (Plenum, New York, 1988), p. 87
84. L. Dunsch, A. Petr, Ber. Bunsenges. Phys. Chem. **97**, 436 (1993)
85. J. Lippe, PhD Thesis. Universität Oldenburg, 1991
86. R. Holze, J. Lippe, Bull. Electrochem. **8**, 516 (1992)
87. for a complete list see: R. Holze, Electroanalysis **5**, 497 (1993)
88. T. Luczak, M. Beltowska-Brzezinska, R. Holze, Electrochim. Acta **38**, 717 (1993)
89. R. Holze, T. Luczak, M. Beltowska-Brzezinska, Electrochim. Acta **39**, 991 (1994)
90. R.E. Malpas, A.J. Bard, Anal. Chem. **52**, 109 (1980)
91. G.H. Brilmyer, A.J. Bard, Anal. Chem. **52**, 685 (1980)
92. S. Yoshihara, M. Ueno, Y. Nagae, A. Fujishima, J. Electroanal. Chem. **243**, 475 (1988)
93. J. Rappich, J.K. Dohrmann, Hauptversammlung der Bunsengesellschaft. Passau, 12.–15.5.1988

94. C.E. Vallet, S. Berns, J.J. Hendrickson, C.W. White, J. Electrochem. Soc. **135**, 387 (1988)
95. M.F. Toney, J. McBreen, Interface **2**, 22 (1993)
96. H. Baumgärtel, H.-W. Jochims, B. Brutschy, Z. Phys. Chem. NF **154**, 1 (1987)
97. A. Michalowicz, J. Huet, A. Gaudemer, Nouv. J. Chim. **1982**, 79
98. M.S. Co, W.A. Hendrickson, K.O. Hodgson, S.A. Doniach, J. Am. Chem. Soc. **105**, 1144 (1983)
99. J.C. Poncet, R. Guilard, Polyhedron **2**, 417 (1983)
100. W.H. Liu, X.F. Wang, T.Y. Teng, H.W. Huang, Rev. Sci. Instr. **54**, 1653 (1983)
101. J. Goulon, P. Friant, J.L. Poncet, R. Guilard, J. Fischer, L. Ricard, Springer Ser. Chem. Phys. **27**, 100 (1983)
102. C. Goulon-Ginet, J. Goulon, J.P. Battioni, D. Mansuy, J.C. Chottard, Springer Ser. Chem. Phys. **27**, 349 (1983)
103. L. Blum, A.D. Abruna, J. White, J.G. Gordon II, G.L. Borges, M.G. Samant, O.R. Melroy, J. Chem. Phys. **85**, 6732 (1986)
104. W.E. O'Grady, J. Electrochem. Soc. **127**, 555 (1980)
105. C. Fierro, R.E. Carbonio, D. Scherson, E.B. Yeager, J. Phys. Chem. **91**, 6579 (1987)
106. D.A. Corrigan, R.S. Conell, C. Fierro, D.A. Scherson, J. Chem. Phys. **86**, 5009 (1987)
107. D.A. Scherson, C. Fierro, D. Tryk, S.L. Gupta, E.B. Yeager, J. Eldridge, R.W. Hoffman, J. Electroanal. Chem. **184**, 419 (1985)
108. D.A. Scherson, S.B. Yao, E.B. Yeager, J. Eldridge, M.E. Kordesch, R.W. Hoffman, J. Electroanal. Chem. **150**, 535 (1983)
109. D. Scherson, S.B. Yao, E.B. Yeager, J. Eldridge, M.E. Kordesch, R.W. Hoffman, Appl. Surf. Sci. **10**, 325 (1982)
110. D.A. Scherson, S.B. Yao, E.B. Yeager, J. Eldridge, M.E. Kordesch, R.W. Hoffman, J. Phys. Chem. **87**, 932 (1983)
111. H. Leidheiser, J. Electrochem. Soc. **135**, C5 (1988)
112. M.V. Ananth, N.V. Parthasaradhy, J. Sci. Ind. Res. **47**, 28 (1988)
113. D.A. Scherson, S.L. Gupta, C. Fierro, E.B. Yeager, M.E. Kordesch, J. Eldridge, R.W. Hoffman, J. Blue, Electrochim. Acta **28**, 1205 (1983)
114. Y.C. Jean, K.C. Cheng, ECS-Meeting. Toronto, 12–17 May 1985, Ext. Abstr. #676
115. A.T. Hubbard, Europ. Spec. **78**, 28 (1988)
116. J.A. Schoeffel, A.T. Hubbard, Anal. Chem. **49**, 2370 (1977)
117. A.T. Hubbard, J. Electroanal. Chem. **168**, 43 (1984)
118. V.K.F. Chia, J.L. Stickney, M.P. Soriaga, S.B. Rosasco, G.N. Salaita, A.T. Hubbard, J.B. Benziger, K.W.P. Pang, J. Electroanal. Chem. **163**, 407 (1984)
119. A.T. Hubbard, Acc. Chem. Res. **13**, 177 (1980)
120. J. Aberdam, in *Spectroscopic and Diffraction Techniques in Interfacial Electrochemistry*, ed. by C. Gutiérrez, C. Melendres. NATO ASI Series C, vol. 320 (Kluwer Academic, Dordrecht, 1990), p. 383
121. H.J. Mathieu, in *Surface Analysis – The Principal Techniques*, ed. by J.C. Vickerman (Wiley, Chichester, 1997), p. 99
122. B. Ratner, D. Castner, in *Surface Analysis – The Principal Techniques*, ed. by J.C. Vickerman (Wiley, Chichester, 1997), p. 43
123. R. Kötz, in *Spectroscopic and Diffraction Techniques in Interfacial Electrochemistry*, ed. by C. Gutiérrez, C. Melendres. NATO ASI Series C, vol. 320 (Kluwer Academic, Dordrecht, 1990), p. 409
124. H.W. Buschmann, S. Wilhelm, W. Vielstich, Electrochim. Acta **31**, 939 (1986)

125. J.C. Vickerman, A. Swift, in *Surface Analysis – The Principal Techniques*, ed. by J.C. Vickerman (Wiley, Chichester, 1997), p. 135
126. S. Wilhelm, W. Vielstich, H.W. Buschmann, T. Iwasita, J. Electroanal. Chem. **229**, 377 (1987)
127. S. Wilhelm, H.W. Buschmann, W. Vielstich, *DECHEMA Monographie*, vol. 112 (VCH, Weinheim, 1988), p. 113
128. E. Alessio, S. Daff, M. Elliot, E. Iengo, L.A. Jack, K.G. Macnamara, J.M. Pratt, L.J. Yellowlees, in *Trends in Molecular Electrochemistry*, ed. by A.J.L. Pombeiro, C. Amatore (FontisMedia/Dekker, Lausanne/New York, 2004), p. 339

Part II

Methods and Applications

In this part the various non-traditional methods employed in electrochemical investigations are described in three chapters. The description follows a very simply outline in order to facilitate readability. Fundamentals of methods are provided in detail only where the author believes that these fundamentals cannot be assumed to be generally known or where a detailed knowledge is necessary for understanding the displayed examples and results.

The term "electrode", according to W. Nernst, refers to the electron (or hole) conducting part of an electrochemical system (a half cell) together with the ion conducting part (electrolyte solution, molten salt, solid electrolyte). Commonly, the term is used frequently with reference to the former part only. Although the Nernstian description is certainly the more precise one, experimental methods mostly serve to investigate the interface (or interphase) between both phases with the former part of the system being the focus of interest. Consequently, in the following text the term "electrode" will be used to designate the former part only. Electrochemical symbols are used as suggested by IUPAC [1].

As already pointed out, the myriad of conceivable combinations of probes and signals and the experimental ingenuity of researchers adapting a given method to an encountered problem has resulted in a considerable wealth of spectroelectrochemical and surface analytical methods established as non-traditional methods of investigation. Organizing these methods in a reasonable way as a means of orientation to help the reader poses a problem almost as large as the breadth of the methods to be described.

Based on the traditional designation of spectroscopy as a method wherein either absorption or scattering of electromagnetic radiation independently of the type and energy of the radiation is the central feature, spectroscopic methods are collected in Chap. 5. Methods wherein diffraction of any kind is the central aspect are collected in Chap. 6. In a still rapidly growing collection of methods, the topography of a surface or an interface is visualized using a scanning probe method. In addition, some of these methods provide additional information on the surface with spatial resolution. These methods are collected together with further techniques, wherein already established methods are adapted in a way enabling the acquisition of localized information in the final chapter. Because these methods are essentially not spectroscopic techniques but surface analytical tools, some additional non-spectroscopic methods used to study surface properties like surface conductivity are added to this chapter.

Collections of reviews of single methods or small selections of methods have been published; as already indicated, they do not provide the broad picture of the complementary use of spectroelectrochemical methods or deal with too much detail [2–10].

References

1. R. Parsons, Pure Appl. Chem. **37**, 499 (1974)
2. E. Yeager, B.D. Cahan, D. Scherson, M. Hanson, Nav. Res. Rev. **37**, 13 (1985)
3. W. Plieth, G.S. Wilson, C.G. de la Fe, Pure Appl. Chem. **70**, 1395, 2409 (1998)

4. H.D. Abruna (ed.), *Electrochemical Interfaces* (VCH, New York, 1991)
5. A. Wieckowski (ed.), *Interfacial Electrochemistry* (Dekker, New York, 1999)
6. M. Hunger, J. Weitkamp, Angew. Chem. **113**, 3040 (2001)
7. J.F. Haw (ed.), *In-Situ Spectroscopy in Heterogeneous Catalysis* (VCH/Wiley, Weinheim, 2002)
8. B.M. Weckhuysen, Chem. Commun. **2002**, 97
9. J.C. Lindon, G.E. Tranter, J.L. Holmes (eds.), *Encyclopedia of Spectroscopy and Spectrometry* (Academic Press, New York, 2000)
10. F. Hoffman, Surf. Sci. Rep. **3**, 107 (1983)

5 Spectroscopy at Electrochemical Interfaces

The following section provides a systematic overview of spectroscopic methods as applied in electrochemical investigations. They are grouped according to the type and energy of the employed probe, i.e. according to the type of electromagnetic radiation used. (Suggestions for a subdivision of the electromagnetic spectrum have been provided elsewhere [1].) In a few cases, the light returning as the signal from the investigated interphase is not analysed with respect to energy, but with respect to some other experimental parameter, e.g. the plane of polarization or intensity. Sometimes the overall intensity is measured only as a function of another experimental parameter. These techniques are, strictly speaking, not spectroscopic ones; they are nevertheless incorporated in appropriately designated sections together with the real spectroscopic methods. Further experimental developments underway when this text was prepared made these assignments preliminary ones, anyway. In the case of ellipsometry, only measurements at fixed wavelengths were done initially. In the meantime, spectroscopic ellipsometry, i.e. measurements of optical constants of the investigated interface as a function of the wavelength of the sample light beam, have become standard. In some cases (e.g. photothermal measurements), investigations with and without taking into account the actual wavelength of the sampling light have been done.

As already pointed out, the number of known and conceivable interfaces and interphases where electrochemical processes may proceed is overwhelmingly large. The following listing of methods deals exclusively with those which are the subject of electrochemical investigations. No methods used exclusively for bioelectrochemical systems and samples are included. Semiconductor electrodes are included despite the fact that the understanding of their spectroelectrochemical investigations requires a deeper understanding beyond the knowledge of the metal/solution interface. (For a broad introduction and overview see [2].)

Obviously the increasing importance of solid electrolytes as employed in solid oxide or polymer membrane fuel cells calls for experimental methods adapted specifically to the needs of these experimental setups, which are considerably different from those employing liquid electrolyte solutions. The number of experimental methods beyond classical electrochemical ones adapted specifically to these requirements was fairly low when preparing this chapter. In most cases standard surface analytical or solid state analytical techniques were employed; for an introductory overview see [3]. Nevertheless, these electrochemical systems are not taken into consideration in the following overview when a standard technique has been modified and adopted substantially.

Besides solid electrolytes of various types, molten salts and ionic liquids[1] [4–7] have attracted considerable interest as electrolyte systems. Although their practical application is currently focused almost entirely on a number of electrolytical processes (production of aluminum, fluorine), advanced systems for electrochemical energy conversion and storage and electrochemical production using both molten salt electrolytes and ionic liquids are under investigation. The specific properties of molten salts, their high viscosity and, in many cases, extreme chemical aggressiveness pose specific challenges in cell design and the selection of appropriate materials. Reported experimental setups are consequently of only limited general applicability. Spectroelectrochemical and surface analytical methods as employed exclusively with molten salt electrolytes are not covered in this book. Fortunately the state of the art has been reviewed previously and the reader will find helpful introductory overviews elsewhere [8, 9]. Liquid/liquid interfaces between immiscible liquid solutions are not treated broadly for basically the same reasons. There have been a considerable number of reports on electrochemical investigations, but spectroscopic methods have been applied infrequently only. An overview has been provided elsewhere [10].

In the following sections spectroscopic methods and related surface analytical methods are treated in detail. They are grouped according to the probe employed by the method. Sometimes this organizational scheme failed and thus a few methods are collected at the end of this section. Most experimental setups reported in the literature are included. Their advantages and limitations are commented upon only briefly; in many cases judgments may differ, anyway. In order to enable the reader and potential user to get an idea of the essentials of the setup and its connection to a spectrometer, simplified and schematic drawings are provided. Detailed pictures containing even the most minute aspects (including sometimes the most obvious) are not included because these additions tend to detract from the essentials. Further information about details can be obtained from the original papers always quoted with the figures.

Growing interest in detailed information related to particular locations at an electrochemical interface (e.g. steps on a single crystal surface or particular features at a modified electrode surface) combined with instrumental developments have resulted in experimental methods providing localized information. In cases where this information is obtained point-by-point (sequentially), the procedure is called a scanning one and the method may be called mapping. In some cases (e.g. with focal plane

[1] The particular advantages of ionic liquids (extremely low vapor pressure, not combustible, physico-chemical properties can be adjusted easily by variation of mixing ratios, good solvation capabilities without specific coordination, broad electrochemical window between anodic/cathodic decomposition, good electrolytic conductivity, high thermal stability, broad temperature range of the liquid state) seem to imply a promising future for their application. These properties may nevertheless turn into serious drawbacks when separation of polar or ionic products from the liquid is required or when low solubility of frequently encountered reactands (methanol) or gases (hydrogen, oxygen) is becoming effective.

detectors in infrared spectroscopy), large parts of the surface under investigation can be imaged onto a detector built of numerous elements. This approach is called imaging.

General information about preparation, purification, etc. of solvents for electrolyte solutions is provided by Mann [11]. Suggestions for further reading are provided only where texts deemed suitable are available.

5.1 Optical Spectroscopy in the Visible Range

Fundamentals. Species involved in electrochemical reactions that show optical absorption caused by electronic transitions in the UV-Vis region of the electromagnetic spectrum can be studied *in situ* with a variety of spectroelectrochemical techniques. The choice of a suitable technique depends on the type of species to be investigated and upon the exact location of the species (e.g. directly adsorbed on the metallic electrode surface, incorporated in a film attached to the electrode surface, dissolved in the electrolyte solution). Basically, two families of spectroelectrochemical techniques have been established so far (see Fig. 5.1):

- Measurement in the transmission mode using optically transparent or semitransparent electrodes
- Measurement in the reflectance mode using either external specular reflection, internal reflection [attenuated total reflection (ATR)] or diffuse reflection

We will describe these techniques in detail. Methods where absorption of light causes emission of species, e.g. photons or electrical currents across the electrochemical interface, are treated in Sects. 5.1.6 and 5.1.9.

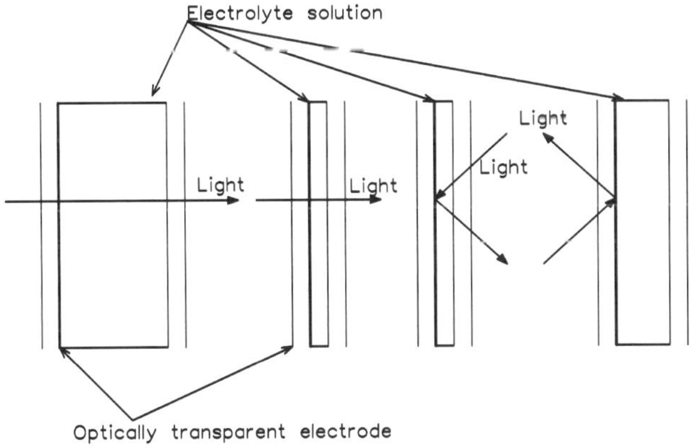

Fig. 5.1. Typical spectroelectrochemical arrangements for *in situ* measurements in the UV-Vis range; from left to right: standard arrangement; arrangement with thin layer of electrolyte solution (TLC) only; external reflection, internal reflection at the backside of the electrode

5.1.1 UV-Vis Spectroscopy with Optically Transparent Electrodes

Fundamentals. A species generated electrochemically or formed in an electrochemical reaction sequence by a chemical reaction with an electronic absorption in the UV-Vis range of the electromagnetic spectrum can be detected by means of its optical absorption provided that the actual extinction (based on the molar extinction coefficient and the concentration) is sufficient. Detection can be achieved very simply by putting the electrochemical cell into the beam of a suitable spectrometer. Species to be detected may interact with the light beam when the beam is passing in parallel just in front of the electrode surface; this requires tedious optical adjustment and is not commonly used. For a development with two-dimensional spectroscopy, see p. 44. A very simple approach is the passage of the beam perpendicular through the electrode and the electrolyte solution. This works only with electrodes that have a sufficient transparency for the used light. Very thin layers of noble metals, metal oxides or carbon will have this property. Fine metal grids or metal meshes that are either made from metal wires or micromachined in various ways (e.g. LIGA technique[2]) will have an optical transparency of zero where the wires are; between them, passage of the beam is unimpeded. Species formed at the transparent layers or at the grids can cause the desired optical absorption. The rather complicated transport processes of these three-dimensional systems can be simplified considerably when the time scale of the electrochemical experiments is long enough to allow the diffusion layer thickness to become larger than the openings in the electrode. In this case the diffusion process feeding the electrode reaction becomes one-dimensional. Further details of investigated electrochemical systems in particular with respect to the use of thin layer cells are provided elsewhere [12–14]. The evaluation of a typical thin layer cell with respect to electrochemical response behavior has been discussed [15].

Instrumentation. Various materials are optically transparent in the UV-Vis range of the electromagnetic spectrum while maintaining a considerable electronic conductivity. Thin sputtered or vapour deposited metal layers (e.g. gold, platinum) and indium-doped tin oxide (ITO) on glass are typical examples. Their absorption spectra depend on the thickness, degree of doping, state of oxidation and other variables. Typical absorption spectra are displayed in Figs. 5.2 and 5.3.

These coated glasses can be used as working electrodes [optically transparent electrodes (OTE)] in standard three-electrode arrangements provided that both glass and coating are chemically and electrochemically stable and inert in the used electrolyte solution and the applied range of electrode potentials. The use of a modified infrared spectroscopy transmission cell equipped with quartz windows for UV-Vis spectroelectrochemistry has been described [18]. Platinum layers deposited onto the quartz served as an optically transparent working electrode and an additional platinum layer served as a pseudo-reference electrode. A counter electrode outside the thin layer zone (in one of the tubes used for solution supply) served as a counter

---

[2] LIGA = lithographic-galvanic; this term refers to the manufacturing process applied for making microstructured honeycomb electrodes.

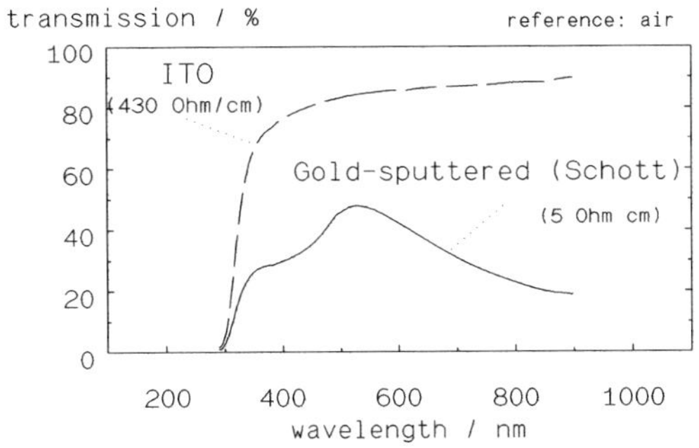

Fig. 5.2. Absorption spectra of ITO-coated glass and of a glass sheet coated with a gold sputter layer [16]

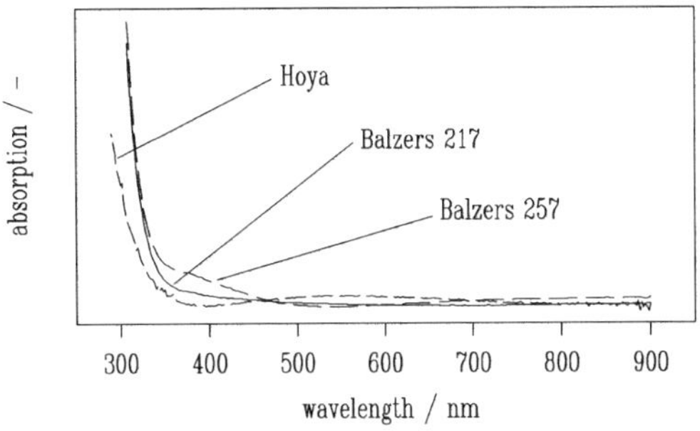

Fig. 5.3. Absorption spectra of ITO-coated glass with coatings of various electrical conductivities [17]

electrode. Chemical modifications of ITO surfaces have been discussed [19] and the electrochemistry of ITO has been investigated in various media [20]. Recently, free-standing synthetic diamond discs that were highly doped with boron have been used as transparent electrodes [21]. Transparent carbon films obtained by pyrolysis of diluted commercial photoresist spin-coated onto glass slides have been described [22]. Compared with ITO-coated glass, diamond shows a significantly higher chemical stability in various environments, has a fairly well-defined surface chemistry, withstands cathodic polarization (where ITO is reduced), has a wide electrochemical potential window and shows a small background current in this range.

The optical arrangement is fairly simple. The working electrode is mounted in an electrochemical cell, in most cases consisting just of a standard cuvet fitted with a tight lid providing feedthroughs for the electrodes and for a gas purge. As a counter electrode, a piece of metal wire (e.g. platinum or gold) is mounted in the cuvet without protruding into the light beam. A reference electrode of any type can be connected to the electrolyte solution via a salt bridge (e.g. a liquid-filled thin plastic tube closed with a porous glass or ceramic plug). The cell is arranged with the working electrode perpendicular to the light beam in the UV-Vis spectrometer; in the case of a two-beam instrument in the sample beam path. A cuvet filled with the

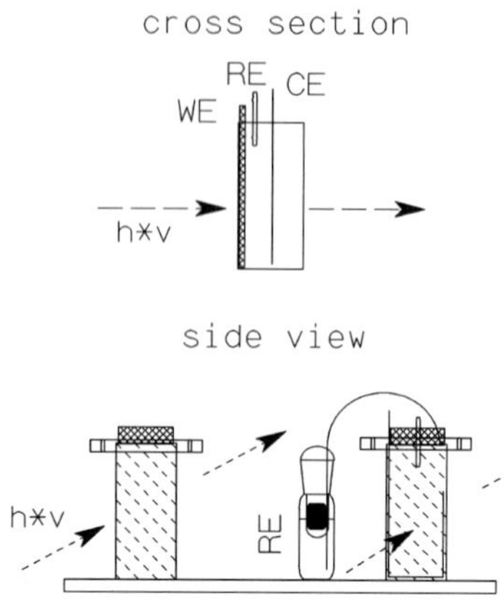

Fig. 5.4. Cell arrangement for *in situ* UV-Vis spectroelectrochemistry with optically transparent electrodes

electrolyte solution to be investigated and an electrode of the same kind as the working electrode (without connection to the potentiostat) are mounted in the reference beam. In a single-beam instrument, the desired UV-Vis-spectra showing changes of optical absorption as a function of electrode potential, time, etc., are obtained by using a defined spectrum (e.g. at an initial potential value, at the spontaneously established rest potential of the electrode, at time zero) as the background spectrum in the standard spectral calculations. A typical arrangement showing the major parts is displayed in Fig. 5.4.

Beyond numerous studies of soluble reaction intermediates and products (as an example see a study of electroreduction of nitrosobenzene [23] or investigations of chromium aryl complexes [24]) this design has also been employed successfully in studies of polymer films deposited onto these electrodes. Films showing redox activity and, in many cases, intrinsic electronic conductivity [intrinsically conducting polymers (ICPs)] have been studied; for an overview see [25, 26]. In a typical set of spectra (Fig. 5.5) obtained with a film of polyaniline, optical absorptions corresponding to the $\pi \rightarrow \pi^*$ transition (around 330 nm) and to further transitions involving species like radical cations (polarons) and dications (bipolarons) formed in the sequence of electrooxidation of the film are observed.

Solid materials showing electrochemical activity can be studied by mechanically attaching them as small particles to an ITO-electrode [27]. The electroreduction of indigo has been studied using this approach.

A considerably different approach employs a mini-grid electrode. This type of electrode can be made of a variety of metals (gold, platinum, etc.) and is basically a very fine mesh. Expanded metal-like structures, finely woven wires [28] or structures prepared via the LIGA-process [29, 30] are currently in use [31–33]. They are inserted into electrochemical cells of basically the same type as that used with OTE. The light beam passes the grid or the honeycomb structure of the LIGA element. Because the wires or the structural LIGA elements are extremely fine, only

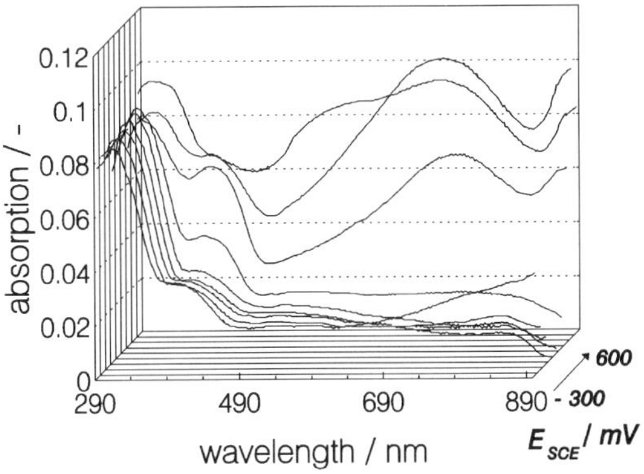

Fig. 5.5. UV-Vis spectra of a polyaniline film deposited on an ITO electrode in contact with an aqueous solution of 1 **M** $HClO_4$, electrode potentials as indicated

a tiny fraction of the light is stopped. The light passing the open parts will interact with species in the solution phase. Consequently, this method is not surface sensitive; rather, it is only sensitive to species present in the solution phase. Nevertheless, these species are essentially close to the electrode surface because of the fine structure. A cell design employing a completely sealed cell with connectors for electrolyte solution flow has been described elsewhere [34]. A cell suitable for work at low temperatures has been described by Hartl et al. [35]. Because carbon, as an electrode material, has some advantages over metallic materials (higher hydrogen overpotential, low chemical reactivity, corrosion resistance), graphite-coatings of metal mesh electrodes have been developed and employed [36, 37].

The length of the optical pathway through electrolyte solution-filled portions of the setup might cause problems when nonaqueous solvents with considerable light absorption in the investigated spectral range are used, when further optically absorbing species besides the species of interest (e.g. educts) are present at considerably concentrations, when electrolyte solutions of poor conductivity are employed or when complete electrochemical conversion of a species is desired. High absorbance might result in a poor signal-to-noise ratio in spectral ranges of low transmission, poor conductivity may cause a sluggish response of the electrochemical cell to rapid potential changes used e.g. in potential scan or potential step electrochemistry. A thin layer cell as depicted in Fig. 5.6 has been suggested [38]; for another design, see [39].

A Pyrex culture tube fitting into a standard 10-mm cuvet for UV-Vis spectroscopy is equipped with two holes. One hole is closed with a fused silica window permitting entrance of the light beam and a piece of NMR-tube closed with an ITO glass disc is fitted into the second one. The latter serves as a working electrode. Contact is made via a platinum wire touching the ITO surface that also serves as a spacer between the working electrode and the silica window. The counter electrode is a platinum wire loop. The cell has shown no distortions of cyclic voltammograms up to scan rates of 25 mV s^{-1}. A similar cell design suitable for NIR spectroscopy

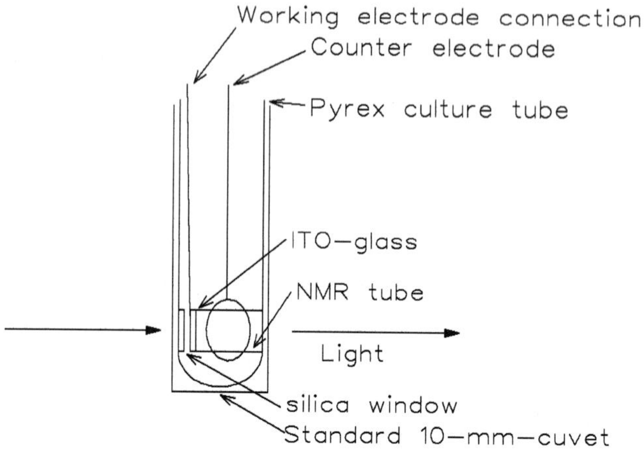

Fig. 5.6. Spectroelectrochemical cell for *in situ* UV-Vis spectroelectrochemistry with optically transparent electrodes [39]

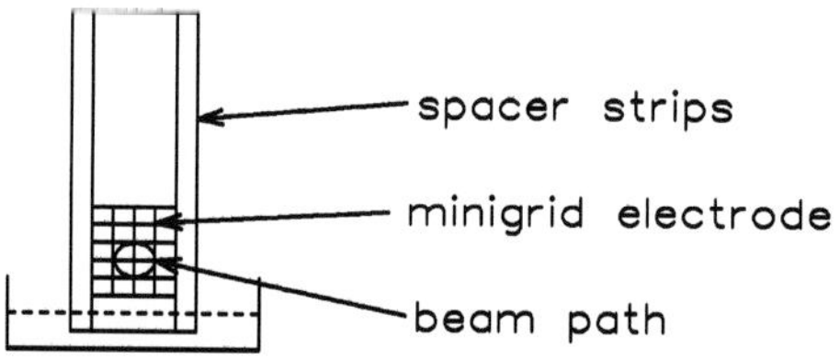

Fig. 5.7. Spectroelectrochemical thin layer cell for *in situ* UV-Vis spectroelectrochemistry with a mini-grid electrode [40]

will be discussed below. A considerably more simple design employing two glass sheets kept at a fixed distance by thin PTFE tape strips has been suggested [40, 41]. A metal wire grid mounted between the plates acts as a working electrode. The setup is immersed with its bottom edge into a beaker filled with an electrolyte solution. A metal wire counter electrode and a reference electrode are also immersed into this reservoir. The solution is sucked into the gap by applying a vacuum at the top edge and capillary action keeps the solution in place. With sufficiently hydrophilic glass surfaces, the capillary action takes place without additional suction. The light beam passes through the glass plate arrangement at the position where the grid working electrode is placed. Further simplification of the design is possible with glass plates (one or both of them coated with ITO) as the working electrode, as was initially suggested elsewhere [42]. For a schematic cell design of this type, see Fig. 5.7.

With only weakly absorbing solvents, low concentrations of the species to be detected or weak effects (i.e. small extinction coefficient), a long optical pathlength might be desirable. In a design described by Zak et al. [43], a cell body holding both the counter and the reference electrode is manufactured from PTFE. The working electrode is attached to the cell body with a spacer of about 0.1-mm thickness. The light beam of the spectrometer passes through the gap between the working electrode and the cell body via quartz glass windows. The pathlength is equivalent to the distance between the windows, i.e. the length of the working electrode (up 1–2 cm). The beam width is given by the thickness of the spacers and the distance between them (e.g. 0.1–7 mm). Very small reactand concentrations can be detected. A long optical path cell employing standard 10-mm cuvets has been described [44]. A gold-coated PTFE-block is positioned 0.5 mm above the bottom of the cuvet. A hole through the block provides access for the counter and reference electrode. The fairly large working electrode surface and the small active electrolyte volume enclosed by the cuvet and the PTFE-block (in the microliter range) allow fast conversion of reactants. A similar design employing optical fibres for coupling a spectroelectrochemical thin layer cell to a spectrometer has been described by Brewster and Anderson [45]. A cell design employing reticulated vitreous carbon and providing a long optical pathlength and simplified handling has been reported [46]. Cell designs for experiments with molten salts have been provided elsewhere [13, 47].

Recording optical absorption at selected wavelengths as a function of electrode potential or recording the first derivative ($\mathrm{d}A/\mathrm{d}E$ vs. E) are useful methods for identification of electrochromic sites in electroactive polymers [48, 49] and for calculating formal Nernst potentials E_0 [50, 51].

UV-Vis spectroscopy with TLC and OTE can be used beyond the identification of species to determine redox potentials, particularly with complicated systems showing multiple electron transfers [52].

5.1.2 External Reflectance Spectroscopy

Fundamentals. UV-Vis spectroscopic measurements with solid electrodes of a thickness beyond that of transparent films (e.g. metal or glassy carbon discs) are possible in the external reflection mode. A beam of light is directed through a window in the electrochemical cell and the electrolyte solution towards the electrode surface. This surface is polished in order to maintain a high reflectivity. The use of rough surfaces is also possible; techniques employing the diffuse reflectance observed in this case are treated in the following section (see p. 57). After a single specular reflection, the beam is directed through the electrolyte solution and the cell window towards the detector. Alternatively, multiple reflections are possible (see also p. 46). No influence of the electrooptical properties of the electrode material itself (e.g. electrode potential or wavelength dependent reflectivity) or of any modification on the electrode (e.g. a polymer film or a thin metal deposit) is assumed to be effective; these effects are studied with electroreflectance spectroscopy (see p. 56 for the former case and the previous section for the latter case). Measurements with films or layers deposited onto the electrode are also possible.

Instrumentation. A straightforward design of an electrochemical cell enabling simultaneous electrochemical and spectroscopic measurements (in both the UV-Vis and the NIR region[3] of the spectrum) with near normal incidence specular reflection has been described by Salbeck [53]. A smooth metal disc acting as a working electrode is pressed against an optically transparent cell window. The electrolyte solution layer of about 15 μm thickness thus formed establishes semi-infinite diffusion conditions. Light from the spectrometer source is guided towards the electrode surface at near normal incidence with fiber optics. The reflected light is picked up with a second fiber and piped to the detector. Combined with fast multichannel detectors (e.g. diode arrays) or intensified multichannel detectors, spectra can be acquired in a rapid sequence, enabling real time experiments for kinetic and mechanistic studies [54–56]. A setup that is similar in design and suitable in a broad range of temperatures (−55 to 80°C) and which has an external reflectance attachment (i.e. without fiber optics) has been described [57].

Besides cell designs that employ metals or other electrode materials in disc shape for external reflection spectroscopy, Robinson and McCreery have successfully employed cylindrical carbon fibers of 12 μm diameter [58, 59]. The carbon fiber was illuminated with a tunable dye laser. Scattered light was collected with fiber optics and guided to a photomultiplier detector. Because no thin layer arrangement and consequently poor electrochemical cell response were involved, fast experiments on a microsecond time scale were possible. Studies of polyaniline films deposited on platinum discs have been described [60].

A combination of transmission and external reflectance spectroscopy resulting in a cell for bidimensional[4] UV-Vis spectroelectrochemistry has been described [61]. With an optically transparent electrode (OTL), the schematic setup shown in Fig. 5.8 illustrates the different pathways of the light. One beam passes through the electrode and the electrolyte solution in front of it and the second beam passes only through the solution in front of the electrode close to it, guided strictly in parallel to the surface. Thus the former beam carries information pertaining to both the solution and the electrochemical interface (e.g. polymer films or other modifications on the electrode surface), whereas the latter beam carries only information about the solution phase. Proper data treatment enables separation of both parts. Identification of

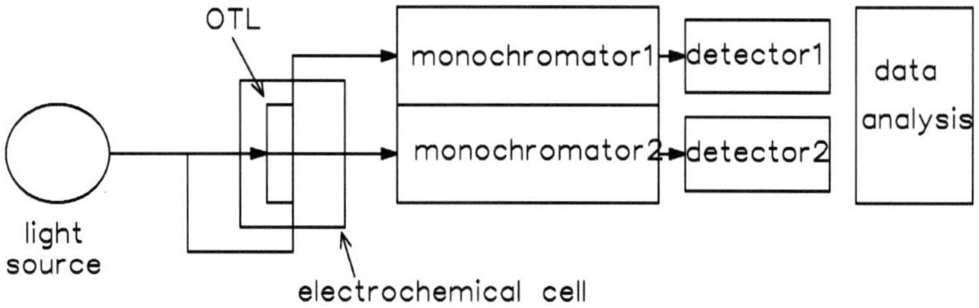

Fig. 5.8. Schematic setup of bidimensional UV-Vis spectroelectrochemistry (based on [62])

[3] The accessible section of the spectrum depends on the actual optical transparency of the employed components (cell window, fiber optics, etc.).

[4] In other spectroscopies, the term "two-dimensional" is substituted for "bidimensional".

insoluble species attached to the electrode and soluble species is easily possible and the setup has been applied to numerous systems (for a review, see [62]).

In a modified setup using a solid platinum electrode modified by depositing an intrinsically conducting polymer, one beam of light impinges on the polymer/electrode surface at almost normal incidence and the second beam is passed along the electrode surface through the electrolyte solution. As a function of the electrode potential, two different UV-Vis absorption spectra are acquired. Because different locations in the electrolyte solution and, in case of the former beam, the deposited polymer are probed, complex processes can be elucidated. In the case of electrochemically controlled release of hexacyanoferrate from a film of polypyrrole, the simultaneous release of pyrrole oligomers could be verified [62].

5.1.3 Attenuated Total Reflectance Spectroscopy[5]

Fundamentals. A plane wave of electromagnetic radiation (i.e. light) impinging upon an interface between two media of different optical density from the medium with the higher refractive index $n_p = \sqrt{\varepsilon_p}$ will be reflected totally (internally) provided the angle of incidence (see Fig. 5.9) is larger than a critical value θ_c [63]. At smaller angles most of the light is transmitted and the reflectivity decreases abruptly (see Fig. 5.10). The actual value of θ_c depends on the refractive indices n of both media according to Snell's law $\sin\theta_c = n_d/n_p$ with d indicating the more dense medium. The intensity of the electromagnetic wave along the z-axis decays at values of θ above θ_c exponentially with a decay length $l = \lambda/(2\pi(n\sin\theta)^2 - 1)$ (see Fig. 5.9) and the wave extending beyond the phase boundary is called an evanescent wave. Only the interfacial range is illuminated and only within this range can light and absorbing species interact. The intensity of the incoming light I_{in} is enhanced by a factor of about 4 at θ_c (Fig. 5.11).

Instrumentation. A setup employing an ITO-coated glass sheet as the working electrode and ATR element as shown in Fig. 5.12 has been proposed for studies of thin films deposited on the ITO layer [64]. Light is piped through an optical fiber to

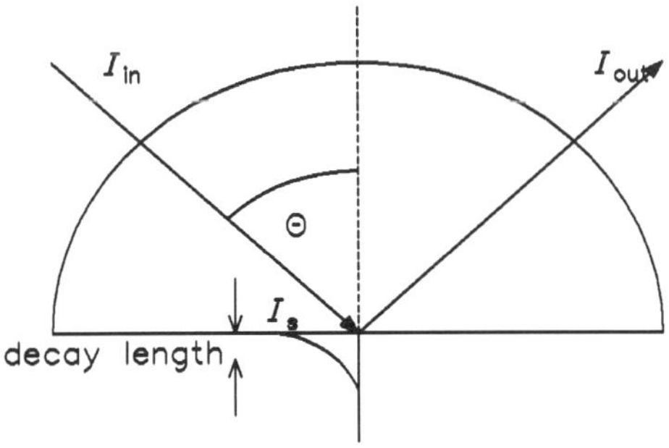

Fig. 5.9. Schematics of total internal reflection; for details, see text

[5] For further details, see the corresponding section in Sect. 5.1.2 on infrared spectroscopy.

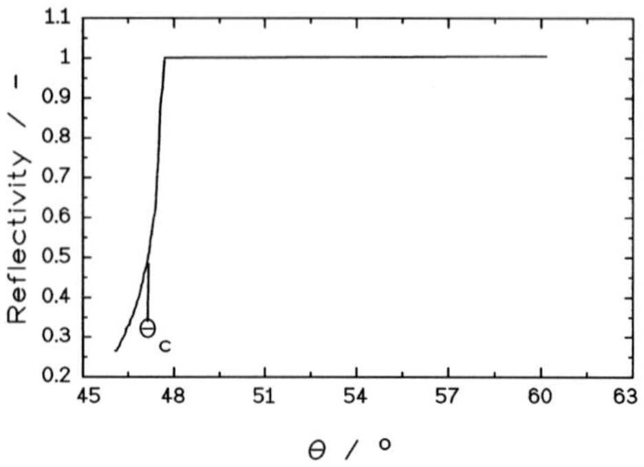

Fig. 5.10. Reflectivity I_{out}/I_{in} as a function of the angle of incidence θ calculated for ε_p = 3.4036 and $\varepsilon_p = 1.778$ (based on data in [64])

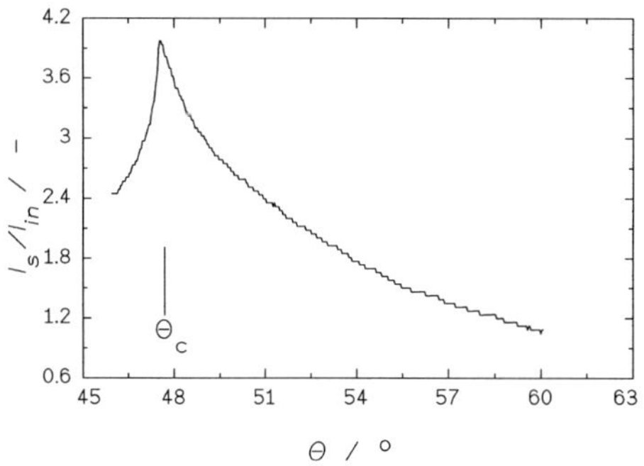

Fig. 5.11. Intensity enhancement I_s/I_{in} at the interface ($z = 0$) as a function of the angle of incidence θ (based on data in [64])

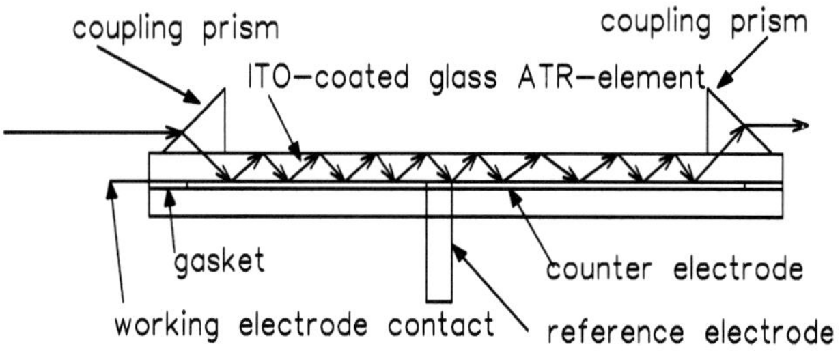

Fig. 5.12. Spectroelectrochemical ATR cell for *in situ* UV-Vis spectroelectrochemistry [65]

a prism that couples the light into the ATR element. A second prism guides the light into an optical fiber connected to the spectrometer.

Spectral information is gathered from species within the range of the evanescent wave penetration length at the ITO-solution interface. The reported cell design allows measurements with polarized light. Results reported so far deal with neutral copper phthalocyanine bilayers [65]. The use of colloidal gold nanoparticles deposited onto ITO as a working electrode has been described [65].

5.1.4 Luminescence Spectroscopy

Fundamental. Luminescence is the emission of electromagnetic radiation, particularly light in the visible range of the electromagnetic spectrum, which is effected by a supply of energy. Numerous forms of luminescence caused by a host of different forms of energy have been observed. Chemiluminescence is the emission of light by energetically excited species formed in the course of a chemical reaction. For example, the reacting species may be radicals generated by electrolysis [66]. In this case it may be called more specifically electrochemiluminescence (ECL). Photoluminescence is caused by the absorption of light, see Sect. 5.1.6. When the emission of light (luminescence) follows rapidly after absorption, the phenomenon is generally called fluorescence [67] (see p. 48). Delayed emission is called phosphorescence. An introduction to the luminescent properties of semiconductors in contact with electrolyte solutions has been provided elsewhere [68].

Instrumentation. A cell design employing reticulated vitreous carbon as the working electrode material that enables both UV-Vis absorption and luminescence measurements has been described [47]. A thin-layer cell with a platinum working electrode has been developed [69]. The luminescence of the electrooxidation products of *o*-tolidine as a function of electrode potential was studied. A simplified flow cell design has been reported [70]. Luminescence spectra and fluorescence intensity[6] for various aromatic compounds and their electrochemical and photochemical reaction products were observed as a function of flow rate, current and time after the potential step. In the latter study the electrooxidation of *p*-phenylenediamine (PPD) was examined. The cyclic voltammogram showed two oxidation peaks; the first one is assumed to be caused by the formation of the radical cation according to

$$\mathrm{PPD} \rightarrow \mathrm{PPD}^{+} + \mathrm{e}^{-}$$

Upon excitation with light of about $\lambda = 245$ nm, a luminescence peak of the parent compound at $\lambda = 391$ nm (trace (a) in Fig. 5.13) was observed. During electrooxidation a different spectrum with a peak around $\lambda = 342$ nm (trace (b) in Fig. 5.13) was seen. The latter spectrum was initially assigned to the radical cation, but no agreement between experimental data and this assumption could be reached. In a second step it was assigned to the dication formed according to

$$\mathrm{PPD}^{+} \rightarrow \mathrm{PPD}^{2+} + \mathrm{e}^{-}$$

This assumption was discarded when experiments with an electrode potential beyond the second voltammetric oxidation peak where the dication is generated

[6] The terms fluorescence and luminescence are unfortunately somewhat mixed up in this report.

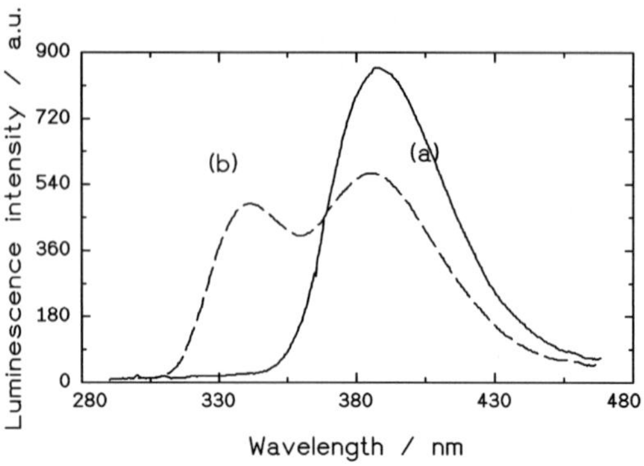

Fig. 5.13. Luminescence spectra of a solution of 0.777 m**M** *p*-phenylenediamine in acetonitrile with 0.1 **M** tetra-*n*-butylammonium perchlorate; (*a*): no electrode potential applied; (*b*): electrode potential positive to the voltammetric peak where the radical cation is generated applied (based on data in [71])

directly via electrooxidation yielded essentially the same spectrum. Based on evidence obtained from literature, the spectrum was finally assigned to a dimeric cation species formed according to

$$2\mathrm{PPD}^+ \rightleftarrows (\mathrm{PPD}^+)_2$$

The photoelectrooxidation of bis(benzylidene)acenaphthene has been studied by Compton et al. [71] using the cell described above [71]; a highly fluorescent product was identified.

Photoluminescence data of numerous semiconductors in contact with various electrolyte solutions have been reviewed [69].

Luminescence generated by chemical reactions [chemiluminescence (CL)] that, in turn, are initiated by electrogenerated species [electrogenerated chemiluminescence (ECL)] has also been studied in simple setups, including flow cells. Reported examples include ECL of acridinium esters reacting with peroxide formed upon reduction of dissolved dioxygen in solution [72]. Electrogenerated chemiluminescence may simply be used to report the rate of electrochemical reactions. As an alternative, the use of light-emitting diodes suitably coupled to the working or counter electrode to measure electrochemical activity without the need to make actual current measurements [73] has been proposed.

5.1.5 Fluorescence Spectroscopy

Fundamentals. Fluorescence is the spontaneous emission of light after excitation of a species with light (see also p. 47). Fluorescence spectra of molecules provide valuable information about structural features [74].

Instrumentation. A universal cell design also useful here has been reported [75, 45]. It is similar to the cell design previously described by Salbeck [76]. The disc-shaped electrode under investigation, which is embedded in an inert sheath, is situated closely behind an optically transparent window. Its surface and the electrolyte solution volume immediately in front of the electrode are illuminated with exciting light from a monochrome source using a fiber optic cable at an angle of approx. 45°. Fluorescence is detected with a second fiber optic cable mounted perpendicularly in front of the window and connected to a fluorometer. The feasibility of fluorescence spectroelectrochemistry has been demonstrated using perylene dipentylimide [76]:

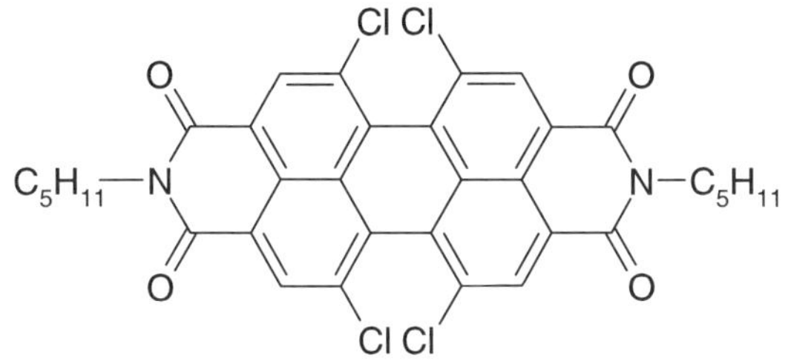

In its neutral state the molecule shows strong fluorescence with bands at 425, 480 and 520 nm when excited with light of $\lambda_0 = 480$ nm. Cyclic voltammetry showed two reversible one-electron reduction processes (Fig. 5.14).

The fluorescence of the molecule disappeared during the first reduction process; this could not be evidenced at the low concentration of the reactand with UV-Vis absorption spectroscopy. The potential-dependent fluorescence of various 5-substituted indole trimer films has been investigated [77].

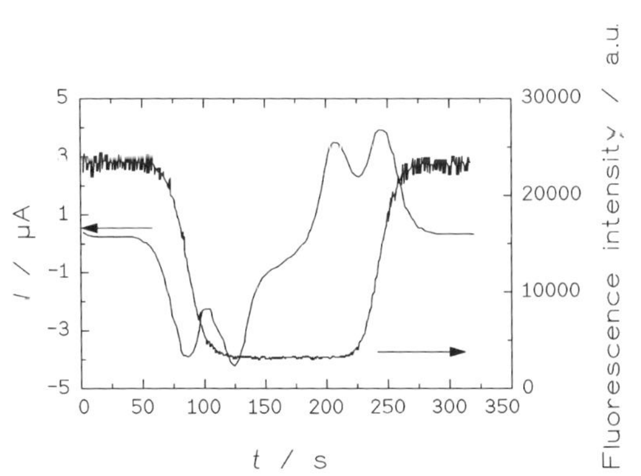

Fig. 5.14. Cyclic voltammogram and fluorescence intensity plotted vs. time of perylene dipentylimide at 0.1 m**M** concentration in a solution of 0.5 M TBAHP in methylene chloride (based on data in [76])

Potential-modulated fluorescence spectroscopy at liquid/liquid interfaces between immiscible liquids has been reported and a cell design has been provided [78]. The dependence of the adsorption of the free, bare or the water-soluble porphyrins at the polarized water/1,2-dichloroethane interface has been studied [79]. Observed spectral differences suggest a solvation structure at the interface that is different from that inside the bulk of the respective solution phases. For further studies with related porphyrins at the same interface, see [79]. Details of the transfer mechanism of the rose bengal dianion across the water/1,2-dichloroethane interface have been elucidated [80].

5.1.6 Electroreflectance Spectroscopy[7] (ERS)

Fundamentals. Beyond probing species that are more or less strongly absorbing in solution or that are adsorbed/attached to the electrode surface UV-Vis spectroscopy can be used to study electrooptical properties of the electrode material itself. (For a thorough introduction, see [81–83]; the fundamentals have been introduced elsewhere [84, 85].) The complex refractive index of a medium containing its refractive index and its extinction coefficient can be used in good approximation to describe the wavelength-dependent response of a medium to an impinging beam of electromagnetic radiation. Further consideration of the complex refractive index leads to the complex dielectric function. Contributions from free and bound electrons as well as from plasmon excitations result in frequency dependencies of the real and the imaginary part of the dielectric constant typical for any given material. The reflectivity of an interface can be calculated using Fresnel's equations. With polarized light (as most commonly used in this type of spectroscopy), the reflectivities for light polarized perpendicular or parallel to the plane of incidence are characteristically different depending again on the wavelength. They can be derived by taking into account the strength of the electric field vectors of the incoming and the reflected beam at the reflecting interface for various angles. In the infrared region, most metals are almost perfect reflectors; only parallel polarized light will interact with species attached to the metal surface (for surface selection rules, see Sect. 5.2). Because of the more complicated situation in the UV-Vis-range, interaction (i.e. absorption) between light and species at the interface may occur even at normal incidence. As already pointed out before, the interface solution/electrode is actually composed of three phases:

1. The bulk solution
2. The bulk electrode
3. The interphase

Thus a complete description of the system under study has to take into account properties of these three phases. Optical properties of the interphase can be obtained from measurements of the reflectivity as a function of wavelength; with systems showing no pronounced interphase, the optical properties of the metal can be studied.

[7] Initially, this method was called the electrolyte electro reflectance (EER) technique.

Instrumentation. The measurement of absolute reflectance values for a given interface is difficult because numerous contributions of components in the optical beam path will contribute. It is more convenient to measure relative changes of reflectivity $\Delta R/R$ as a function of wavelength (or energy). This result is directly related to $\Delta I/I$, with I being the intensity of light. Typical values of $\Delta R/R$ are 10^{-1} to 10^{-5}. The change in reflectivity (i.e. in intensity) has to be effected by some well-defined electrochemical manipulation. Most frequently this is a modulation of the electrode potential. Combined with lock-in amplification, the necessary signal-to-noise ratio can be obtained. Measurements in a range of electrode potentials wherein no faradaic reactions occur and the change in reflectivity is assumed to be caused directly by the change of the modulated electrode potential are called electroreflectance spectroscopy (ERS). The experimental setup includes a light source (frequently a xenon lamp) coupled with a monochromator and a polarizer. The light is guided to the interface under investigation at a selected angle of incidence; the reflected light is detected by a photomultiplier. The signal from the photomultiplier is processed using the lock-in amplifier, yielding the desired electroreflectance spectrum (for a typical setup, see [86]).

The design of spectroelectrochemical cells is strongly simplified by the fact that the most commonly used electrolyte solutions are highly transparent in the UV-Vis range. Cells with the working electrode placed in the center of a cylindrical cell body provide perfect electrochemical response; the use of bent quartz windows permits investigations at various angles of incidence by just rotating the cell. Measurements with single crystal electrodes in the dipping (hanging meniscus) technique are also possible. Typical examples are shown in Fig. 5.15 and a review of experimental setups is available [84].

The spectral range accessible for measurements is limited by the transmission properties of the electrolyte solution and the optical components, the intensity distribution of the used light source and the spectral sensitivity of the detector to values of about $220 < \lambda < 1400$ nm.

The broad variety of investigated systems includes single- and polycrystalline metal electrodes (for an overview, see [84]), adlayers of metal atoms, ions and molecular species and films like metal oxides of nanometric thickness. Overviews of studies of anodic oxides (including corrosion, passivation and electrochromic layers) on various metals have been provided [84, 87]. Thick corrosion layers on 304 stainless steel have been investigated [88]. Polarization anisotropy with a polycrystalline gold electrode was observed by Zhao et al. [89] and data pertaining to a Ag(110) electrode surface have been presented and reviewed in a broader context by Furtak and Lynch [90, 91]. Semiconductor/electrolyte solution interfaces have been studied by Shaklee et al. [92]. In a study of the n-GaAs/electrolyte solution interface, changes in the potential distribution in the semiconductor depletion layer were identified [93]. The possible advantages of a simultaneous (double beam) application of electroreflectance spectroscopy and second harmonic generation have been pointed out [94].

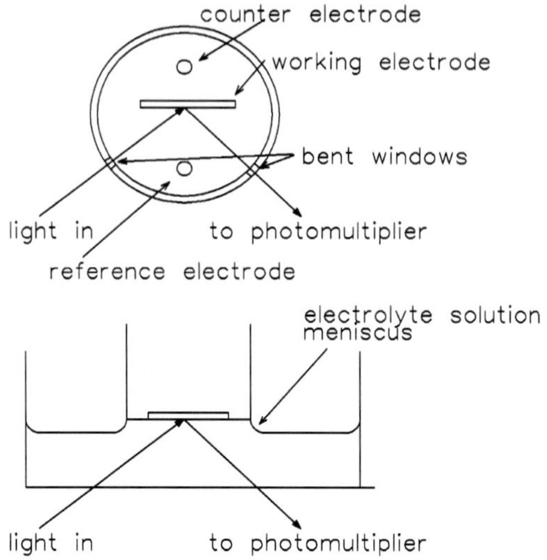

Fig. 5.15. Schematics of spectroelectrochemical cells for electroreflectance spectroscopy. *Top*: Arrangement for measurements at various angles of incidence; *bottom*: Cell for measurement with electrodes in the dipping technique

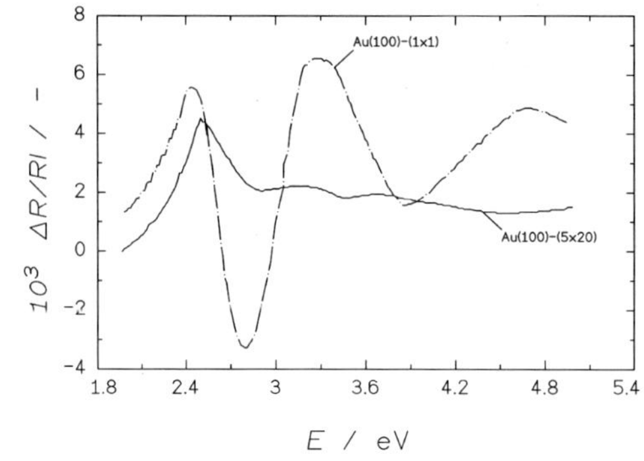

Fig. 5.16. Normal incidence electroreflectance spectra of two reconstructed Au(100) surfaces in contact with a solution of 0.01 **M** $HClO_4$, potential step from $E_{SCE} = -0.2$ V to 0.3 V (based on data in [98])

The optical properties of a Au(100) surface in its reconstructed state differ markedly because of the participation of electronic surface states in the optical excitation; these states depend on the crystallographic surface structure [95]. Surface band structure calculations have revealed the existence of empty surface states [96]. These surface states can be shifted in their energy by the electrode potential (Stark shift) [97]. Optical transitions into these states thus become potential dependent.

In Fig. 5.16 the electroreflectance spectrum of the Au(100)–(1 $\times$ 1) surface shows two derivative-like features around 3 and 4.2 eV assigned to transitions from the bulk d-band into the aforementioned unoccupied surface states [98]. These fea-

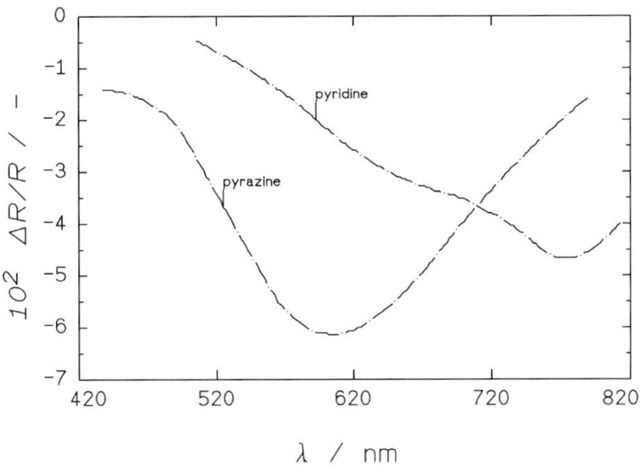

Fig. 5.17. Reflectance spectra of Ag(111) single crystal electrodes in contact with a solution of 0.1 **M** KCl, potential step from $E_{SCE} = -0.2$ V, reference spectrum recorded before oxidation-reduction cycle (based on data in [98])

tures are absent in the ER spectrum of a Au(100)–(5 × 20) surface. This difference can be used to monitor structural transitions between these reconstructions. The influence of the surface anisotropy of single crystal silver electrodes has been studied with ERS [87]; for an introductory overview, see [84]. Since the electric field vector of the impinging light is perpendicular to the (111) direction, the effect is twice as large as that of the orientation parallel to this direction. A somewhat less metallic character of the surface, when it is oriented in the former direction, can be deduced from surface plasmon dispersion curves measured with an air gap.

Optical properties of gold nanorod arrays have been studied and were found to be dominated by surface plasmon modes superimposed on interference effects [99]. The formation of a surface complex composed of molecules (e.g. pyridine or pyrazine) strongly bound to silver formed during oxidation-reduction electrode potential cycles is evident in ER spectra shown in Fig. 5.18. They show pronounced absorptions not found with the respective molecules in solution.

This approach works also with adsorption/desorption-induced changes of reflectivity, provided the sorption kinetics are fast enough to follow the potential modulation. In this case absorption features in the obtained spectra are attributed to changes in coverage (no influence of the electrode potential is presumed). Precisely speaking, these spectra should not be called electroreflectance spectra (see also [88] concerning this classification).

The derivative-like feature around $\lambda = 600$ nm in Fig. 5.18 indicates an electrode potential induced shift of this absorption.

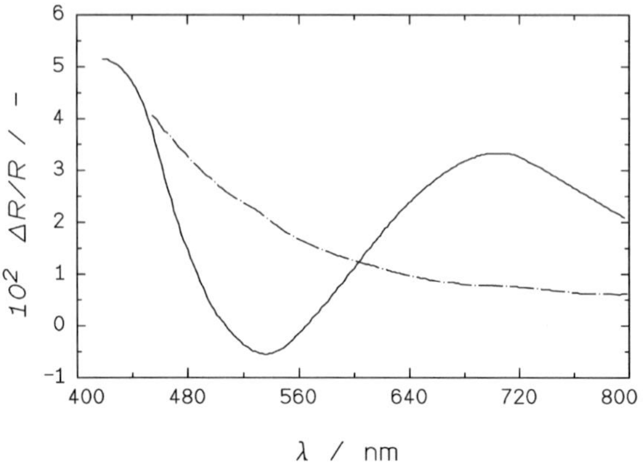

Fig. 5.18. Electroreflectance spectra of a Ag(111) single crystal electrode in contact with a solution of 0.1 **M** KCl + 0.05 **M** pyrazine before (- - -) and after (——) oxidation-reduction cycles, potential step from $E_{SCE} = -0.5$ V to 0.0 V (based on data in [83])

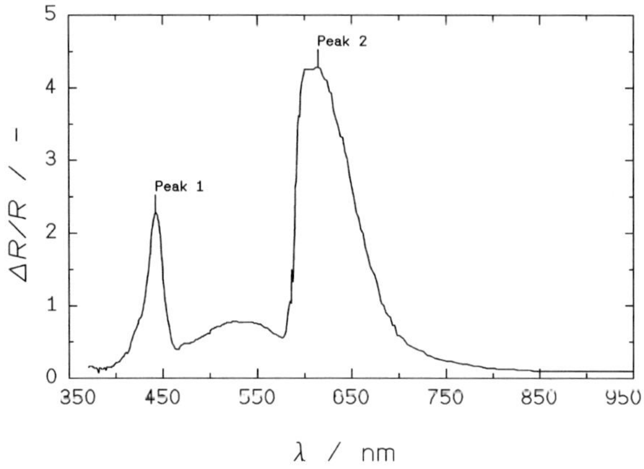

Fig. 5.19. Electroreflectance spectra of a polycrystalline platinum electrode in contact with a solution of 0.5 **M** $Na_2SO_4 + 5 \times 10^{-5}$ **M** *p*-nitroaniline, $E_{MSE} = 440$ mV, $\Delta E = 100$ mV, (based on data in [101])

The structure of the electrochemical double layer in the presence of adsorbed molecules can be investigated with ERS.[8] In a typical study, Schmidt and Plieth [100] have investigated the adsorption of *p*-nitroaniline on a polycrystalline platinum electrode from sulfuric acid solution. The ER spectrum as displayed in Fig. 5.19 shows two distinct bands when light polarized parallel to the plane of reflection is used (with light polarized perpendicular, only a flat baseline was found).

[8] This should not be confused with UV-Vis spectroscopy of dissolved species in the electrolyte solution more or less close to the surface, but not necessarily adsorbed onto the surface, as discussed above.

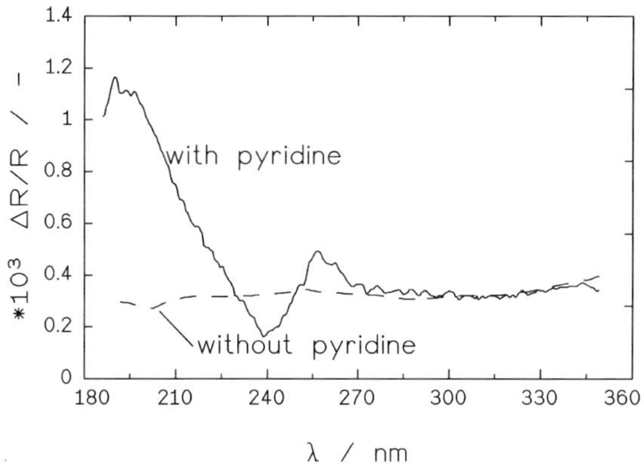

Fig. 5.20. Electroreflectance spectra of a Au(110) electrode in contact with an aqueous solution of 0.1 **M** $NaClO_4$ and 10^{-3} **M** pyridine, potential modulation between $E_{SCE} = -0.3$ V and 0.1 V, modulation frequency 18 Hz (based on data in [102])

Peak 1 is assigned to the $\pi \rightarrow \pi^*$ transition of the adsorbed molecule, peak 2 is caused by intermolecular interactions in the adsorbate layer. A perpendicular orientation of the molecule with the nitro-group interacting with the platinum surface was concluded.

The adsorption of pyridine on a Au(110) electrode surface has been studied with ERS [101]. Spectra obtained in the absence and in the presence of pyridine in solution show marked differences (see Fig. 5.20). The electrode potential was modulated between two values ($E_{SCE} = -0.3$ V and 0.1 V) where the coverage was constant (i.e. in the flat range of the adsorption isotherm). Thus the ER spectra do not show any coverage effect; rather, they show only electrode potential induced changes of the adsorbate layer. In the absence of pyridine, a rather flat spectrum is obtained. With pyridine in solution, the spectrum shows a derivative-like feature around 257 nm and a further absorption around 197 nm collides with the onset of water absorption.

Both bands are caused by electronic transitions (absorptions) of pyridine. The derivative-like structure indicates a small electrode potential induced shift of the transition energy assigned to a Stark effect. From the polarity of the lobes (when going from longer to shorter wavelength: first an increase, then a decrease in reflectivity) a redshift can be concluded. The orientation of pyridine adsorbed on a Au(110) electrode surface has been deduced from ER spectra measured subsequently with light of p- and s-polarization (see Fig. 5.21) [102]. When the plane of polarization is switched from p to s the band at 257 nm shows a dramatic increase. This band is z-polarized and is associated with the molecular dipole moment along the molecular axis intersecting the nitrogen atom and the *p*-C atom. Consequently, the molecule is adsorbed in a perpendicular orientation as already tentatively deduced from the comparison of spectra obtained with p-polarized light in the presence/absence of

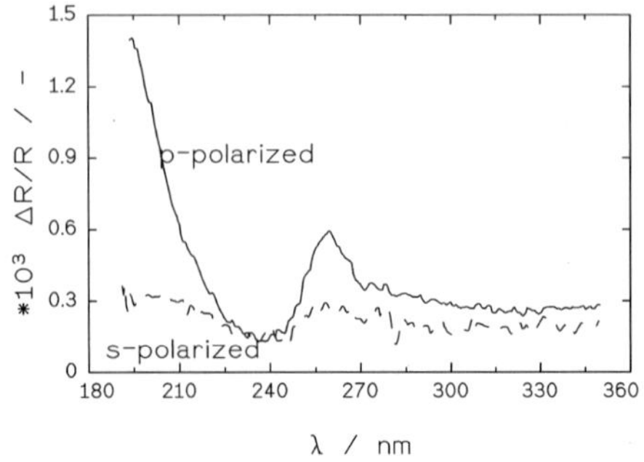

Fig. 5.21. Electroreflectance spectra of a Au(110) electrode in contact with an aqueous solution of 0.1 **M** $NaClO_4$ and 10^{-3} **M** pyridine, potential modulation between $E_{SCE} = -0.3$ V and 0.1 V, modulation frequency 18 Hz (based on data in [102])

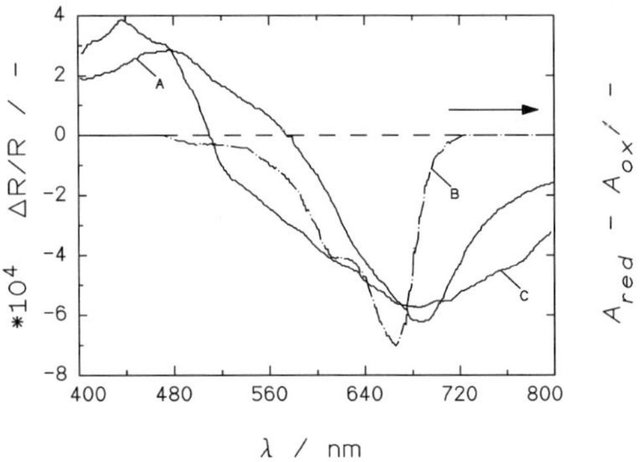

Fig. 5.22. Electroreflectance spectra of a polycrystalline gold electrode in contact with a buffer solution; A and B: absorption spectra of MB_{red}; C: after electrode transferred from MB-containing solution into solution without MB, $E_{Ag/AgCl} = -190$ mV, $\Delta E = 80$ mV; (based on data in [104])

pyridine in solution (see p. 55). A similar study pertaining to the Au(100) surface has been reported [102].

In a study of the redox spectroelectrochemistry of methylene blue (MB),[9] ER spectra as displayed in Fig. 5.22 were recorded [103].

The bipolar structure of the ER spectrum (A) cannot be explained by considering the optical absorption of MV in its reduced and oxidized forms. The reduced form (leuco-MB) is colorless. The spectrum displayed in Fig. 5.22 (trace B) shows only a single-sided band attributed to the oxidized form MB_{ox}. Thus the ER spectrum A is not caused simply by species in solution phase in front of the electrode. This assumption is confirmed in spectrum C, which was obtained after transfer of the gold electrode from the MB-containing solution into a solution containing no MB. The observation of an ER spectrum implies that irreversibly adsorbed MB is still present

[9] MB = 3,7-bis(dimethylamino)phenothiazin-5-ium chloride.

on the gold surface and its surface state differs from the state in the presence of MB in solution as indicated by the changes in the observed ER spectra A and C. Redox properties of polypyrrole-modified electrodes doped with metalloporphyrines have been studied [104]. Results permitted the localization of the proceeding redox process and determination of their redox potential. Gold nanorods embedded in a porous alumina matrix have been investigated with simple reflectance measurements and ERS [105]. Depending on the plane of polarization of the incident light, the transverse or both the transverse and the longitudinal plasmon mode could be identified.

With slow kinetics of the involved processes or with interfaces where electrode potential modulation might be detrimental because of crystallographic changes in the metal surface, other spectroscopic techniques have to be used. The whole spectrum of interest can be scanned or registered within a few milliseconds with a rapid scan spectrometer or a multichannel (diode array) spectrometer. Repeated acquisition provides the required signal-to-noise ratio. After a potential step, the acquisition is repeated and spectral calculation yields $\Delta R/R$. This single potential step procedure allows investigation of systems where repeated potential modulation has failed.

Using pyrolytically prepared graphite films, the intercalation of Li^+ and K^+ has been monitored with ERS [106]. For a review on ERS, see also [107].

5.1.7 Diffuse Reflectance Spectroscopy

Fundamentals. Measurement of the radiation diffusively (i.e. not specularly) reflected from a surface as a function of wavelength results in a diffuse reflectance spectrum. This method is applied frequently in the various ranges of the electromagnetic spectrum for investigations of powders or other poorly reflecting substances (see also Sect. 5.1.2). Combining a fiber optics spectrometer with an optical microscope, this method can be applied to spectroelectrochemical investigations of small amounts of electrochemically active colored materials deposited on a soft electrode surface by mechanical embedding (abrasive stripping voltammetry technique) [108]. Results will be displayed as Kubelka–Munk functions of the wavelength.

Instrumentation. A suitable electrochemical cell is displayed in Fig. 5.23. The material to be investigated is embedded in the front surface of the working electrode (e.g. a soft graphite electrode). Visual inspection of the surface is feasible via an optical microscope attached to the cell bottom via a transparent window. Illumination of the surface is affected by a halogen lamp incorporated in the microscope. The working electrode is connected to a potentiostat, counter and reference electrode, yielding the standard three-electrode arrangement. The fiber optics spectrometer is also coupled with the microscope and great care must be exercised in order to keep specularly reflected light from the microscope illumination lamp from entering the spectrometer.

Figure 5.24 shows the Kubelka–Munk function of silver octacyanomolybdate(IV) and (V). Upon oxidation of the molybdenum ion, a silver ion is released into solution and the optical absorption of the solid compound increases drastically [110].

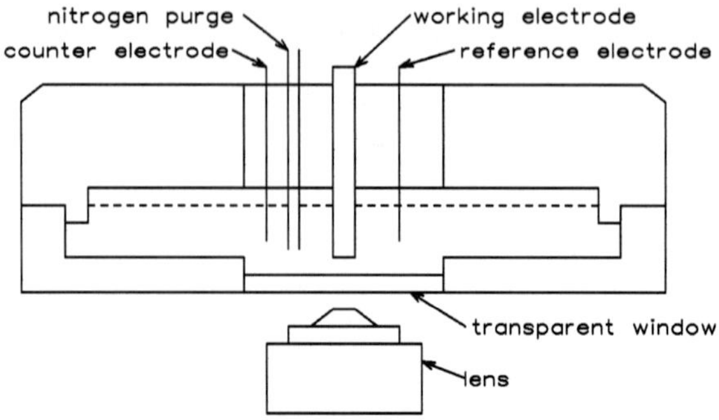

Fig. 5.23. Spectroelectrochemical cell for diffuse reflectance spectroscopy [109]

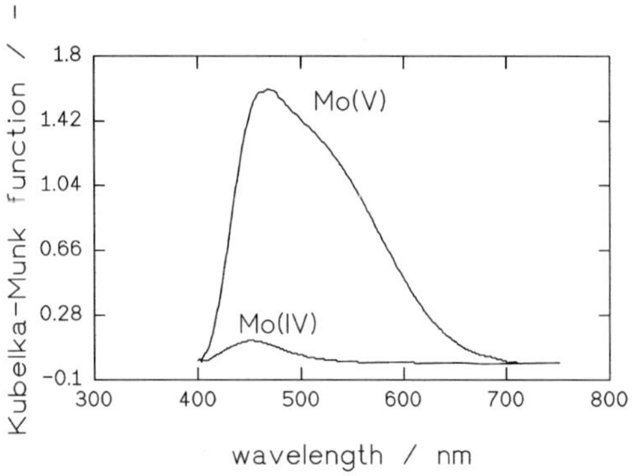

Fig. 5.24. Kubelka–Munk function of silver octacyanomolybdate(IV) and (V) immobilized on a graphite electrode in contact with an aqueous solution of 0.1 **M** $AgNO_3$ [110]

The change of the oxidation state of the central ion can be effected electrochemically, as evident in the cyclic voltammogram displayed in Fig. 5.25. The corresponding change of reflectance indicating the change of the Mo(IV) complex into the Mo(V) complex is included, too.

The considerable hysteresis between the reflectance values recorded in the cathodic- and anodic-going scans already implies more complicated features of the ongoing process beyond a mere change of the state of oxidation. Displaying the first derivative thus provides further information about kinetics and topology of the electrochemical processes.

5.1.8 Reflection Anisotropy Spectroscopy[10]

Fundamentals. When the reflectivity at normal or near normal incidence is measured relative to various crystallographic directions, anisotropies may be observed.

[10] This method is sometimes called reflectance difference spectroscopy (RDS) and, because of considerable overlap, this method is sometimes also considered to be a variation of electroreflectance spectroscopy (see p. 50 for further details).

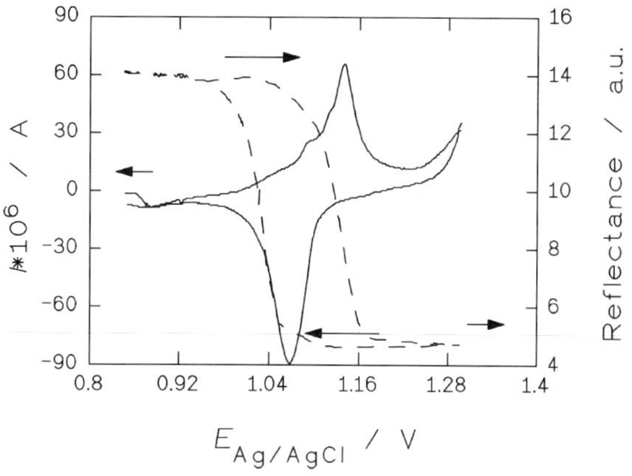

Fig. 5.25. Cyclic voltammogram of silver octacyanomolybdate(IV) and (V) immobilized on a graphite electrode in contact with an aqueous solution of 0.1 **M** $AgNO_3$ at a scan rate of 1 mV/s; change of the reflectance at $\lambda = 550$ nm (*dotted line*) [110]

For a bulk isotropic crystal, this reflects anisotropies of the surface optical response. Various theories and models describing the physical fundamentals of this observation have been reported; for an overview, see [110]. For a silver single crystal, a simple explanation was proposed [111] for the differences between the dispersion relations of surface plasmons on different crystal faces and along the two principal symmetry directions on Ag(110). A relationship between the energy of the surface plasmon and the packing of silver atoms was found. This was explained by assuming that the free electrons have a smaller optical effective mass along the denser surfaces. This is closely related to different free-electron conductivities along the crystallographic orientations. Based on a simple examination, this explanation was found to be insufficient [111]. Further developments, including interband and intraband electronic transitions [112, 113], have been described [111].

Instrumentation. An RDS setup (as commercially supplied [111, 114]) is equivalent to a normal incidence ellipsometer. It directly delivers the real and the imaginary parts (or their ratio) of $r_{1\bar{1}0}$ and r_{001}. These are the complex reflectances along the respective crystallographic directions.

Results are displayed as

$$\mathrm{Re}\left\{\frac{\Delta r}{r}\right\} = \mathrm{Re}\left\{\frac{r_1 - r_2}{r}\right\}$$

with the indices referring to orientation of the light detection with respect to crystallographic orientation. Sometimes the real and the imaginary parts are displayed separately.

Various possible optical configurations have been compared elsewhere [115]. In order to record reflectance-difference spectra beyond a light source (e.g. a Xe

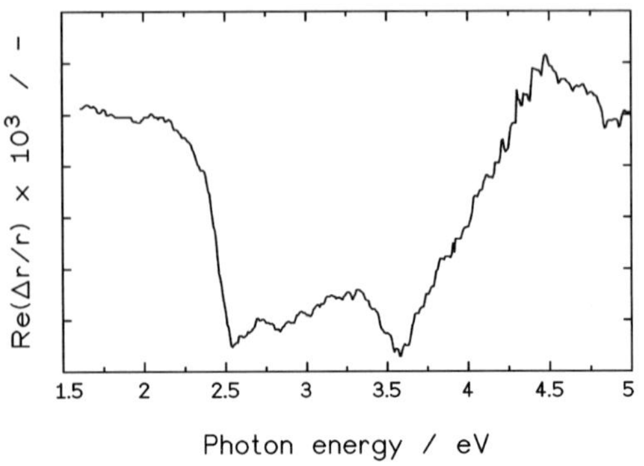

Fig. 5.26. Reflectance anisotropy spectrum of a Au(110) surface in contact with an aqueous solution of 0.1 **M** Na_2SO_4 at $E_{Ag/AgCl} = 0.6$ V (based on data in [119])

lamp), a monochromator (if wavelength dependencies are to be studied), a detector (a photomultiplier), a polarizer and a photoelastic modulator or a rotating polarizer are needed [116]. A multichannel spectrometer design-enabling rapid acquisition of RA spectra has been described [117].

RA spectra obtained with a Au(110) surface exposed to an aqueous solution of 0.1 **M** Na_2SO_4 [118] are displayed in Fig. 5.26.

At the applied electrode potential, the gold surface shows the 1×1 arrangement. The observed features around 2.5, 2.8, 3.5 and 4.5 eV have been assigned to interband transitions from the filled d-band into empty states above the Fermi level; for further details, see [119].

Further results reported so far deal with surface reconstruction [119], metal deposition [118, 120] and adsorbate layers (including molecular orientation) [121]. The electrochemical oxidation of a Au(110) surface has been studied with RAS [122]. Evidence for surface oxidation and adsorption of hydroxyl ions was obtained and both processes destroy surface states of the Au(110)(1×2) in a reversible manner. No significant kinetic barriers for any of these processes were found.

5.1.9 Photoacoustic Spectroscopy[11] (PAS)

Fundamentals. Illumination of a surface with electromagnetic radiation will result in absorption of this radiation provided the surface is capable of interacting with the radiation in a way that results in absorption of energy. Recording the absorption as a function of the wavelength of the incident radiation will provide an absorption spectrum that, in turn, may yield information about structural, electrooptical or other properties of the surface and of regions of the material close to the surface. When light in the visible range of the electromagnetic spectrum is used, the spectrum will give information about optical transitions of species adsorbed on the surface or optical transitions in surface layers (e.g. in semiconducting oxide layers). Measuring the

[11] See also Sect. 5.9.4, p. 190.

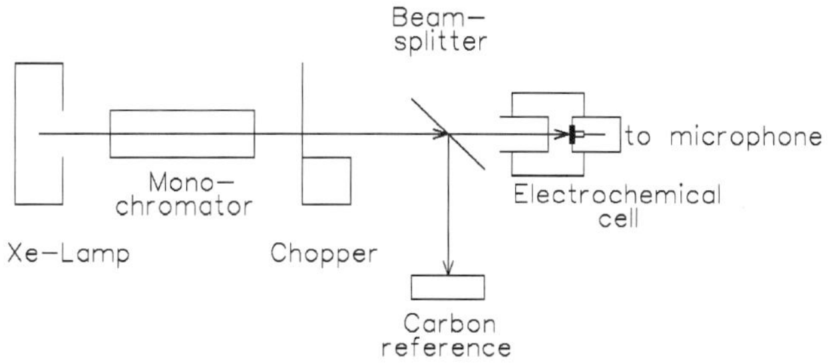

Fig. 5.27. Principle of photoacoustic spectroscopy according to [123]

amount of absorbed radiation can be accomplished in various fashions. In various techniques that employ reflection or attenuated total reflection setups, the intensity of the light before and after the interaction is measured. These techniques cannot be used with rough or dull surfaces because these surfaces tend to scatter the incident radiation in a diffuse way. With these surfaces, photoacoustic detection will provide the required information, as already established in infrared spectroscopy (see also Sect. 5.1.9, p. 60).

Instrumentation. The light of a Xe-lamp is passed through a monochromator and a mechanical chopper. Part of the monochromatic light is directed towards a carbon black sample fitted with a microphone that serves as a reference sample.

The other part of the radiation passes the beamsplitter and enters the electrochemical cell via a quartz glass window. The electrode surface to be investigated is mounted close to the window in order to minimize absorption of radiation in the electrolyte solution. The electrode and the microphone head enclose a small gas volume. Changes in the pressure of the enclosed gas, which is affected by absorption of electromagnetic radiation that is converted into thermal energy, are detected by this microphone. Since detection of the absorption proceeds on the back side of the investigated surface in a manner contrary to that employed with conventional photoacoustic spectroscopy, the applicability of this mode had to be checked. This was done successfully by Vallet et al. [124]. The signal of both microphones was passed via a lock-in amplifier and displayed as a function of incident light wavelength. In a simplified setup, a piezoelectric detector was used instead of a microphone [124]. In this arrangement, the electrode material under investigation was glued directly onto the piezoelectric detector. As an example, the photoacoustic response obtained during electroreduction of diheptyl viologen (DHV^{2+}) at a platinum electrode in an aqueous solution of 0.3 **M** KBr is shown in Fig. 5.28.

Upon reduction of DHV^{2+} a deeply colored film of the salt DHVBr is formed on the electrode surface. This becomes clearly evident in the photoacoustic spectrum, when the electrode potential is switched from an initial value positive to the reduction potential to a value negative of the reduction potential. The recorded spectrum was found to be in good agreement with spectra of HVBr films previously recorded. For further details, see [125].

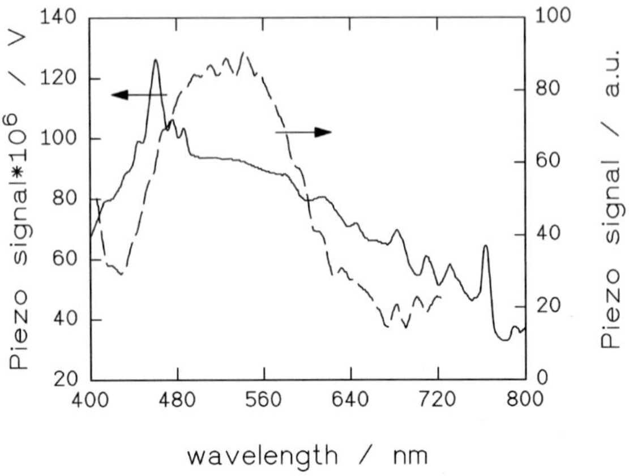

Fig. 5.28. Piezoelectric signal of a platinum electrode in contact with an aqueous solution of 10^{-2} **M** DHV^{2+} and 0.3 **M** KBr at an electrode potential positive (1) and negative (2) to the reduction potential (based on data in [125])

5.1.10 Photothermal Spectroscopy (PTS)

Fundamentals. As already described in the preceding section (p. 23), absorption of electromagnetic radiation that illuminates an electrode surface by a surface layer or in particular by adsorbates on the electrode surface can result in thermal effects. These can be detected by various means as described below.

Instrumentation. A rather simple approach is the measurement of the temperature change of the investigated electrode. Using a differential thermistor setup, this was accomplished in a very simple fashion by Brilmyer and Bard [126]. Two identical thermistors were used. One thermistor was glued to the backside of the working electrode and the other was inserted into the electrolyte solution close to the working electrode but without thermal contact with the electrode. Both thermistors were incorporated into differential amplifier circuits based on standard operational amplifiers. The output signal was passed through a lock-in amplifier connected to the chopper and displayed as a function of incident light wavelength. As an example, the photothermal spectrum of a film of DHVBr deposited by electroreduction on a platinum electrode is displayed in Fig. 5.29.

It closely resembles corresponding spectra recorded with other methods, including photoacoustic spectroscopy.

Measurements of the deflection of a light beam caused by local heating at the solid/solution interface as a function of the wavelength of light illuminating this interface are an alternative possibility; the resulting method is called photothermal deflection spectroscopy (PDS) [127, 128]. The experimental setup as depicted in Fig. 5.30 closely resembles the arrangement for probe beam deflection studies treated in Sect. 5.1.9.

Local heating of the electrode surface that is caused by the impinging monochromatic light results in density gradients in the electrolyte solution in front of the

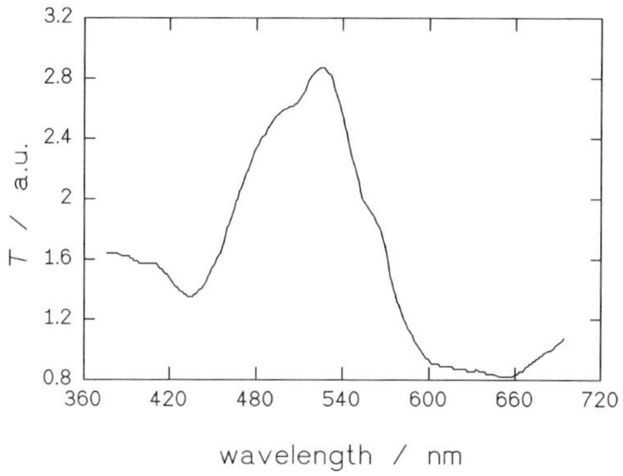

Fig. 5.29. Photothermal spectrum of a platinum electrode in contact with an aqueous solution of DHV^{2+} at an electrode potential negative to the reduction potential [127]

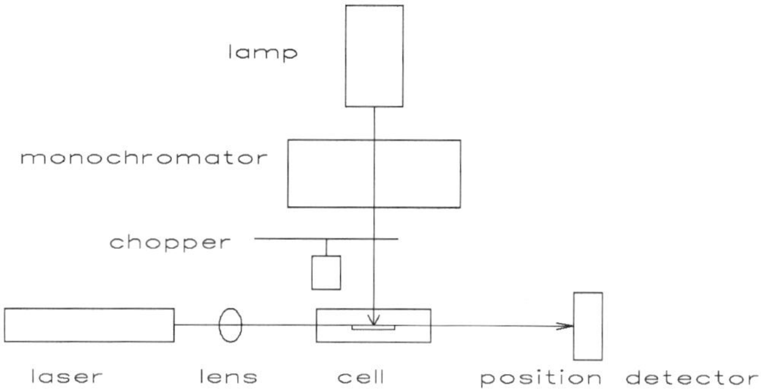

Fig. 5.30. Probe beam deflection setup for photothermal spectroscopy [129]

electrode surface. A laser light beam passing close to the interface through this part of the solution will be deflected ("mirage effect"). Its actual position can be detected with a position sensitive detector. Recording the deflection as a function of the illuminating wavelength results in the desired absorption spectrum of the electrode surface. Submonolayer studies of adsorbate layers on a platinum surface in contact with a solution of perchloric acid [129] and on copper electrodes [129] as well as results on photocorrosion of semiconductors [130] have been reported. A slightly modified procedure known as differential photothermal deflection spectroscopy (DPDS) has been suggested by Barbero et al. [131]. It is based on the acquisition of photothermal deflection spectra at two different states of the electrode/solution interface (e.g. two different electrode potentials). Subsequent spectral ratioing results in spectra independent of the intensity fluctuation of the employed lamp, spectral intensity distribution of the lamp, etc. Physical parameters of conducting polymer layers have been studied with various types of photothermal spectroscopies [132].

5.1.11 Circular Dichroism (CD)

Fundamentals. Optically active molecules show different extinction coefficients for circularly polarized light ε_L and ε_R as a function of wavelength (circular dichroism (CD) [133, 134]). The observed difference $\Delta\varepsilon = \varepsilon_L - \varepsilon_R$ is small compared to the extinction coefficient ε—generally about $\Delta\varepsilon/\varepsilon \leq 10^{-3}$ [135]. The name of the effect derives from the fact that plane-polarized light becomes elliptically polarized upon passage through a sample showing CD. Together with optical rotary dispersion (ORD, anomalous dispersion), CD is grouped under the term "Cotton effect".

Instrumentation. The standard optical device for transforming light into circularly polarized light is a quarter-wave plate. Unfortunately this device produces the desired kind of light only at a single wavelength—at all other wavelengths, elliptically polarized light is produced. Circular dichroism spectroscopy became possible only with the advent of the Pockels cell. Its optical properties can be adjusted properly at every wavelength. More recently the Pockels cell has been replaced by photoelastic modulators (PEM). With suitable crystals used as PEM, the accessible range of wavelengths could be extended into the infrared; with synchrotron radiation, the vacuum UV range can be studied.

A first spectroelectrochemical setup employing an optically transparent (ITO-coated) electrode in a thin layer three-electrode arrangement was described by Daub et al. [136–138]. Two ITO-coated glass sheets were mounted at a distance of 0.1 mm at a PTFE body. This body was used as the head of an electrochemical glass cell. The counter electrode, reference electrode and the purge gas inlet and outlet were also attached to the cell head. The volume of electrolyte solution in the cell was adjusted to maintain an upper level just reaching the lower edge of the two ITO-electrodes. Capillary action sucked the solution into the tiny gap. The cell was placed in the beam of an UV-Vis spectrometer with the ITO-electrodes positioned perpendicularly in the beam. The CD spectroelectrochemistry of the optically active esters depicted in Fig. 5.31 has been studied [139].

The electrochemical reduction proceeds via a radical anion, a subsequently formed diradical dianion and a tetraanion formed after intramolecular rearrangement. The CD spectra of the two enantiomeric forms of the diradical dianion, as shown in Fig. 5.32, are distinctly different.

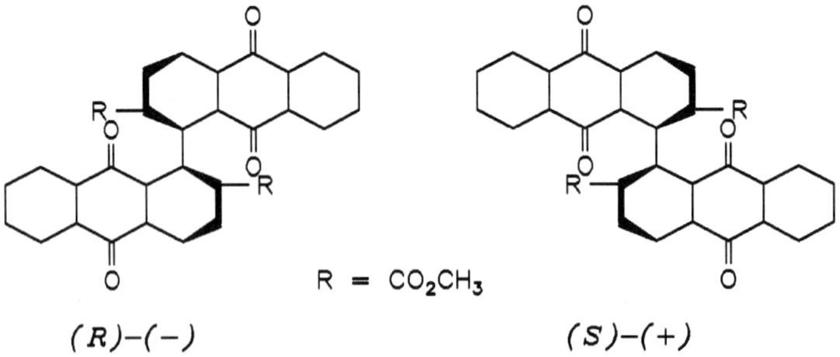

Fig. 5.31. The two forms of an optically active bianthraquinone ester

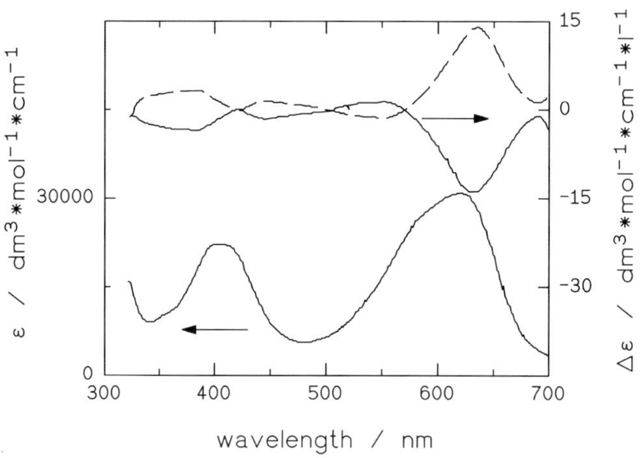

Fig. 5.32. *Bottom*: Conventional UV-Vis absorption spectrum of the (*S*)-(+)-diradical dianion of the bianthraquinone ester (see Fig. 5.31). *Top*: CD spectrum of the (*S*)-(+)-diradical dianion (—) and the (*R*)-(−)-diradical dianion (—) (based on data in [139])

No racemization was observed when the electrode potential was scanned only to a value where the dianion is formed. Upon formation of the tetraanion, subsequent chemical reactions were found. With a slightly different electrolyte salt (Me_4NBF_4 instead of Bu_4NF_6), reversibility without racemization was found even up to the tetraanion formation. Further examples include the spectroelectrochemistry of vitamin D_2 [139], which has been studied with a long pathlength cell similar to the design described by Zak et al. [44]. Optical rotary dispersion and CD of optically active polybithiophene that has been electropolymerized in a cholesteric electrolyte have been studied [140]. The optical rotation of this chiral polymer could be controlled via the electrode potential.

5.1.12 Near Infrared Spectroscopy

Fundamentals: Besides optical absorption in the UV-Vis region of the electromagnetic spectrum, absorption in the near infrared region can be employed to study electrochemical processes. This part of the electromagnetic spectrum[12] between ca. 900 nm and 3000 nm (10000 cm^{-1} to 3333 cm^{-1}) shows predominant overtone bands of electronic absorptions already seen in the UV-Vis region. In addition, absorptions caused by mobile charge carriers are seen [26, 27]. Thus, this part of the spectrum has not attracted much attention. Investigations of intrinsically conducting polymers and transition metal complexes showing large absorptions in the NIR

[12] The actual limits, particularly the lower wavelength one, seem to be subject to arbitrary definition. Spectra showing absorptions around 12000 cm^{-1} have been designated NIR spectra, whereas sometimes the range from 4000 cm^{-1} to 14000 cm^{-1} has been called NIR.

because of intramolecular charge transfers have caused an increased interest in NIR spectroelectrochemistry. For an introductory overview see [141, 142].

Instrumentation. Numerous spectrometers for the NIR range are basically either UV-Vis or infrared spectrometers with suitable additional optics and light sources. They come with two-beam arrangements or single beam setups with internal sample chambers or with fiber optics used for coupling the spectrometer with a cuvet, i.e. a spectroelectrochemical cell. Depending on the type of instrument and the spectral range actually to be investigated, optical components and materials used in the construction of the spectroelectrochemical cell have to be selected. For investigations extending only slightly into the NIR range with an electrolyte solution containing only weakly absorbing solvents, most components (i.e. cuvets and optically transparent electrodes) as used with *in situ* UV-Vis spectroelectrochemistry can be used. Optics made of fused silica with a low hydroxyl group content will cause no significant optical absorption. An overview of the NIR absorption of common solvents used in spectroelectrochemical investigations in this spectral range is shown in Figs. 5.33–5.35.

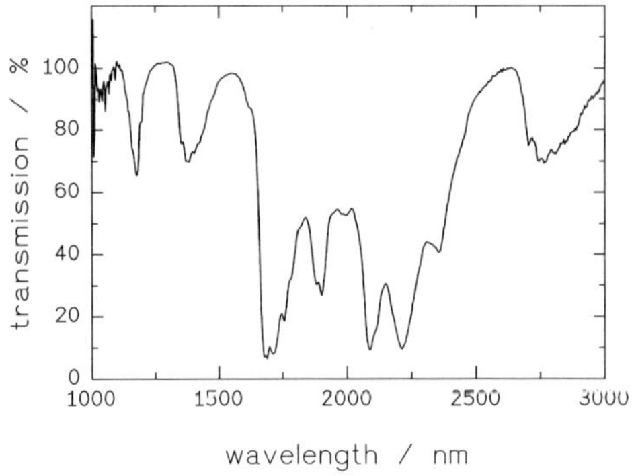

Fig. 5.33. NIR transmission spectrum of propylene carbonate

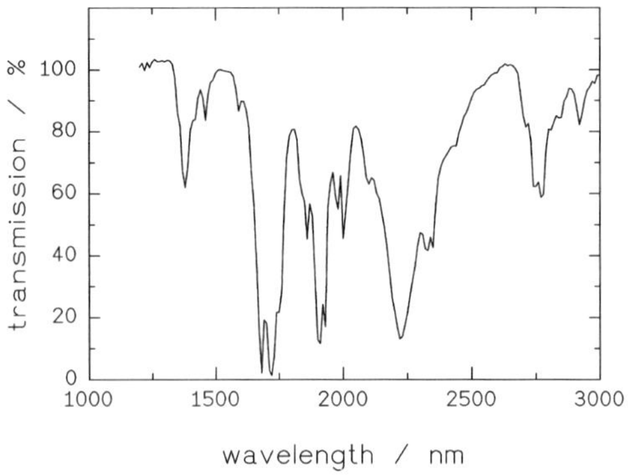

Fig. 5.34. NIR transmission spectrum of acetonitrile

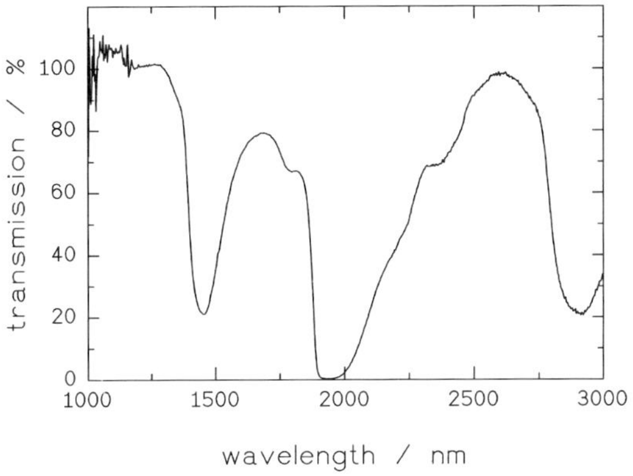

Fig. 5.35. NIR transmission spectrum of water

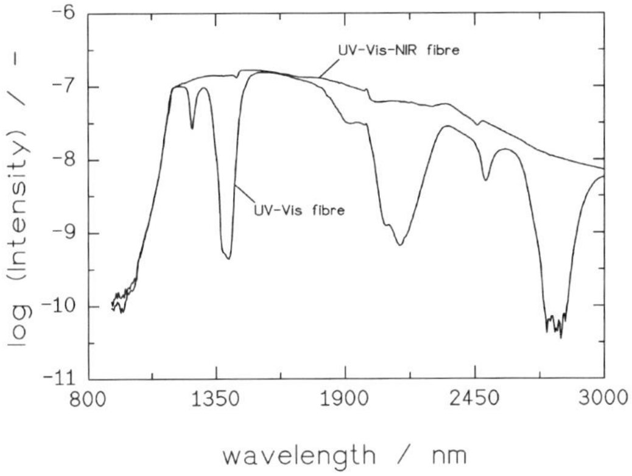

Fig. 5.36. NIR transmission of various fused silica fibers

A cell suitable for measurements at low temperatures that employs a platinum gauze working electrode inside a cuvet that is used as a cell with counter and reference electrode placed in a tubular extension on top of the cuvet has been described [143]. Measurements extending up to the limit of a lead sulfide detector (i.e. 3000 nm) require particularly careful selection of materials and cell design. Because of the strong and very pronounced optical absorption of glass fibers used in the UV-Vis and lower NIR range, these fibers are not suitable for investigations at higher wavelengths. Figure 5.36 provides information on the optical transparency of typical fibers with very low (UV-Vis–NIR fiber) and slightly higher hydroxyl (UV-Vis fiber) content. Near infrared spectrometers with fiber optics allow a simpler experimental setup.

The absorption of commonly used solvents in this spectral range requires the use of thin layer cells in order to avoid spectra with poor signal-to-noise ratios.

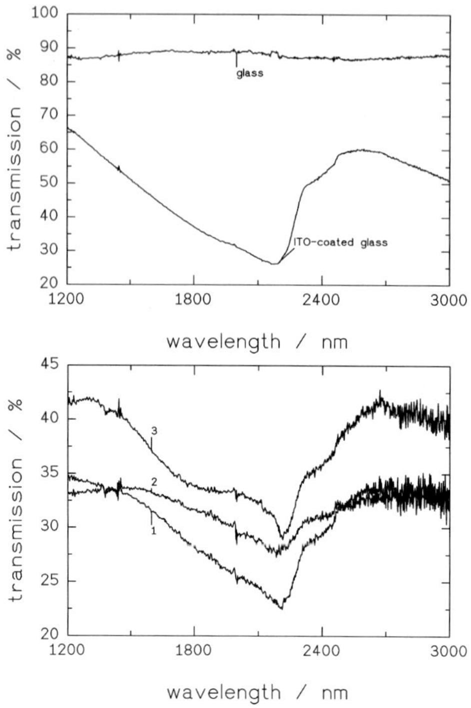

Fig. 5.37. NIR-transmission of standard glass and ITO-coated glass; *top*: reference air; *bottom*: reference glass; 1, 2 and 3: different coating conditions, resulting layer thickness, conductivity, etc.

The design described above for use with UV-Vis spectroelectrochemistry can be used here provided that both the glass used as cell window (cuvet) and the ITO-coated glass have sufficiently low NIR absorption. As shown in Fig. 5.37, the optical absorption of ITO-coated glass shows some major bands where the signal-to-noise ratio might be worse than in other parts of the spectrum.

Highly absorbing electrolyte solutions require thin layer arrangements. In the simplest setup, two glass sheets (one of which is coated with ITO and is slightly longer) are clamped together with narrow PTFE strips acting as spacers at the edges. The bottom edge of the assembly is immersed into the electrolyte solution. Most solutions will penetrate into the gap because of capillary action. A more sophisticated setup, including a cell body and provisions for reference and counter electrode, is depicted in Fig. 5.38.

A setup suitable for work with highly reflective solid electrodes (e.g. platinum, gold or glassy carbon discs) has been described by Salbeck [145]. As shown in the cross section in Fig. 5.39, the polished electrode surface is mounted close to the NIR-transparent cell bottom, leaving only a thin layer of electrolyte solution in the narrow gap. Connection to the NIR spectrometer is accomplished with a fiber optic

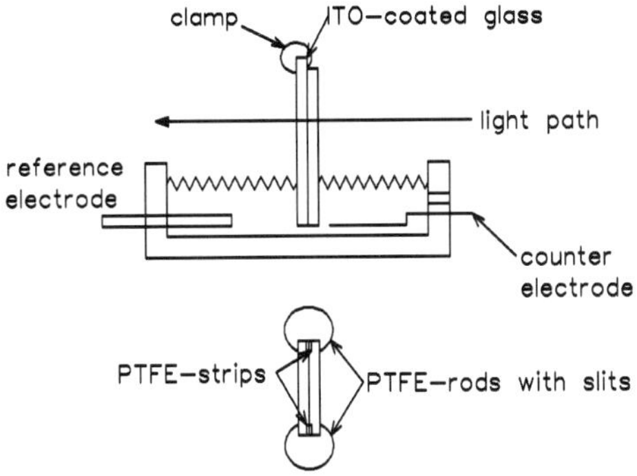

Fig. 5.38. Thin layer cell for NIR spectroelectrochemistry in the transmission mode according to [142, 144]

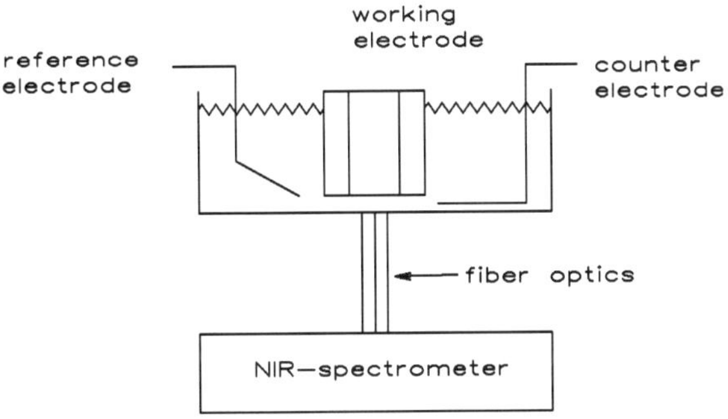

Fig. 5.39. Thin layer cell for NIR spectroelectrochemistry in the external reflection mode according to [145]

cable containing bundled cables that extend from the light source to the detector. The light coming via one fiber from the source passes through the window and the solution. It is externally reflected at the electrode surface back into the fiber that leads to the detector. The cell can be thermostatted down to −40°C and it shows both thin layer and semi-infinite diffusion behavior.

Investigated systems include a broad variety of dissolved organometallic species, electrochemically active organic molecules and redox active polymers like polyaniline (for a review, see [142]). Both dissolved species and species attached by adsorption, covalent bonding or film-forming deposition have been studied. Dissolved polyaniline dispersions as prepared by chemical oxidation [145] show various transitions in the NIR, as depicted in a set of NIR spectra in Fig. 5.40. The bands around $\lambda = 1490$ nm and 1950 nm are overtones of the N–H stretch mode of an aromatic amine, whereas the band around $\lambda = 2300$ nm is caused by the oligomer itself, which presumably indicates the presence of mobile charge carriers.

NIR spectra of polyaniline films (see Fig. 5.41) that are deposited on optically transparent ITO-coated glass sheets and recorded in the same spectral range show

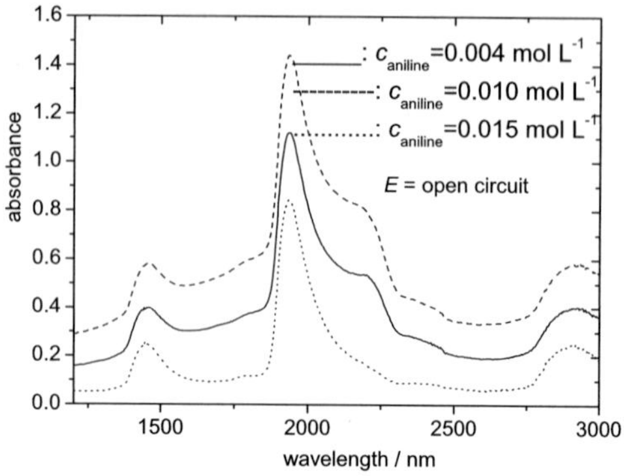

Fig. 5.40. NIR spectra of a polyaniline dispersion prepared with various concentrations of aniline (dispersion prepared according to [146]; for further details, see [146])

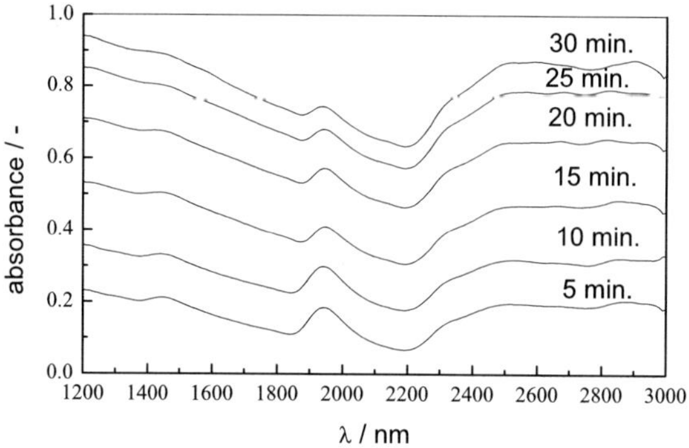

Fig. 5.41. NIR spectra of a polyaniline film in a solution of 0.5 **M** sulfuric acid at $E_{RHE} = 1$ V at various times of electropolymerization (for further details, see [147])

similar features. Depending on the reference (or background) spectrum, the observed features can be assigned to the film and its changes (with the film in the spectroelectrochemical cell at a fixed electrode potential taken as a reference), to the growing film (with a spectroelectrochemical cell filled with supporting electrolyte solution that contains the monomer) or to the aniline in the system being present as dissolved monomer or as a building block in the polymer film (with a spectroelectrochemical cell that is only filled with supporting electrolyte solution).

The bands discussed above are again present, although very weak. A distinction between absorption caused by the aniline in solution and the aniline in the polymer is impossible. With the electrolyte solution containing the monomer as a background, a different result is observed.

The broad absorption already seen above is present again, implying that it must be caused by species in the film. These are most likely mobile charge carriers (for a detailed discussion, see [147]).

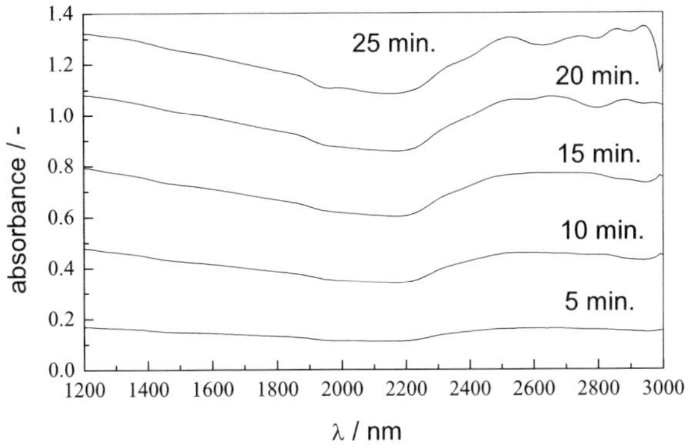

Fig. 5.42. NIR spectra of a polyaniline film in a solution of 0.5 **M** sulfuric acid and 1 **M** aniline at $E_{RHE} = 1$ V at various times of electropolymerization (for further details, see [147])

5.2 Optical Spectroscopy in the Infrared Range

Rotational and vibrational modes of molecules and polyatomic ions are probed with infrared and Raman spectroscopy. Energies needed to promote a species into a higher vibrational or rotational state are generally in the mid-infrared (approx. 400–4000 cm^{-1}) and in the far infrared (10–400 cm^{-1}). Transitions can be studied by supplying the needed energy and measuring the absorption by the species under study; this is done with infrared absorption spectroscopy. Alternatively, the species under investigation can be exposed to monochromatic light. The inelastically scattered light is shifted in wavelength (Stokes and anti-Stokes shift) and the shift again yields the desired information about the energies of rotational and vibrational transitions. Although this method, known as Raman spectroscopy, works with light in the visible or NIR range, it is nevertheless treated in this chapter because the results provide information about rotational and vibrational properties of species at the interface under investigation. Furthermore, both spectroscopies can be employed in a complementary manner, i.e. results of one spectroscopy can be enhanced and complemented (e.g. with respect to the accessible range of wavelength or the observed vibrational modes that might show up with one spectroscopy only because of symmetry reasons) by the other one. In addition, results of one spectroscopy can be used to support and confirm data obtained with the other one. This complementary approach is also expressed in monographs treating both spectroscopies almost simultaneously [147]. A general overview of symmetry, selection rules and nomenclature in various surface spectroscopies has been provided by Bradshaw and Richardson [148].

Beyond intramolecular modes of species present in the electrolyte solution or interacting with an electrode surface, modes caused by adsorptive interaction between the adsorbed species and the electrode surface can be studied (e.g. the silver-halide stretching mode with surface enhanced Raman spectroscopy (SERS); for details, see p. 104). These modes that appear mostly at fairly low wavenumbers provide particularly useful information about mode and strength of interaction.

Fundamentals, applications and experimental details of both infrared and Raman spectroscopy are treated in numerous textbooks and handbooks [149–157]; thus only a few general details of these spectroscopies, which are of particular importance for applications in electrochemistry, are treated here.

Infrared spectroscopy, as applied in numerous different experimental setups for electrochemical investigations, is at first glance seriously hampered by the fact that most electrolyte systems (including both the solvent and the dissolved supporting electrolyte) tend to absorb infrared radiation strongly. Thus the detection of absorption of radiation of species present in an adsorbate layer or dissolved in this electrolyte solution is difficult. Various approaches have been described to overcome this problem.

In thin layer cells similar to those used in UV-Vis spectroscopy, the absorption by the electrolyte solution is kept low by keeping the electrolyte volume that is passed by the beam of light thin. Unfortunately this also limits the number of species under investigation that can be probed by the light. Strongly absorbing species and solvents with supporting electrolytes that show absorptions in a range of the spectrum far away from absorption bands of interest can nevertheless be studied successfully.

Species adsorbed on surfaces can be probed with radiation in external or attenuated total reflection mode. In the former mode, a beam of infrared light is directed at the surface of interest. The reflected light is measured and attenuated by absorption caused by the species under investigation. In an *in situ* study in the external reflection mode, the light beam has to pass through the electrolyte solution in contact with the surface under investigation and the window of the electrochemical cell; thus the light also contains information about radiation absorption by these components. This information is mostly unwanted. Unfortunately even a very thin electrolyte solution film and an almost completely transparent cell window will absorb more energy than the rather small number of adsorbed species. This poses a problem in the detection of only these absorptions. Modulation or differential techniques have to be applied. Recording spectra with the electrode at different electrode potentials yields only bands influenced by the electrode potential after spectral ratioing. All unwanted absorptions listed above are eliminated. A similar approach employs the surface selection rules, i.e. the differences in interaction between light of various types of polarization with adsorbed species. Details of these methods are discussed below (p. 76).

With attenuated total reflection spectroscopy, the light absorption by the electrolyte solution and the cell window is no obstacle. The probe beam enters a crystal transparent for infrared light. It is directed to the outer surface of the crystal, which is coated with a thin layer of the electrode material under investigation. The beam is reflected, but a small part (the evanescent wave) penetrates the surface and thus can probe species located immediately on the electrode surface. The returning beam contains exactly this information. As discussed below (p. 91) in detail, this approach shows also serious limitations.

Spectrometers used for this investigation are in most cases commercial instruments equipped with accessories for the desired mode of spectroscopy. Because of the strong absorptions already mentioned, instruments with powerful light sources and high optical throughput are preferred and sensitive detectors are desirable. Besides commonly known infrared light sources, synchrotron radiation provides a source of high intensity polarized light of variable frequency [158]. Both dispersive and Fourier transform instruments can be used. In the former case custom-built spectrometers have been used in practically all reported studies, whereas in the latter case mostly commercial instruments can be used after only minor adaptations.

The design of a spectroelectrochemical setup for *in situ* infrared studies and for all methodologies described below is rather straightforward. The light provided by a glowbar or a Nernst glower is collected and guided towards the electrochemical cell. The transmitted or reflected light (depending on the method) is sent towards the entrance slit of the monochromator. At the exit slit an infrared detector connected to data processing electronics is positioned. In most cases, a chopper in the light beam and a lock-in amplifier are used in combination to enhance the sensitivity of the method when applying a modulation technique light; e.g. electrode potential modulation. A typical experimental setup with computer-based data processing is described in [159].

The use of a Fourier transform infrared (FTIR) spectrometer has provided a completely different methodological approach to this spectroscopy [160, 161]. The central component of any FTIR spectrometer is an interferometer. No dispersive elements are used. Based on the construction, various advantages in comparison with dispersive instruments are achieved:

- The Jacquinot advantage: Light from the source does not pass through a slit, but only through a large circular hole (the Jacquinot stop) at the entrance of the interferometer optics. Thus a much larger fraction of the light from the source is used.
- The Felgett advantage:[13] Light of all wavelengths emitted from the source is passed through the spectrometer simultaneously and reaches the detector, resulting in a much stronger signal. In a dispersive instrument, only light of the selected wavelength reaches the detector. Thus, in the former case, the signal level is way above the detector noise level.
- The Connes advantage: Because of the way data points (intensities) are measured at any given position of the moving mirror in the interferometer, superposition of two subsequently recorded interferograms (i.e. coaddition of measured signals) is easily possible. This results in a greatly improved signal-to-noise ratio.

Raman spectroscopy is not hampered by most of the problems described above. Water, as the most frequently used electrolyte solvent, is a very weak Raman scatterer. Both incident light and scattered light are in the visible or NIR range, wherein cell windows of common glass are transparent. Unfortunately the yield of scattered

[13] This advantage is also called the multiplex advantage.

light is very low (10^{-6}) [162] and this makes detection of Raman light a challenging task, even with liquid or solid samples in standard Raman spectroscopy. The number of scatterers adsorbed on a metal surface is extremely small by comparison; thus detection of light that is scattered by them seems to be almost impossible. Enhancement of scattered intensity by surface effects outlined (p. 104) and significant advances in detector and data processing technology allow detection of even very weak scattered light [163, 164].

Localized vibrational spectra can be obtained with various microspectroscopic setups that employ near field optics and other tools (see Sect. 7.3). Tunneling spectroscopy performed with a scanning tunneling microscopy can also be used (see Sect. 7.2.11).

Electron energy loss spectroscopy (EELS) and high resolution electron energy loss spectroscopy (HREELS) can also provide vibrational information of adsorbates on surfaces [165]. Because these methods employ electrons instead of electromagnetic radiation, surface selection rules (see p. 76) are not effective; this allows investigation of modes not observed with infrared spectroscopy. Unfortunately the use of electrons both as probe and signal prevents *in situ* application. Studies of electrode surfaces are feasible with these methods after emersion of the electrode from the solution, but they have been reported only infrequently.

5.2.1 Infrared Transmission Spectroscopy with Thin Layer Cells

Fundamentals. Dissolved species formed or changed during electrochemical reactions that show infrared absorption can be studied in solution using a thin layer cell that only has optically transparent components in the spectral range of interest. By recording the spectral absorption at various electrode potentials or at time intervals and subsequent spectra ratioing changes of infrared absorption and the corresponding composition of solution as a function of electrode potential or time can be identified. This method provides information about species in solution and thus it is not surface sensitive. The same information can be obtained in an external reflection setup operated in a non-surface sensitive manner (see p. 76).

Instrumentation. Construction materials that are optically transparent in the mid-infrared range are limited to various salt crystals (KBr, NaCl, ZnSe, etc.) and semiconductors (Si, Ge). These materials have to be used as cell windows. Aqueous electrolyte solutions are obviously unsuitable. Electrodes can be made of opaque materials and, in this case, have to be placed out of the light beam. This results in a sluggish cell response, i.e. optical absorption will be detected only after considerable time lags caused by slow diffusion of products from the electrode into the beam. Optically transparent electrodes can be made from metal minigrids or fine metal mesh. Handling of these fragile metal materials can be simplified by embedding them into melt-sealed polyethylene sheets. A design reported by Krejcik et al. [166] is shown in Fig. 5.43.

A gold or platinum wire mesh or minigrid acting as the working electrode and located in the beam path, an additional metal grid serving as the counter electrode

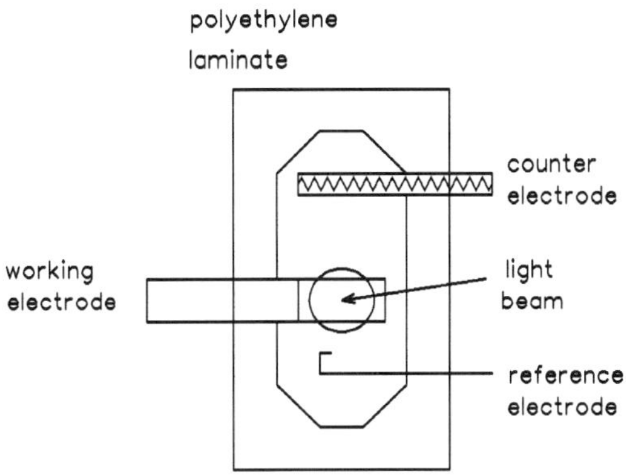

Fig. 5.43. Schematic design of the three-electrode arrangement embedded in polyethylene melt-seal laminate according to [169]

and a silver chloride coated silver wire as reference electrode are embedded into the melt-seal polyethylene laminate. Using PTFE seals, this laminate is mounted between two KBr-plates as cell windows with suitable steel pressure plates and screws. The whole assembly is placed into a standard FTIR spectrometer. By spectral ratioing single beam spectra, changes caused by electrochemical processes can be detected as functions of time or electrode potential. A cell employing a mesh electrode has been used in studies of the spectroelectrochemistry of hydrogenase enzymes and related compounds [167]. A cell design with adjustable thickness of the electrolyte solution layer has been presented [168]. Alternative designs employing preferable standard components for transmission infrared spectroscopy have also been described [169]. In a typical study, the electrochemical conversion of substituted phosphine complexes was monitored [170]. A thin layer cell micromachined from silicon has been developed [171]. Because of limitations in the use of ATR crystals (see p. 91) in *in situ* investigations of lithium ion batteries, a transmission cell for studies of $LiFePO_4$ as a cathode material for these batteries has been developed and applied [172, 173].

5.2.2 Infrared Reflection Spectroscopy

Any species showing infrared active vibrational modes adsorbed on a reflecting surface can be studied with infrared spectroscopy. The beam of light will interact absorptively with the species when passing through the adsorbate layer before and after the point of reflection. This enables studies of all kinds of adsorbates on many surfaces. Of particular interest in electrochemistry are surfaces of metals and semiconductors employed as electrodes. Thus the following text deals only with reflection at these surfaces; other surface and interfaces are not treated. Attempts to record infrared spectra of emersed electrodes (i.e. *ex situ* measurements) have been reported infrequently in studies of adsorption of hydroquinone and benzoquinone on a polycrystalline platinum electrode [174–177]. Further development of this approach has obviously stopped due to the need for large electrodes, a cumbersome experimental setup and the lack of any particular advantage in comparison to spectroscopies employing working electrodes immersed in the electrolyte solution.

5.2.3 External Reflection Spectroscopy

Fundamentals. Infrared light impinging at an angle of incidence θ (according to convention and as depicted in Fig. 5.44, this is the angle between the surface normal and the beam of light) onto a reflecting surface (for example, at the interface metal/dielectric) will be reflected specularly at the same angle θ. The impinging and reflected waves form a node at the point of reflection. At small values of θ, the resulting amplitude of the electric field vector will be zero and absorption of radiation by species showing infrared active vibration (modes with a change of the dipole moment different from zero during a normal vibration) is not possible. At larger angles of incidence the behavior depends strongly upon the polarization of the incoming light with respect to the plane of reflection, which is defined by the incoming and the reflected beam of light. The phase shift of the electric field vector is of particular interest. It suffices to discuss the case of p-polarized light with the vector of the electric light in the plane of reflection and s-polarized light with this vector being perpendicular to the plane. With s-polarized light the phase shift is an almost constant 180° at all angles of incidence and the effective field vector (which would have been parallel to the reflecting surface) is zero, as shown in Fig. 5.44.

With p-polarized light the phase shift varies from 0 to 180° depending on the actual value of θ. Vectorial addition of the electric field vectors of the impinging and reflected rays yields the effective electric field at the point of reflection. In the case of s-polarized light, the phase shift of approx. 180° will result in practical zero field strength at all values of θ. At a 90° phase shift the effective field will be

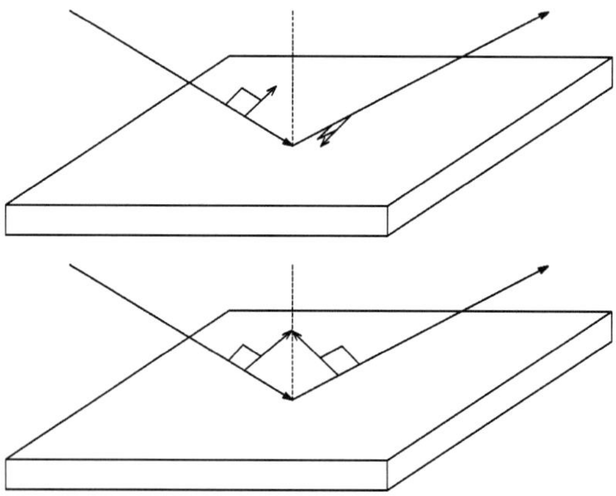

Fig. 5.44. Schematic illustration of phase shift of the electric field vector of polarized light reflected at a surface; *top*: s-polarized light; *bottom*: p-polarized light

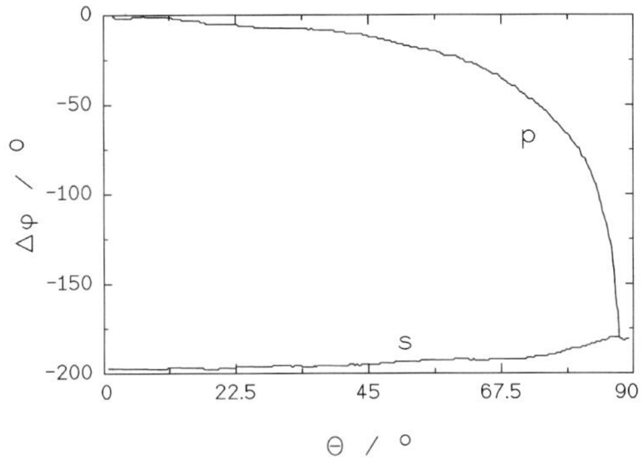

Fig. 5.45. Calculated phase shift for s- and p-polarized light at a reflecting surface (based on data in [181])

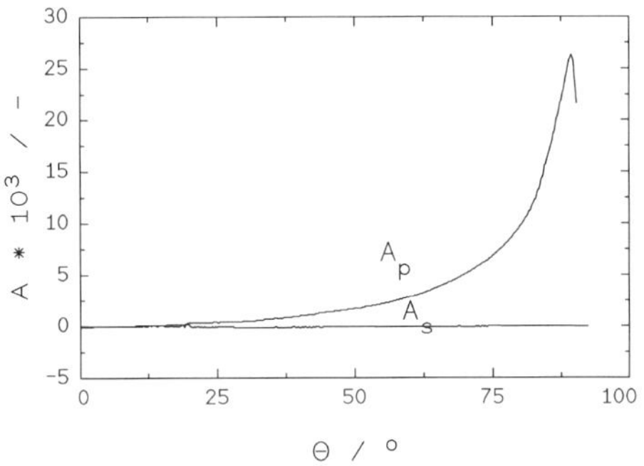

Fig. 5.46. Calculated absorption coefficients A_s and A_p for a layer of 1 nm thickness of acetone on a reflecting surface as a function of the angle of incidence θ (based on data in [179])

maximal for p-polarized radiation; at lower and higher values the phase shift will differ [178]. The calculated phase shift for both cases is depicted in Fig. 5.45.

At angles close to 90° the phase shift is at the value yielding a maximum effective field strength. Calculation of absorption coefficients A_p for p-polarized and A_s for s-polarized light based on the Maxwell equations, assuming a layer of 1 nm of acetone on a reflecting surface, show basically the close relationship between angle of incidence and phase shift (see Fig. 5.46).

In *ex situ* studies of CO adsorbed on palladium sheets at grazing incidence, considerable discrepancies between the behavior predicted by Greenler and the actual results were observed [180, 181]. The actual plane of polarization of the incoming light was without influence on the obtained spectra. This contradiction was explained by invoking surface microroughness in the subnanometer range. Calculations have indeed indicated that a roughness of 2.5 to 5 nm is enough to provide statistically distributed orientations of the adsorbed CO dipoles. This again recommends perfect polishing of the electrode surface. Further details have been reviewed previously [182–184].

Utilization of the differences in interaction between infrared light of the two possible types of polarization with species at the interface (or in the interphase) allows the dedicated design of setups for measurements of surface species or of solution phase species. The latter possibility has already been discussed and will be considered again briefly at the end of this section (p. 90); the former case will be treated below.

Surface enhanced infrared absorption (SEIRA) has been observed in external reflection spectroscopy [185], for further details, see Sect. 5.2.5.

Instrumentation. The subsequent discussion focuses initially on surface sensitive setups; the use of external reflection spectroscopy for solution phase studies is reviewed afterwards. Because the size of the illuminated surface area is given by the—mostly elliptically deformed—infrared beam and the reflected beam contains information averaging surface properties, no particular spatial resolution is possible. With infrared microscopy this would be possible; the geometric arrangement for external reflection based on an FTIR spectrometer and an attached infrared microscopy as applied to studies of a variety of individually addressable arrays (combinatorial chemistry) of nanostructured platinum as electrocatalyst for methanol oxidation has been described [186]. A study has been reported of interfacial processes and species at the tin/organic solvent-based electrolyte solution as employed in lithium batteries using an infrared microscope [187] and the results indicate alloying/dealloying of lithium with tin.

Taking into account the general considerations outlined above the spectroelectrochemical cell has to be of a thin layer design. A typical example is shown in a cross section in Fig. 5.48; another view is presented in Fig. 5.49, showing major features of the cell. In cases where a standard spectrometer is used, the cell has to be mounted in the optical beam using an external reflection attachment as schematically depicted in Fig. 5.47.

A fundamental problem of thin layer cells is the fairly high ionic solution resistance even when aqueous solutions of strong electrolytes are used. (Values of layer thickness vary considerably; Parry et al. [189] have estimated a thickness of less than 10 μm, whereas Bae et al. [190] have claimed a value below 1.5 μm.) This can result in ionic migration, which in turn might cause spurious bands in the observed spectra not caused by adsorbed species but by migrating dissolved ones. A quantitative treatment has been reported [191]. The sensitivity of the method can be improved considerably, as already suggested by the calculations of Greenler [192], by using infrared radiation impinging on the electrode surface at a large angle of incidence (grazing angle). The use of hemispherical or prismatic windows allows such arrangement [193]. A cell mounted perpendicularly in the sample chamber has been described [194]. For a general review, see [195], an early overview of experimental setups including data acquisition and processing has been provided [196]. The influence of interference effects in the thin layer arrangement, particularly on the sensitivity, has been discussed thoroughly [197]. A cell suitable for measurements in a wide range of temperatures from $T = -100$ to 25°C with rather precise temperature control (±0.5°C) has been described [198]. Its capabilities were

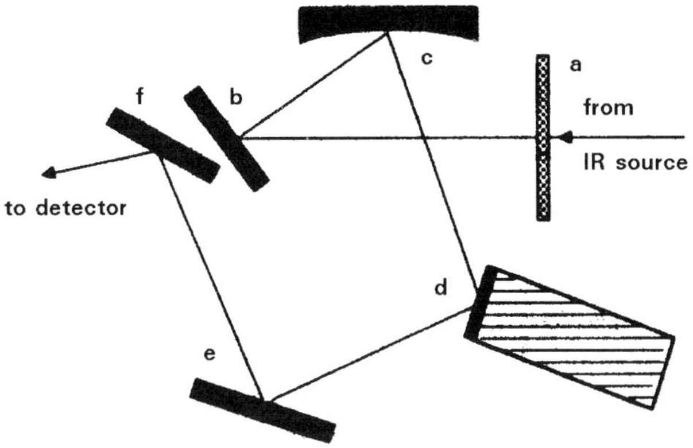

Fig. 5.47. Optical path of IR light in an external reflection attachment equipped with an electrochemical cell; *a*: polarizer; *b*, *e*, *f*: plane mirrors; *c*: concave mirror; *d*: cell with working electrode

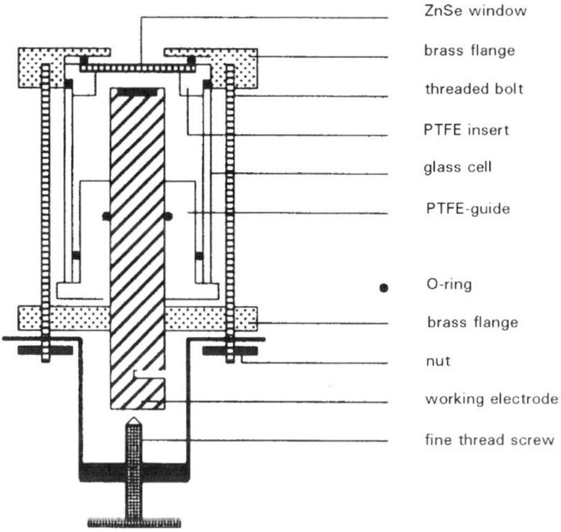

Fig. 5.48. Axial cross section of a spectroelectrochemical cell for external reflection measurement according to [188]

demonstrated with the identification of the electrooxidation reaction intermediate of 1,4-bis(2-ferrocenylvinyl)benzene.

A spectroelectrochemical cell has been described that allows investigations of electrochemically active species in an electrolyte solution that interact with gaseous reactands dissolved in this solution at elevated pressure [199]. Results obtained with organometallic nickel complexes interacting with dissolved CO imply the formation of various carbonyl adducts [202]. Special requirements of measurements at active materials for lithium ion batteries have been considered in a cell described elsewhere [200] for SNIFTIRS studies. The incorporation of a gold electrode that

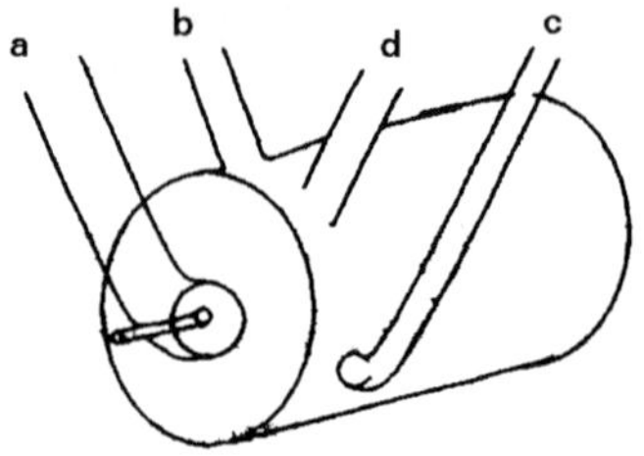

Fig. 5.49. Glass body of the cell depicted in Fig. 5.48; *a*: reference electrode compartment with Luggin capillary; *b*: counter electrode compartment; *c*: gas inlet; *d*: gas outlet, according to [191]

is evaporated onto a piezoelectric quartz crystal as employed in a quartz crystal microbalance in the spectroelectrochemical cell in an FTIR spectrometer has been described [201]. Differences in the formation process and in the properties of the reaction product of aniline electropolymerization and aniline/*o*-aminophenol copolymerization could be identified.

Two modes of operation of the spectroelectrochemical setup (including both the infrared spectrometer and the devices for electrode potential control) are possible in order to obtain the requested surface sensitivity and to remove unwanted absorption contributions from solution, gas phase in sample chamber, etc.:

1. Modulation of the electrode potential
2. Modulation of the plane of polarization of the infrared light

In the first case, the electrode potential is switched between two values previously identified as being suitable with cyclic voltammetry. By calculating the difference between measured surface reflectivities (i.e. spectral intensities recorded at the selected electrode potential as a function of wavenumber) as observed at both potentials, the change of infrared absorption by species being adsorbed is detected because absorption by all other species in the path of light will be the same at both electrode potentials and they will cancel out each other. Depending on the behavior of the adsorbate at the respective electrode potential, various combinations and results are conceivable; the most likely major cases are:

1. The adsorbed species is present only at one of the electrode potentials and it is completely desorbed (absent from the surface) at the other. Assignment of the electrode potentials is arbitrary. In this example the former potential may reasonably be called measurement potential (E_m) and the latter potential may be called reference potential (E_r). Infrared absorption by the adsorbate will thus be observed only at the former potential. In the case of the latter potential no absorption will be effective. In the case of a FTIR spectrometer, measurement of the reflected intensity R as a function of the wavenumber at these electrode potentials is equivalent to recording a single beam spectra. Beyond the infrared absorption by the adsorbate, both single beam spectra contain all the other absorptions of infrared active species in the optical beam path. Since these absorp-

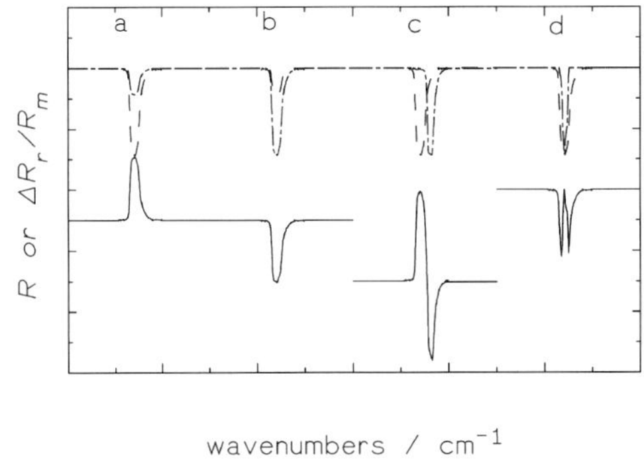

Fig. 5.50. Schematic infrared reflection absorption spectra, —·——·— reflectance at E_r; —·—· at E_m; —— resulting spectrum; for details, see text

tions should in no way be influenced by the change in electrode potential, they should be equal in both cases. Spectral calculation will thus cancel out these contributions and, as a result, only the absorption spectrum of the adsorbate will be obtained:

$$\frac{\Delta R}{R} = \frac{R_m}{R_r} - \frac{R_r}{R_r}.$$

In resulting spectra, $\Delta R/R$ without any indices is displayed. When the spectrometer software is run in a mode resulting in transmission spectra (with a percent abscissa), calculation of the simple intensity ratio will result in a spectrum around the 100% line. Subtraction of the second term causes a shift with the spectrum now displayed around a 0% line. Schematically this is shown in Fig. 5.50 (a); for an experimental example, see Fig. 5.51.

2. Adsorbate coverage and band position are independent of electrode potential. When the adsorbate is present at both electrode potentials and neither the degree of coverage (i.e. the number of infrared radiation absorbing species) nor the band position of observable infrared modes changes as a function of electrode potential, the absorptions at both electrode potentials in the respective single beam spectra will completely cancel out. Although this is not impossible, it is at lcast highly unlikely.
3. Band position and/or coverage depend on electrode potential. In a more common situation, at least one property will change. If the band position changes because the strength of metal-adsorbate interaction changes as a function of electrode potential, the position of the absorption in the single beam spectra will be different, spectral calculation will result in an s-shaped spectrum (which looks like a differential spectrum[14]) when the change is small. In cases of larger changes, two bands pointing in opposite directions will be observed (see Fig. 5.50). If the band position does not change as a function of electrode

[14] This similarity is indeed justified, because the band position is measured as a function of electrode potential: $d\bar{\nu}/dE$.

potential, but the coverage does, the resulting spectrum is somewhat ambiguous, as shown in Fig. 5.50. In most cases both properties will change and the observed spectra will be somewhere between the two extreme possibilities outlined above.

In Fig. 5.50, selected examples are illustrated schematically. For more variations, see [160, 202]. In case a, the absorption at the measurement potential E_m is smaller than at E_r (because of a lower degree of coverage), the result is an upward pointing band in the display of $\Delta R/R$. Generally, the infrared reflection absorption spectra (not single beam spectra, as available with some methods) obtained with any of the methods described here are displayed in a way that results in bands above the baseline that indicate higher absorption of any infrared-active species at E_r and downward pointing bands that indicate higher absorption at E_m. In case b, the situation is just reversed, resulting in a downward pointing spectrum. In both cases, determination of the band position is easy. Case c shows the major problem of the experimental approach: The obtained band shows the typical "differential shape", making determination of the band position almost impossible. A rather rare case with a considerably narrower band at E_r is displayed in case d.

Obviously the absolute band position can only be extracted easily from the obtained spectra in cases a and b. In most investigations, one electrode potential is kept constant and subsequently called E_r, whereas the other potential is changed and designated E_m. The applied spectral calculation amounts to a division of the measured infrared intensities arriving at the detector, which correspond directly to the reflectivities of the electrode surface. For traditional reasons, this procedure has been called spectral subtraction and the method outlined above is thus called SNIFTIRS: subtractively normalized interfacial Fourier transform infrared spectroscopy. Because the electrode potential is switched (modulated), albeit very slow (after several interferograms have been acquired), this method has also been called square wave Fourier transform infrared reflection spectroscopy (SW-FTIRS) [203]. At this point, the Connes advantage typical of FTIR spectroscopy can be usefully employed. At both potentials E_r and E_m interferograms are collected and stored separately. Subsequently, the electrode potential is switched repeatedly between both values; fast Fourier transformation is applied only at the end of the sequence. Any fluctuations will thus be eliminated and a significantly improved signal-to-noise ratio results. This procedure is unfortunately only applicable when the changes at the electrochemical interface that occur as a function of the switched electrode potential are entirely reversible, i.e. no chemical reaction interferes. In cases of irreversible processes (e.g. dissociative chemisorption), only a single potential switch is possible. This methodology has been called single potential alternation infrared reflection spectroscopy (SPAIRS). When single beam spectra are recorded during a very slow electrode potentials scan with only one reference spectrum recorded initially or finally at E_r, the single beam spectra recorded at slowly changing electrode potentials are assigned to the average value of E_m during the time interval of interferogram acquisition. The method is called linear potential scan infrared reflection spectroscopy (LPSIRS). The spectroelectrochemical setup always includes an FTIR spectrome-

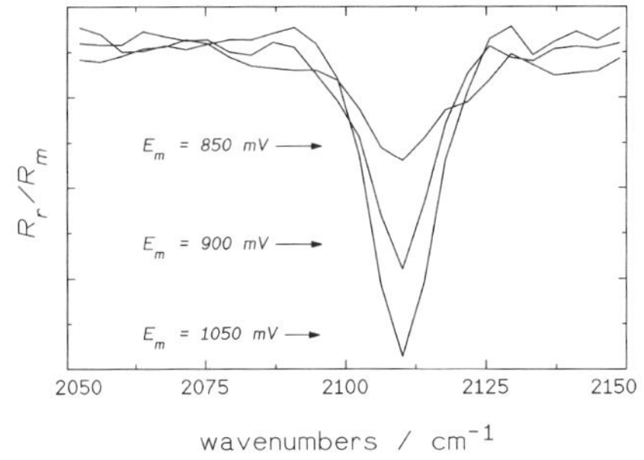

Fig. 5.51. IRRA spectra of CO_{ad} on polycrystalline copper in contact with an aqueous solution of 0.1 **M** $KClO_4$, $E_{r,RHE} = -0.1$ V, $E_{m,RHE}$ as indicated

ter, an external reflection accessory, a polarizer and the spectroelectrochemical cell as depicted above in Figs. 5.47–5.49.

The first case (a) discussed above is nicely demonstrated in a study of carbon monoxide adsorption on copper in contact with an aqueous solution of 0.1 **M** $KClO_4$ [204]; only a single-sided band of the CO-stretching vibration is observed at 2110 cm^{-1} (Fig. 5.51).

The absence of an electrode potential induced shift of the band position has been attributed to a relatively weak metal–CO interaction. Carbon monoxide is completely desorbed at E_{ref}, thus none of the typical problematic features of differential bands appear.

Electrode potential modulation[15] can also be applied with dispersive infrared spectrometers (EMIRS). In this case, potential modulation is done by using a square wave potential function, switching the electrode potential between E_r and E_m. The frequency is in the range of a few Hertz up to about 20 Hertz. The infrared intensity arriving at the detector changes as a function of the changing electrode potential provided that the adsorbed species show any of the potential dependent changes discussed above. The absorption of all other components in the infrared beam does not change and results only in a constant intensity at the detector. Assignment of the observed intensity (i.e. reflectivity) to the respective electrode potential is achieved by using a lock-in amplifier controlled from the potential modulation device. The use of this technique results in an improved signal-to-noise ratio. However, this advantage can only be practically employed if the studied range of wavenumbers is limited to a few hundred wavenumbers. A comparison of EMIRS with other infrared techniques, particularly SNIFTIRS, has been reported [205]. The infrared intensity measured at the detector is recorded as a function of the slowly changed wavenumber of the infrared radiation passing through the employed monochromator. As there is no commercially available dispersive standard infrared spectrometer suitable for this purpose, the required optical components are arranged as needed. The basic optical layout has been described frequently (see [199, 208]). The need for a rather

[15] The acronym EMIRS somewhat misleadingly refers to electrochemically modulated infrared reflectance spectroscopy.

fast potential modulation to operate the lock-in amplifier combined with the electrochemical thin layer cell arrangement causes some specific problems. Thin layer cell arrangements—as pointed out repeatedly already—are prone to sluggish response to potential changes. This is mostly caused by the large potential (*iR*) drop along the poorly conducting thin electrolyte solution film that is trapped between the cell window and the electrode surface. Thus electrode potential values can only be approximate because the tip of the Luggin capillary is at the edge of the electrode (whereas the infrared beam probes mostly the central part of the circular electrode surface). The use of EMIRS in investigations of electrocatalysts has been reviewed [208].

In the fundamentally different second approach, the surface selection rules as discussed above [181] are employed in a different manner. A species adsorbed on the electronically conducting surface of the electrode with a change of its dipole moment perpendicular to the surface during a vibrational mode will interact only with infrared radiation with its plane of polarization oriented parallel to the plane of reflection (p-polarized, for further details see above). With s-polarized light no absorption will occur. Absorption by anything else in the optical path of light will affect light of both planes of polarization in exactly the same way. Provided that the light in both planes shows exactly equal intensity, spectral subtraction should yield a spectrum with ΔR as a function of wavenumber showing only the absorption by species at the interface; all other absorptions should cancel each other out. Various experimental approaches to obtain and process the spectra have been described [206–208]. In the most frequently employed spectroelectrochemical arrangement (based on the initial setup described in [209]), the light from the infrared radiation source is intensity-modulated by a mechanical chopper at a frequency ω_c and passed through a polarizer. Using a photoelastic modulator (PEM) (for the mode of operation of a PEM, see [209]) the plane of polarization with respect to the plane of reflection at the electrode surface is modulated between s- and p-polarization at a frequency $\omega_m = 2\omega_c$. The modulated infrared intensity R arriving at the detector is demodulated with lock-in amplifiers at these frequencies. As a result of further data treatment

$$R = \frac{R_p - R_s}{R_p + R_s}$$

is obtained. The modulation scheme also cancels out sample emission effects and thus can also be applied with heated samples. Measurements are performed at a fixed electrode potential, i.e. the obtained spectrum pertains to a single electrode potential because no electrode potential modulation is involved. This in turn results in spectra that are considerably more easy to interpret as compared to those obtained by potential modulation: Sometimes these spectra are also called "absolute spectra" (as compared to "differential spectra"). The method has been reviewed in detail [210] and its limitations have been pointed out [211]. Detailed considerations of the optimization of a low-noise spectrometer have been discussed [212].

Because of the advantages of FTIR spectrometers and their broad availability, attempts have also been made to perform polarization measurements on such spectrometers. An experimental setup for polarization modulated Fourier transform in-

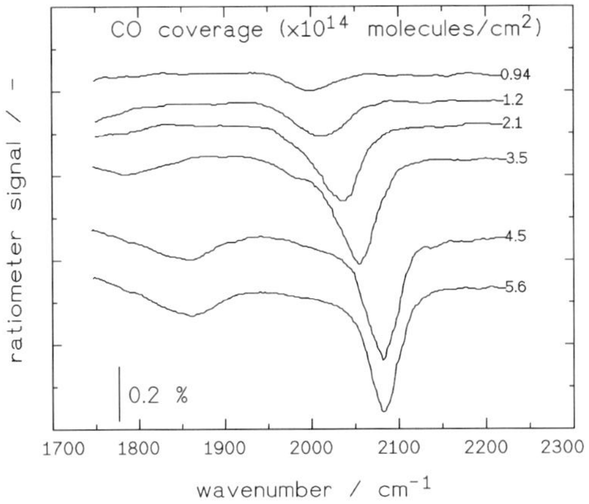

Fig. 5.52. IRRA spectra of CO_{ad} on polycrystalline platinum in 0.5 **M** K_2SO_4 electrolyte solution, $E_{m,SCE} = -0.35$ V; based on data in [222]

frared reflection absorption spectrometry (PMFTIRRAS) has been described [213, 214] based on a spectrometer of the Genzel–Happ design [215, 216]. This type of interferometer shows an almost perpendicular incidence of infrared radiation on the beamsplitter and has rather equal transparency for all planes of polarization of the incoming light. Beyond the addition of a polarizer and a photoelastic modulator, extensive additions to the electronic data processing hardware and software of the spectrometer are required. Results reported so far pertain to numerous different adsorbates, interphases and species of interest, including dye films [217]; for selected examples (see p. 85 ff.). The described modified PMFTIRRAS methodology results in imperfect suppression of contributions from solution phase species [218].

Results of a typical application that nicely demonstrate the advantage of polarization modulation pertaining to CO adsorption on a platinum electrode as a function of the degree of CO coverage are shown in Fig. 5.52 [219].

The shift of the CO absorption band appearing 1978 cm^{-1} at the lower coverage limit of $0.94 \cdot 10^{14}$ molecules$\cdot cm^{-2}$ to 2063 cm^{-1} at the upper coverage limit of $0.5.6 \cdot 10^{14}$ molecules$\cdot cm^{-2}$ has been attributed to electronic coupling between the adsorbed molecules, which is obviously more effective at higher coverages [220, 221]. The mode around 1800 cm^{-2} assigned to CO adsorbed in "bridging" positions (i.e. interacting with two platinum atoms instead of one as in the "on-top" position) appears only when the "on-top" adsorption sites are filled; in addition, these species have a lower infrared absorption cross section [222]. Carbon monoxide adsorbed onto palladium layers growing on a Pt(111) electrode surface was used as a probe to monitor the mode of layer growth and palladium deposition [223]. Further details of CO adsorption on metal surfaces as studied *ex situ* have been reviewed elsewhere [224]. The use of theoretical methods, particularly density functional theory (DFT), to the understanding of the metal–CO bonds (including some other related adsorbates) has been reviewed [225].

A Michelson interferometer (the type most frequently employed in commercial FTIR spectrometers) has largely different transparencies for s- and p-polarized light and is actually an effective polarizer. Attempts to use this type of spectrometer and the spectral mathematics outlined above, which would be simple even without extensive modification of the spectrometer, have failed. Applying various normalization procedures as initially suggested by Golden et al. [217], the single beam spectra (i.e. spectra obtained at a selected plane of polarization and electrode potential) were also normalized and ratioed without success. In a procedure described by Westerhoff et al. [226–230], a somewhat different approach was tried. In the sample chamber of the spectrometer, a polarizer mounted on a motor-driven holder was added to the electrochemical cell fixed on the external reflection accessory. By an electrical signal synchronized with the interferogram acquisition, the motor could move the polarizer into positions that resulted in only p- or s-polarized light reaching the electrode surface. Because of the significant differences in signal intensity reaching the detector (mostly due to the polarizing effect of the interferometer, but also due to the polarizing effect of the flat ZnSe cell window), normalization of the spectra was required. It was performed by measuring single beam spectra for s- and p-polarized light both in the absence and the presence of the adsorbate to be studied. The former state can be achieved by exchange of electrolyte solution, oxidative desorption or other means. The measured reflectivities are called R_{p} with and R_{p0} without adsorbate for p-polarized light and R_{s} with and R_{s0} for s-polarized light, respectively. The final spectrum is calculated according to[16]

$$\frac{\Delta R}{R} = \frac{R_{\mathrm{p}} - R_{\mathrm{p0}}}{R_{\mathrm{p}} + R_{\mathrm{p0}}} - \frac{R_{\mathrm{s}} - R_{\mathrm{s0}}}{R_{\mathrm{s}} + R_{\mathrm{s0}}}$$

A typical set of results of a study of the adsorption of CO on oxide-free nickel that employs this methodology is displayed in Fig. 5.53.

The positions of both the end-on and the bridged adsorbate can be easily identified as a function of electrode potential. For comparison, SNIFTIR spectra were recorded in the same spectroelectrochemical setup as shown in Fig. 5.54.

The shift of both bands to higher values with increasingly positive electrode potential is obvious, but because of the bipolar nature of the recorded bands less obvious and well defined when studied with SNIFTIRS. Recording the spectra of the electrode free of any adsorbate sometimes causes difficulties. In the worst case the electrode has to be pulled back from the electrode and, after solution exchange, purge, etc., it has to be pushed against the window again. In most cases, this will result in slightly changed optical adjustments and, consequently, spectral artifacts. A modification of the method combining potential and polarization modulation has thus been proposed [231]. As a reference state, the electrode at a potential E_{r} is taken. The measured reflectivities are called R_{pm} and R_{pr} for p-polarized light and

[16] Choice of signs and sequence are important when assigning observed bands, which point upwards and downwards with respect to the baseline, to adsorbates being present at the two electrode potentials E_{m} and E_{r}.

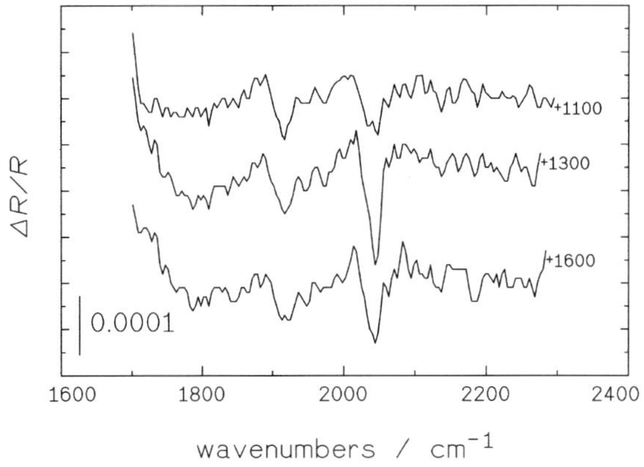

Fig. 5.53. PMFTIRRA spectra of CO_{ad} on a polycrystalline nickel electrode in contact with an aqueous 0.1 **M** $KClO_4$ electrolyte solution, E_{RHE} as indicated

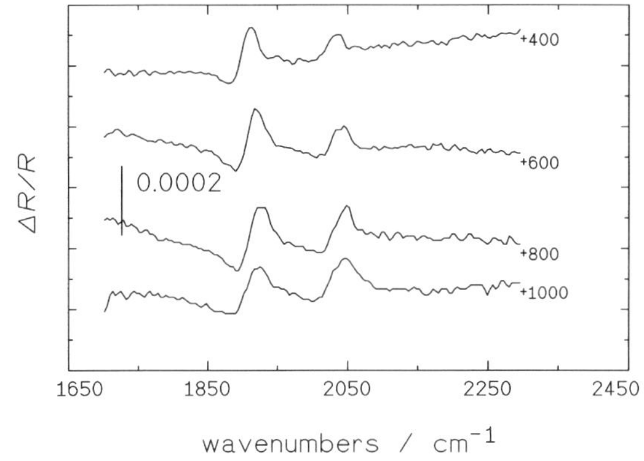

Fig. 5.54. SNIFTIR spectra of CO_{ad} on a polycrystalline nickel electrode in contact with an aqueous 0.1 **M** $KClO_4$ electrolyte solution, $E_{m,RHE}$ as indicated; = 0.1 V (top spectrum), 0.2 V lower spectra; 8*100 or 16*100 interferograms

R_{sm} and R_{sr} for s-polarized light, respectively. The final spectrum is calculated according to

$$\frac{\Delta R}{R} = \frac{R_{pr} - R_{pm}}{R_{pr} + R_{pm}} - \frac{R_{sr} - R_{sm}}{R_{sr} + R_{sm}}.$$

Assuming again that reversible changes occur between both electrode potentials, repeated switching is possible, which results again in considerably improved signal-to-noise ratios after data treatment.

In the following overview of further typical examples, no attempt is made to distinguish between results obtained via potential and polarization modulation techniques. An early overview of investigated systems is available [199].

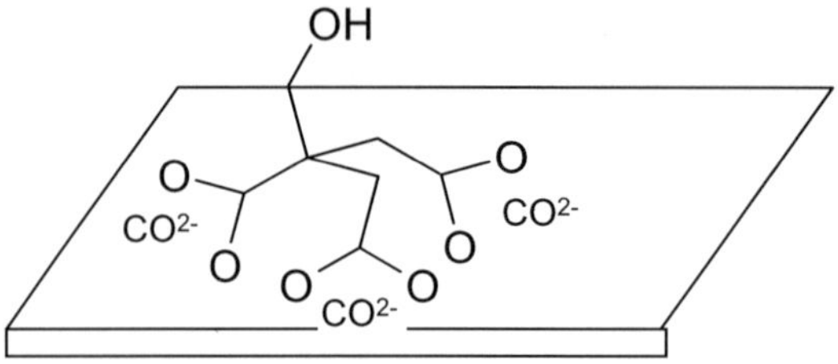

Fig. 5.55. Model of citrate anion adsorption on a Au(111) surface (based on data in [235])

In a typical study of the adsorption of an organic molecule, the interaction between the anion of citric acid and a Au(111) surface has been investigated with SNIFTIRS [232]. Evaluation of the observed bands together with spectral simulation resulted in a model of the adsorbate structure, as depicted in Fig. 5.55, where the carboxylic groups fully deprotonated and interacted with the metal surface. The carboxylate groups are slightly tilted.

Water orientation as a function of electrode potential at the solution/platinum electrode interface has been studied with SNIFTIRS [233]. A change of the water dipole as implied by electrostatic considerations was deduced. Differences in the behavior of H_2O and D_2O were interpreted in terms of differences in hydrogen bonding. A flat orientation of the water molecule around the E_{pzc} and with the electrode potential shifting gradually in anodic direction an orientation with only the oxygen atoms directed to the surface at electrode potentials shifted gradually in anodic direction was derived. The adsorption of phosphoric and trifluoromethanesulfonic acid on platinum surfaces has been studied with SNIFTIRS and the radiotracer method [234]. Coverage data obtained with both methods agree and deviations at very positive potentials were attributed to chemical reactions. Structural models of the anions in the electrochemical double layer were proposed. Sulfate adsorption on Au(111) from solutions of various pH-values as a function of electrode potential has been studied; results were compared with those obtained with STM [235]. Results imply no marked influence of the adlayer ordering observed with the scanning probe microscopy on the vibrational spectroscopy features. Adsorption of 2-halo-benzoic acids on a Au(100) surface has been investigated with SNIFTIRS [236]. Results indicate a change of orientation from flat at negative potentials to vertical with both oxygen atoms oriented towards the gold surface at positive potentials. In the case of 2-fluorobenzoate, evidence of the formation of an ordered adsorption layer was found.

Beyond adsorption/desorption studies, including both the dynamics of the process and the structure at the interface/of the interphase, measurements of various deposits showing infrared active moieties have been reported. They include redox processes at lead electrodes proceeding during charging/discharging processes in lead acid batteries [237]. A quantitative determination of the amount of lead sulphate formed at the lead/solution interface was made by integrating the area under the absorption band assigned to the asymmetric vibrational ν_4 mode of the sulphate mode. Contributions from the solution were removed by spectral subtraction of a solution spectrum recorded under suitable conditions. Numerous other molecular adsorbates have been investigated. In a typical study [238], the angle of the C2-axis and the direction normal to the surface could be calculated ($64 \pm 4°$) for 2-mercaptobenzimidazole adsorbed on a Au(111) electrode. Deposition of gold and silver and their alloys on glassy carbon and silicon substrates has been monitored by measuring the infrared bands of cyanide ions present in the electrolyte solution. Two different silver complexes $[Ag(CN)_2]^-$ and $[Ag(CN)_3]^{2-}$ and one gold complex $[Au(CN)_2]^-$ were identified with SPAIRS [239]. In electrocatalytic studies, the oxidation of ammonia at various platinum single crystal surfaces in contact with an alkaline electrolyte solution was investigated [240]. NH_2 was found as an intermediate. Atmospheric corrosion of zinc caused by organic constituents in thin water films on the metal surface has been studied using SNIFTIRS, with time as the variable instead of the modulated electrode potential [241]. In the presence of formic acid in the water film, zinc formate was observed as the dominant corrosion product.

The use of an infrared microscope enables the investigation of the surface of rather small electrodes. The resulting miniaturization of the necessary electrochemical cell allows its operation as a flow cell in thin layer arrangement [242]. Combined with a rapid-scan FTIR spectrometer, acquisition of infrared spectra during electrode potential scans at a rate of $\mathrm{d}E/\mathrm{d}t = 200\ \mathrm{mV \cdot s^{-1}}$ are possible. The time resolution is equivalent to one complete spectrum recorded every 2.6 mV. The formation of various reaction intermediates of methanol oxidation in alkaline solution at a platinum electrode could be assigned to specific electrode potential ranges.

A drawback of polarization modulated IR spectroscopy using a FTIR spectrometer is the slow scan rate of the moving mirror in the interferometer, which is necessary to allow separation of the signals using a lock-in amplifier, and the increase of noise by this amplifier [243]. Digitally sampling the interferogram at positions corresponding to the two positions of the polarizer (i.e. to the two planes of polarization of the incident light) has been reported as a significant improvement [244, 245]. Applications of this improved system in investigations of imidazole films deposited on copper electrodes [246] and of dye films on polycrystalline electrodes have been described [247].

Infrared reflection absorption spectroscopy [17] (polarization modulated spectroscopy with a dispersive spectrometer) has been used in studies of the kinetics of the formation of CuSCN multilayer films [248]. An advantage of this method—absolute spectra showing no differential bands, which are difficult to interpret—became obvious in a study of CO adsorption on platinum [222]. With the various neutral electrolyte solutions that are used, significant differences were observed in the results obtained with acidic solutions. The mode of CO adsorbed in the bridged position grows considerably in intensity and the position of the "on-top" CO_{ad} is slightly shifted (see Fig. 5.56).

[17] The terms PM-IRRAS, IRRAS and even RAIRS seem to be used in the literature with the same meaning.

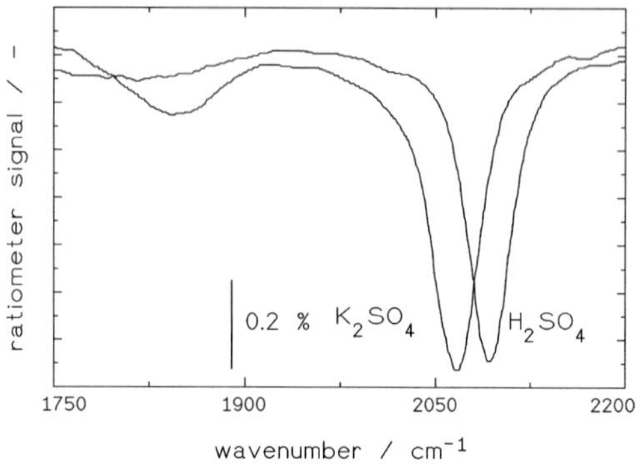

Fig. 5.56. IRRA spectra of CO_{ad} on polycrystalline platinum in 0.5 **M** electrolyte solutions (as indicated), $E_{m,SCE} = 0.15$ V for H_2SO_4 solution; $E_{m,SCE} = -0.35$ V for K_2SO_4 solution; (based on data in [222])

Structural changes in Langmuir–Blodgett bilayers of DMPC[18] that are caused by electrode potential changes have been studied with PMIRRAS [249]. Physical state and molecular arrangement formed at the air/water interface are mostly preserved when bilayers are formed according to the Langmuir–Blodgett and Langmuir–Schaefer method.

In a simplified experimental approach, no selection of a plane of polarization is attempted; thus solution phase species dominate the spectrum. Setups suitable in a broad range of temperatures (−55 to 80°C) with an external reflectance attachment (i.e. without fiber optics) have been described [250–252]. In a standard accessory for liquid samples with two CaF_2 windows, one window is replaced by a Teflon disc of the dimensions of a window. A brass rod with a platinum foil welded on top as the working electrode is pressed into a hole in this disc and a silver wire is pressed into a groove around the circular working electrode as a reference electrode. In one of the filler tubes attached to the back of the Teflon disc, a platinum wire is inserted as counter electrode. The setup is mounted in a standard external reflection accessory. Reported examples of applications include iron carbonyl complexes [253] and bridged ruthenium cluster complexes [254] showing evidence of intramolecular electronic coupling between the ruthenium ions in the infrared range. In a study of a trinuclear ruthenium cluster with isocyanide ligands, unusually broad infrared bands were observed in a selected combination of redox states of the participating ruthenium atoms [255]. As an explanation, Fermi resonance between various modes of the ligand system was suggested. Based on the low wavenumber transitions observed with the Creutz–Traube ion in aqueous solution, a complete delocalization of the odd electron between the two Ru centers on the vibrational time scale has been suggested in good agreement with the results of other spectroscopies [256]. In a series of trinuclear nickel complexes, the bridging behavior of the capping isocyanide or carbonyl ligand upon single electron reduction was de-

[18] 1,2-dimyristoyl-*sn*-glycero-3-phosphocholine.

rived from the infrared measurements [257]. The electrocatalytic reduction of CO_2 with binuclear copper complexes enabled the identification of CO as a reduction intermediate [258]. In a similar setup, one plate in a thin layer flow cell was a quartz disc coated with a platinum layer as the working electrode. The other plate was a CaF_2 disc with platinum counter and pseudo-reference electrode deposited onto acting as cell window [259]. The setup was mounted in an external reflectance accessory. In a study of arylisocyanide chromium complexes using this cell, changes in internal bonding inside the ligand and between the ligand and the metal ion were deduced from changes in band position and intensity. Structural features of electrochemically generated products of iron cyclopentadienyl carbonyl complexes have been elucidated [260]. In a study of a selection of complexes containing various metal ions (besides tin iron, molybdenum, manganese) and phenyl as well as carbon monoxide ligands, structural changes upon electrochemical conversion were identified [261]. Carbon dioxide reduction products formed in the presence of iridium 1,8-diisocyanomethane complexes as catalysts have been identified [262].

The structure of the interphase between a salt melted at room temperature and a gold surface has been studied with SNIFTIRS [263].

5.2.4 Attenuated Total Reflection Spectroscopy[19]

Fundamentals. If an infrared beam passing through a medium is reflected at an interface, it impinges back into the medium provided the angle of incidence is larger than an angle of incidence typical of the medium (Brewster angle, angle of total reflection) [264]. The electric field extends a very small distance beyond the phase boundary (evanescent wave). For a given wavelength λ_1 this depth of penetration d_p can be calculated according to

$$d_p = \frac{\lambda_1}{\sqrt{2\pi(\sin^2 a - (n_p/n_c)^2}},$$

assuming refractive indices n_p and n_c for sample and crystal material, respectively, at an angle of incidence α. Any species present within this distance (about 200 to 300 nm or even higher with materials showing extremely large values of n_K; the actual value of this decay length depends upon the mentioned parameters[20]) may absorb infrared radiation provided it has infrared active modes of vibration with a change of the dipole moment perpendicular to the interface. The reflected beam shows these corresponding absorptions. With a suitable optical arrangement, as outlined below, the reflected beam can be directed once more (or even more often) towards the interface. Consequently the absorption bands will increase in intensity. Another way of enhancing absorption (surface enhanced infrared absorption) is discussed on p. 94.

[19] This method is infrequently called internal reflection spectroscopy.

[20] The actual depth of information also depends on the modulation technique employed; the reported value refers to electrode potential modulation.

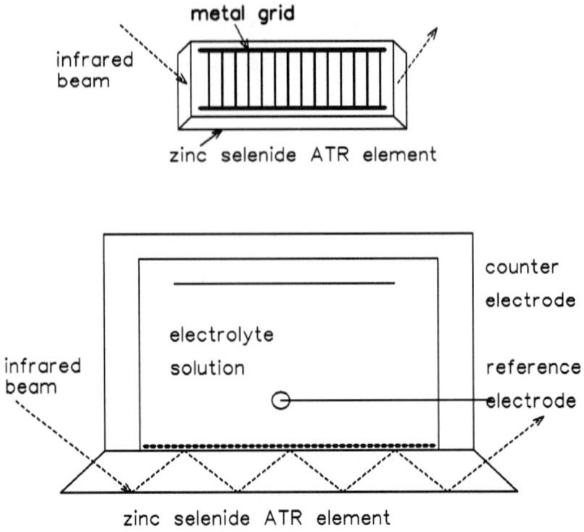

Fig. 5.57. Schematic design of an ATR element coated with a metal grid as conducting support for a polyaniline layer (*top*); three-electrode cell arrangement for FTIR-ATR spectroelectrochemistry [268] (*bottom*)

Instrumentation. An attenuated reflection element (ATR element), i.e. a crystal of a material like zinc selenide or silicon with suitable transparency in the infrared range cut into a shape as depicted below (Fig. 5.57) is coated on one side with the electrode material to be investigated (gold, silver, platinum, iron, etc.) or with a metal grid, which in turn is coated with the material to be studied (e.g. a gold grid coated with a layer of polyaniline [265]). The coated side is attached to a vessel of glass that is used as electrochemical cell. The coating is used as the working electrode. A reference electrode and a metal wire that is used as counterelectrode are immersed in the electrolyte solution inside this cell. The fundamental advantage of this method as compared to external reflection approaches is obvious. The electrolyte solution in contact with the working electrode provides an almost unlimited supply of reactands and there are no limitations caused by thin layer arrangements.

The construction of a cell permitting both FTIR measurements and electrochemical impedance measurements at buried polymer/metal interfaces has been described [266]. Ingress of water and electrolyte, oxidation (corrosion) of the aluminum metal layer, swelling of the polymer and delamination of the polymer were observed. A cell suitable for ATR measurements up to 80°C has been described [267]. The combination of a cell for ATR measurements with DEMS (see Sect. 5.8.1) has been developed [268]. It permits simultaneous detection of stable adsorbed species and relatively stable adsorbed reaction intermediates (via FTIR spectroscopy), quantitative determination of volatile species with DEMS and elucidation of overall reaction kinetics. An arrangement with a gas-fed electrode attached to the ATR element and operated at $T = 60°C$ has been reported [269]. In this study, the establishment of mixed potentials at an oxygen consuming direct methanol fuel cell in the presence of methanol at the cathode was investigated. With infrared spectroscopy, carbonate-containing species formed at the cathode as a result of methanol crossover were detected. Complete methanol oxidation prevailed at elevated temperatures.

Fourier transform infrared spectrometers were used almost exclusively in reported applications. Thus a single beam spectrum designated as a "reference spectrum" is initially recorded. Subsequent single beam spectra recorded at different electrode potentials or as functions of time are ratioed versus this "reference spectrum". Bands appearing in these differential spectra reveal changes in the coating or in the adsorbate or surface layers on the coating that are affected by the changed electrode potential or that occur as a function of time. Accordingly, this method has been designated as a potential dependent attenuated total reflectance Fourier transform infrared spectroscopy (PDATRFTIRS). The agreement of results obtained with this method and with external reflection spectroscopy PDFTIRRAS has been pointed out [270]. Time resolved measurements for kinetic studies are possible with a sufficiently fast instrument [271]. Deposition of the metal film on a hemispherical silicon (or any other suitable material for an ATR element) allows wide variations of the angle of incidence, although obviously only a single reflection is possible. With a platinum film working electrode the refractive index of the metal was found to be different from the bulk value [296]. This was ascribed to the surface corrugation of the evaporated film with only nanometric thickness as observed with scanning tunneling microscopy. Considerable changes in reflection intensity as a function of angle of incidence were observed when resonant absorption in the solution medium occurred. The angle dependent infrared absorption band intensity (not to be confused with the reflection intensity) showed considerable changes in magnitude up to inversion for CO adsorbed on platinum. The excitation of resonant surface plasmon waves at the interface was assumed to be the cause; a modelling of the three-layer system, silicon/platinum/solution, supported the experimental observations and their explanation.

A cell design employing attenuated total reflection without depositing the working electrode as a thin film on top of the ATR element has been described by Visser et al. [272]. In this setup the three-electrode cell with a platinum wire mesh working electrode is mounted on top of an ATR element made of KRS-5 that has a sensing window made of diamond. Consequently the composition of the electrolyte solution in front of this element is probed. A wide spectral window ranging from 250 cm^{-1} to 16700 cm^{-1} is accessible. This arrangement obviously provides no surface sensitivity. Moon et al. [273] have reported on an optical arrangement, where the ATR-crystal is pressed against the electrode surface and the arrangement is surrounded by the electrolyte solution, which also contains counter and reference electrode. This circumvents the need to deposit a metal film as a working electrode and allows measurements on carbon surfaces or finely dispersed metal particles that otherwise cannot be studied with ATR spectroscopy. Although the surface selection rules are not completely maintained, there is nevertheless a large difference between spectra collected with p- and s-polarized light, enabling deduction of adsorbate orientation.

Attenuated total reflection spectroscopy can also be employed in solid state electrochemistry, i.e. in electrochemical systems without a liquid electrolyte solution/melt. These systems usually contain mixed-valence sites (e.g. Fe(III) and Fe(II) sites in hexacyanoferrate systems) between which electron exchange is possible. In addition, mobile counterions may be present. Changes in the spectroscopic properties of the material in the infrared range can be monitored with an ATR setup as described by Kulesza et al. [274]. The material to be investigated is sandwiched between a germanium ATR element, which also acts as a working electrode, and an additional germanium disc used to fix the material to be investigated on the ATR-element; in addition it serves as both counter and reference electrode. The probed thickness of the material was about 1 μm.

5.2.5 Surface Enhanced Infrared Absorption Spectroscopy (SEIRAS)

Fundamentals. Hartstein et al. [275] observed for the first time that infrared absorption of organic films evaporated onto a silicon substrate was significantly stronger when a thin layer of silver was evaporated onto the adsorbate layer (metal overlayer geometry). A similar enhancement was observed when the organic adsorbate was deposited onto silver-coated substrates like CaF_2. Enhancement of infrared absorption of species in close proximity to thin metal layers caused by surface plasmon polaritons (transverse collective electron resonance) causing a locally increased electromagnetic field has been observed frequently thereafter [276–278] and has been treated theoretically [279, 280]. The abnormally high absorption of IR active modes has been found mostly on structured surfaces that show surface features of a nanometric size (e.g. metal islands of about 30-nm diameter and 10-nm height). Obviously, similarities to surface enhancement as observed with Raman spectroscopy can be found (although SEIRAS is not limited to certain metals, particularly coinage metals). This includes the importance of electromagnetic contributions [281]. Other contributions similar to the charge transfer or chemical enhancement mechanism proposed for SERS are discussed [282–284]; for an overview, see [285–287]. Chemisorbed molecules tend to show larger enhancements than physisorbed ones. Chemisorption based on a stronger surface-adsorbate interaction tends to align molecules at the interface. Thus more molecules will interact with the infrared light provided that the vibrational modes under investigation have a significant component perpendicular to the metal surface. In cases of randomly oriented adsorbate molecules, this will be less pronounced. Further contributions by donor-acceptor interactions that result in large enhancement for modes involving strongly polarizable groups are conceivable [285]. Theoretical calculations predict SEIRA if the size and shape of particles as well as their proximity meet certain criteria [285]; for further fundamental considerations, see [288, 289]. Enhancement applies for all polarization states of the employed radiation [290]. Typical enhancement values for CO adsorbed on gold film electrodes prepared by vapour deposition are about 20; flame annealing results in a preferential 111 orientation of the gold surface and an increase of the enhancement factor to about 40 [291]. Surface selection rules apply, as discussed for unenhanced reflection spectroscopy. At first glance,

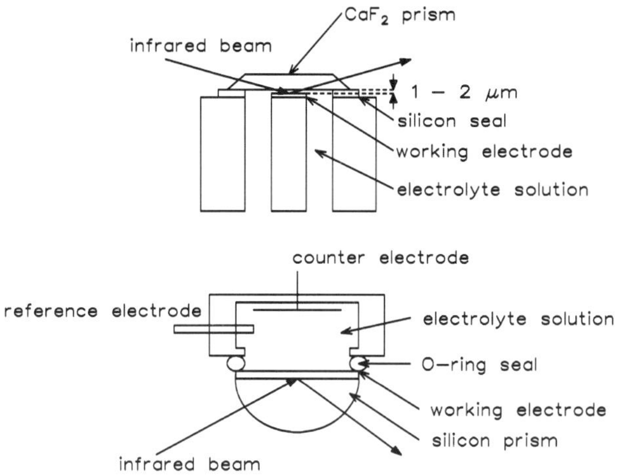

Fig. 5.58. Spectroelectrochemical cells and optical arrangements for SEIRAS in external reflection (*top*) and attenuated total reflection (*bottom*)

this might look surprising because of the many different surface orientations of randomly deposited metal island films. Assuming that the electric field exciting the adsorbate vibrations is always perpendicular to the local metal surface, it seems reasonable. The range of the enhancement (i.e. the distance wherein surface enhancement is effective) is about 5 nm [285]. Observed band intensities depend on the angle of incidence and the plane of polarization of the employed light. An alternative explanation of the enhancement that assigns the enhanced radiation absorption not to the adsorbed molecules themselves but to a change in the effective dielectric function of the metal island film at the frequencies of the molecular vibrations has been proposed elsewhere [292]. The excitation of resonant surface plasmon waves at the metal/solution interface has been invoked as an explanation of the angle dependency of the magnitude and—as observed under certain circumstances—the inversion of infrared absorption bands [293]. Further unusual features of infrared absorption bands generally called AIRE (abnormal infrared effects) like inversion of absorption bands, change from single-side to pseudo-differential bands and increased band width have been discussed elsewhere [294].

Instrumentation. In order to employ local enhancement of infrared absorption by surface plasmon polaritons that cause locally enhanced surface electromagnetic fields, a suitable optical arrangement is needed [295]. Surface enhanced infrared absorption spectroscopy can also be observed in the transmission mode [285, 296]. However, since no application of this approach in spectroelectrochemistry has been reported so far, it is not discussed further.

The Kretschmann design (see also Sect. 5.9) employs a hemicylindrical or triangular prism of germanium coated with a thin layer ($\approx$10 nm) of the metal to be investigated (e.g. silver). In this ATR arrangement, the slight enhancement of the incoming light intensity as described elsewhere is also operative [267]. The preparation of suitable electrode layers fulfilling the morphological requirements mentioned above by deposition of various metals on ATR crystals has been described [297]. Chemical deposition and sputtering of silver layer electrodes for SEIRAS in the Kretschmann mode have been studied extensively [298]. Optimum deposition conditions with respect to maximum intensity could be identified. This coated surface is brought into contact with the electrolyte solution. A standard three-electrode arrangement can be used. The infrared beam enters and leaves the crystal from the backside. The polarization-modulation technique can be used. As a background spectrum, the polarization-modulated spectrum obtained with the empty cell is used as a check for the shape of the baseline [299]. Absorption bands caused by adsorbed species are obtained as a function of electrode potential.

In several studies, adsorption phenomena of both solvents and dissolved species have been investigated. The potential dependent orientation of water molecules at the Au(111) interface in the presence of sulfuric acid has been reported [300]; for more recent results, see [301]. Suggested spatial arrangements of both water molecules and electrolyte anions at the electrochemical interface are depicted in Fig. 5.59.

At low sulfate ion coverage the ice-like structure is not disturbed by the sulfate anions (they occupy water molecule positions) and, at higher degrees of coverage and more positive electrode potentials, water molecules bridge adjacent sulfate anions in various conceivable positions, resulting in slightly different spacings. Structural details of the water layer adjacent to a silver electrode have been reported in a study of acetate anion adsorption [302]. Interaction of the acetate ion via its two oxygen atoms was concluded.

The high sensitivity of SEIRAS (see above) allows measurements in real time during a slow electrode potential scan [303–305]; for particularly fast acquisition, step-scan interferometers may be used [306]. A series of time-resolved SEIRA spectra recorded during reduction of heptyl viologen HV^{2+} to $HV^{\bullet+}$ at a silver electrode in an aqueous solution of 0.3 **M** KBr is displayed in Fig. 5.60 [274].

Spectra were recorded at 100 μs intervals and were displayed in 1 ms intervals. Bands caused by the formed radical developed as a function of time. It precipitated onto the electrode as a film by association with supporting electrolyte anions according to $HV^{\bullet+}Br^{-}$. Band intensities correlated well with coulometrically determined charge consumption and they were proportional to $t^{3/2}$ for $t < 12$ ms and to $t^{1/2}$ for $t > 12$ ms. These time dependencies imply a film formation via initial nucleation and subsequent diffusion-controlled growth.

A significantly improved time resolution on the picosecond scale has been reported [307, 308]. The method has been applied to study the potential jump at the electrochemical interface platinum/aqueous solution of 0.1 **M** $HClO_4$ saturated with carbon monoxide induced by pulse irradiation with visible light. Position and intensity of the infrared absorption caused by the CO stretching mode were measured as a function of time after the 532 nm pump pulse.

The red shift of the stretch mode observed with a delay of about 200 ps is attributed to a negative shift of the electrode potential and the heating of the in-plane

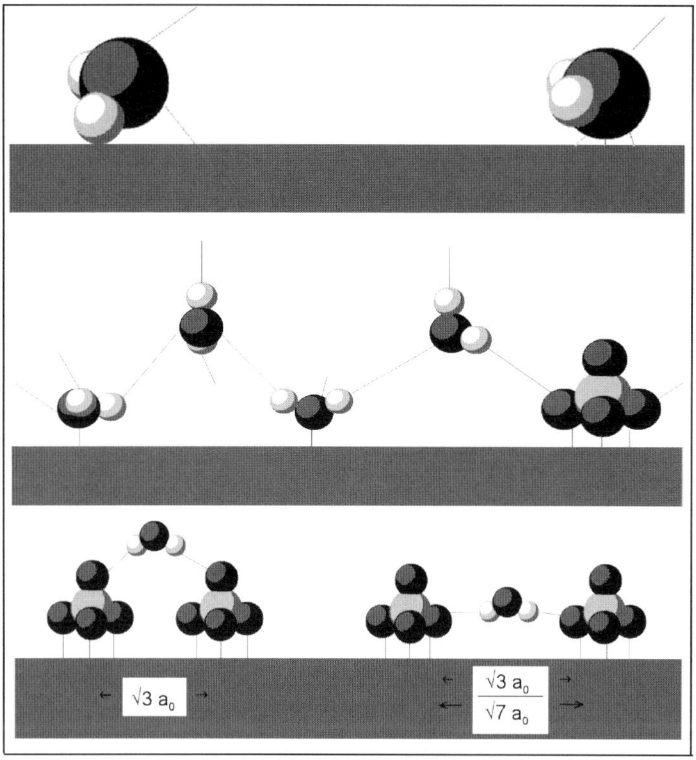

Fig. 5.59. Orientations of water molecules on the Au(111) surface at electrode potentials negative to the E_{pcz} (*top left*), around the E_{pcz} (*top right*), icelike water structure with small coverage of coadsorbed sulfate ions (*middle*) and interfacial structure with large sulphate ion coverage (*bottom*) with water molecules in various conceivable positions (based on [303])

frustrated translational mode of the adsorbed CO (see Fig. 5.61). The established potential jump is caused by the heating of water layers near the metal surface. Measurements of the change of metal reflectivity as a function of time enabled calculation of changes of both surface and solution temperature. The laser light of the pump pulse does not interact with adsorbed CO or water and is only absorbed by the electrode metal. Thus thermal energy transferred will diffuse into metal and solution and any change of the respective temperatures will be delayed with respect to the pulse. Calculations indicate an almost instantaneous temperature change at the metal surface by about 70 K, whereas the temperature change in water layers several nanometers away from the metal amount to a few tens of Kelvins with a delay of about 200 ps. Thus the temperature jump in the water layers adjacent to the metal contributes to the infrared band shift. Separation of the band shift into contributions attributed to the temperature jump of the metal surface and to the potential change in solution caused by the temperature change in the water layers and the associated orientational change of water molecules close to the interface was possible.

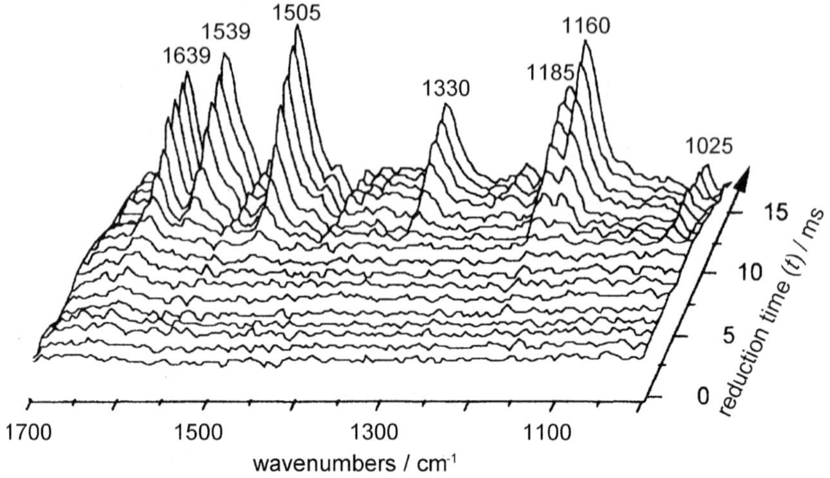

Fig. 5.60. Time-resolved SEIRA spectra of the reduction of HV^{2+} to $HV^{\bullet +}$ at a silver electrode in an aqueous solution of 0.3 **M** KBr (based on data in [274])

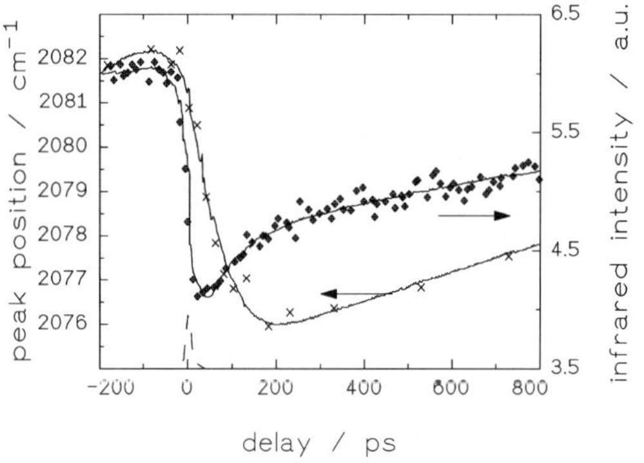

Fig. 5.61. Time-resolved SEIRAS. Pump pulse 532 nm, 35 ps duration, 3 mJ cm^{-2}, temporal profile indicated by *dotted line*; position and infrared absorption intensity of CO stretch mode plotted as a function of time, platinum electrode, aqueous solution of 0.1 **M** $HClO_4$ (based on data in [310])

The transition between physisorption of cytosine on a gold electrode at negative and chemisorption at the positive electrode potential was deduced from changes in SEIRAS spectra in agreement with the charge transfer observed in cyclic voltammograms [309]. In the case of pyridine adsorbed on a Au(111) surface, spectral changes observed as a function of electrode potential were attributed to a reorientation; band intensities were not found to correlate with surface coverage of the gold electrode with pyridine [310]. In a comparative study of pyrazine adsorbed on polycrystalline gold with SEIRAS and SERS end-on adsorption on the gold surface was deduced from SEIRAS data. Comparison with SERS data provided arguments related to the surface enhancement mechanism in SERS [311]. Self-assembled monolayers of benzenethiol on a Au(111) surface have been studied [312]. A symmetry of the well-ordered adlayer commensurate with the supporting gold surface was observed and differences between the amount of adsorbed molecules determined from SEIRAS and cyclic voltammetry were tentatively attributed to coadsorbed sulfur atoms and vacancies.

Further reported examples include electrocatalytic processes and their intermediates [313, 314]. Formate could be identified as an active intermediate of methanol electrooxidation at a polycrystalline platinum electrode [315]. Water molecules coadsorbed during methanol adsorption on platinum were identified as those species that react subsequently with CO_{ad} that was formed as a result of methanol chemisorption [316]. The high sensitivity of SEIRAS allows mapping of two-dimensional spectra (for selected examples, see [285]). Finally, two-dimensional correlation analysis of electrochemical reactions becomes possible [317].

The applicability of SEIRAS to metals other than coinage metals was shown in a study of CO adsorption on platinum deposited chemically as an electrode on an ATR crystal [318]. The extraordinary sensitivity also allowed time- and electrode potential-resolved studies of the oxidation of methanol. Carbon monoxide adsorption on platinum particles that were finely dispersed by electrodeposition into a NAFION® film coated onto a gold electrode was studied [319]. The calculated particle size was approx. 10 nm, thus it is likely that the obtained spectra were surface enhanced. They indicated the presence of CO_{ad} in two chemically different environments. In a comparative study, the adsorption of pyridine at platinum group metal electrodes was studied with SEIRAS [320]. Results revealed electrode-potential dependent adsorbate orientation that was also dependent on the identity of the electrode material. The structure and orientation of water molecules adsorbed on a Au(111) surface [321] was investigated and elsewhere a potential dependent reorientation [322] showing a weakly hydrogen-bonded water below the E_{pzc} and a strongly hydrogen-bonded ice-like structure at electrode potentials slightly above the E_{pzc} was reported. At even more positive electrode potentials the ice-like structure was broken up by specific adsorption of perchlorate anions from the supporting electrolyte. Band shapes and intensities could be correlated to the deduced structural changes. A study of first- and second-layer adsorbates employing the improved sensitivity and spatial resolution has been reported [323]. The distinction between first- and second-layer adsorbed anions could be based on the potential dependency of vibrational modes of anions in the former layer; this dependency was absent in the latter case.

Mechanism and reaction intermediates of the hydrogen oxidation reaction on platinum have been studied with SEIRAS [324]. A band observed around 2090 cm^{-1} was assigned to adsorbed hydrogen species and the band height dependency on electrode potential and hydrogen overpressure was found to match values predicted on the basis of the Volmer–Tafel mechanism.

The interface platinum/NAFION® monomer in the presence of perchloric acid was investigated with SEIRAS [325]. A band showing electrode potential dependence was assigned to the symmetric vibration of the sulfonic acid substituent of the polymer. Changes of band position and intensity suggested an electric field-driven orientation and adsorption.

To prevent denaturation of redox proteins and enzymes upon immobilization on an electrode surface that is necessary for electrochemical investigations, electrodes may have to be modified with suitable biocompatible membranes. Studies of redox processes and associated non-electrochemical processes of bound species with both SEIRAS and SERS have been reviewed [326].

Although in all reported examples spectra were obtained by spectral subtraction of a reference or background spectrum from a sample spectrum, the method has been simply called SEIRAS. An attempt to designate this approach by a different term, SEIDAS (surface enhanced infrared difference absorption spectroscopy), has been reported. In it, the influence of the electrode potential on the spectroscopically relevant changes in cytochrome *c* adsorbed on a chemically modified granular gold surface was asserted [327]. Changes of the internal conformation induced by redox reactions of the adsorbate were observed; they were not disturbed by the interaction between cytochrome *c* and the various sulfur-containing layers used or chemical modification.

5.2.6 Diffuse Reflectance Infrared Spectroscopy (DRIFT)

Fundamentals. Infrared radiation interacting with a surface can be absorbed or reflected. At a more or less ideally reflecting (smooth and planar, mirror-like) surface, most of the reflected light will be observed at an angle identical with the angle of incidence. With rough and corrugated surfaces, a considerable portion of light will be reflected in various directions—basically all positions on a hemisphere on top of the investigated surface will receive some reflected light. This is called diffuse reflection [151]. The spectral information carried by the reflected light is basically the same as with specularly reflected light.

Instrumentation. The interface within a suitably constructed electrochemical cell to be investigated is placed in the sample position of a standard DRIFT accessory for an infrared spectrometer; for a typical design, see [328, 329]. Examples reported so far deal with solid polymer electrolyte fuel cells where the surface of the anode layer exposed to a mixed gas atmosphere containing both water and methanol is separated from the environment via a CaF_2 window [331, 332]. Various oxidized species and penetrating methanol were observed.

5.2.7 Photothermal Deflection Spectroscopy (PDS)

Fundamentals. Absorption of electromagnetic radiation illuminating an electrode surface, particularly the surface layer, or by adsorbates on the electrode surface, can result in thermal effects.[21] These can be detected by various means, as described

[21] The closely related photoacoustic method (wherein no wavelength dependency is taken into consideration) is described in Sect. 5.9.4.

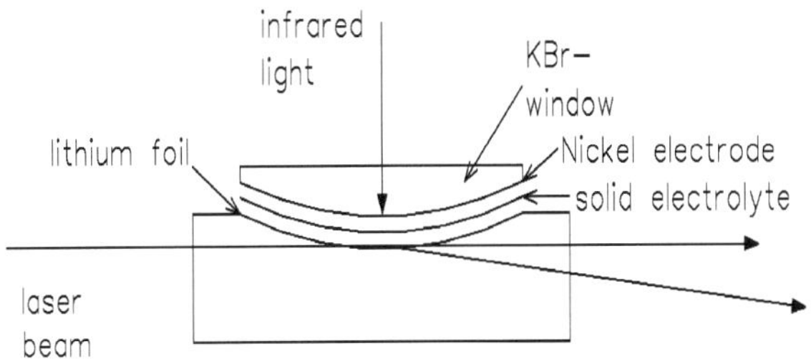

Fig. 5.62. Spectroelectrochemical cell for IRPDS measurements according to [330]

below. Illumination of an interphase in the infrared range causes absorption by vibrational modes of species both in the interphase (e.g. polymer or passivation layers, adsorbed species) and in the solution in front of the electrode surface. The latter absorption is of no interest in most cases and the associated loss of intensity arriving at the electrode surface is an inherent drawback. The actual IR absorption by the species under investigation is rather small compared to this unwanted absorption, causing further experimental difficulties. Ways to overcome these problems in various infrared reflection spectroscopies have been described above. An approach proposed by Deng et al. [330] employing PDS addresses both drawbacks. Illumination of the interphase is performed with a tunable infrared diode laser of much higher intensity as compared to conventional infrared sources. As no reflection is involved, rough and non-reflective surfaces can be studied. Wavelength-dependent infrared absorption in the interphase causes local heating, which in turn generates thermal gradients in the adjacent electrolyte solution. These gradients are measured by passing a laser beam in close proximity to the interphase through this gradient.

Instrumentation. The surface/interphase under investigation is illuminated with a tunable infrared laser. The setup reported by Deng et al. [330] is adapted to investigations of surface layers on lithium metal in contact with thin solid electrolyte films. Thus the thermal gradient has to be detected on the backside of the lithium electrode. As depicted in Fig. 5.62, this is done by attaching the lithium foil electrode to a piece of acrylic coated with a thin evaporated copper layer. The curved electrode surface makes alignment of the laser beam easier; the thermal diffusivity of acrylic is small, resulting in short thermal diffusion length.

The availability of tunable infrared diodes limits the accessible spectral range; in the study by Deng et al. [330], data for the range 600–1000 cm^{-1} could be obtained. LiOH was identified as the dominant species in the lithium/solid polymer electrolyte interphase.

5.2.8 Infrared Emission Spectroscopy

Fundamentals. Emission of radiation by a body at $T > O$ K is described by the laws of Kirchhoff, Stefan–Boltzmann, Planck and Wien with the sum of emission, reflection and transmission being unity [151]. In the mid-infrared range good emission spectra are already obtained with a sample at a temperature around $T = 300$ K provided that the detector is cooled to a lower temperature. Otherwise sample and detector might be in thermal equilibrium without a detectable signal; because liquid nitrogen detectors are used, this requirement is easy to fulfill. According to Planck's law, nitrogen cooled emission decreases sharply at higher wavelengths, thus emission spectra in the CH-region are weak. Emission spectra are equivalent with transmission spectra with respect to band position; with respect to intensity, they are considerably different. Because of the reabsorption of emitted radiation, only species on the surface or very close to it will cause emission bands.

Instrumentation. Measurements are possible with an infrared spectrometer where the emitting sample is placed instead of the radiation source. In FTIR-instruments without an emission port (where the emitting sample is mounted in order to introduce the emitted radiation properly) the sample can also be placed in the sample chamber. Back reflection of the Michelson interferometer (about 50%) provides the needed spectral separation. Reference (background) single beam spectra are obtained with a standard black body; further spectral correction might be necessary [151]. Investigations reported so far are limited to solid electrolytes of solid oxide fuels (SOFC) operating at elevated temperatures (e.g. 550°C), as is typical for SOFCs. The surface of the electrolyte to be investigated, which is coated with the electrocatalyst as used for the electrode reaction, is located in the sample chamber of the FTIR spectrometer under a suitably designed KBr-dome in the sample position of a diffuse reflectance accessory. Emitted radiation is collected, spectrally separated and guided to a liquid nitrogen cooled MCT detector.[22] Single beam spectra (actually, spectral intensity vs. wavenumber relationships) are collected at selected electrode potentials and ratioed vs. a spectrum collected at open circuit electrode potential. Spectral differences showing increased or decreased emittance as a function of wavenumber are indicative of changes in the concentration of the emitting species. Using this technique, various dioxygen species could be identified on a working cathode surface in a solid oxide fuel cell [331].

5.2.9 Far Infrared Spectroscopy

Fundamentals. Absorption of infrared radiation in the range between a few wavenumbers (approx. 10 cm^{-1}) and about 600 cm^{-1} can provide valuable information about interactions between adsorbed species and the electrode surface. In addition, internal molecular modes of metal complexes or inorganic films on the electrode may become observable. The low intensity of most laboratory light sources and the strong absorption by water and numerous gases in this spectral region has hindered investigations considerably. Only one study of lithium deposition on a mercury surface with a nonaqueous electrolyte solution using a conventional light source has been reported [332]. The advent of synchrotron radiation as an intense light source in this region has somewhat changed this situation [333, 334], consequently the corresponding spectroscopy has been named SFIRS (synchrotron far infrared spectroscopy).

[22] MCT: Mercury cadmium telluride.

Instrumentation. An experimental setup with a thin layer electrochemical cell using a reflecting electrode surface in the external reflection mode as described above is used in a standard infrared spectrometer with a detector particularly sensitive in the FIR region [335]. The sample chamber is purged with dry nitrogen in order to minimize light absorption by water vapour. Because of radiation absorption by lattice modes, particularly those involving heavy atoms, most crystalline window materials are not suitable. Instead polyethylene film may be used. Systems that have been investigated so far are adsorbed hydroxyl species on a platinum electrode [336], surface films on a copper electrode [337, 338], oxoanions on gold surfaces [339] and pseudohalide ions on a silver electrode [340].

Alternatively, SFIRS can be done in the attenuated total reflection mode [341]. A gold film is evaporated onto a silicon hemicylinder. Vibrational modes of adsorbed halide ions have been detected [344].

5.2.10 Raman Spectroscopy

Although Raman spectroscopy does not employ absorption of infrared radiation as its fundamental principle of operation, it is combined with other infrared spectroscopies into a joint section. Results obtained with various Raman spectroscopies as described below cover vibrational properties of molecules at interfaces complementing infrared spectroscopy in many cases. A general overview of applications of laser Raman spectroscopy (LRS) as applied to electrochemical interfaces has been provided [342]. Spatially offset Raman spectroscopy (SORS) enables spatially resolved Raman spectroscopic investigations of multilayered systems based on the collection of scattered light from spatial regions of the samples offset from the point of illumination [343]. So far this technique has only been applied in various fields outside electrochemistry [344]. Fourth-order coherent Raman spectroscopy has been developed and applied to solid/liquid interfaces [345]; applications in electrochemical systems have not been reported so far.

Fundamentals. Provided enough substance is present at the electrochemical interface in a fairly thick film of corrosion products or in a modifying layer, identification of the constituent matter is possible with (normal) Raman spectroscopy (NRS). Although this approach is not exactly surface specific, it is included because it has been applied frequently. The same argument applies to Raman spectroscopy applied with the help of a microscopy (Raman microscope).

Instrumentation. An experimental setup suitable for studies of corrosion films formed at the metal/solution interface that employs a microscope has been described [346] and an overview pertaining to corrosion and passivation of metals is available [345]. Measurements at elevated temperatures (up to 200°C) and pressures of 6.9 MPa, resembling the conditions in operating plants, are possible with specifically designed cells described in [345]. Studies of layer formation, particularly on lead and mercury, are favored by the combination of these high atomic number elements, forming large molecules (e.g. Hg_2Br_2) with highly polarizable electronic structures and showing large cross sections for Raman scattering. Green rust corrosion compounds were formed when 1024 mild steel was exposed to phosphate and bicarbonate solutions containing chloride and sulfate ions. A number of further corrosion products was observed during carbon steel corrosion [347]. Stress corrosion cracking of iron in the presence of carbonate-containing solutions has been investigated [348]. Various iron-containing corrosion products were identified. Surface Raman spectra of magnetic nanoparticles composed of maghemite (γ-Fe_2O_3), a surface layer of FeOOH and a positively charged non-stoichiometric Fe(III) oxy-hydroxide layer have been reported with a smooth silver electrode [349]. Presumably the claimed surface enhancement (if any) is caused by the large area of the nanoparticles and their surfaces, which are rich in active sites (micro-roughness). Raman microscopy has been employed in a study of electrochemical intercalation of tetraethylammonium (Et_4N^+) and tetrafluoroborate ($[BF_4]^-$) ions into microcrystalline graphite [350]; a suitable cell setup is described. From changes of the G-band (around 1578 and 1600 cm^{-1} [351]) upon intercalation of tetrafluoroborate into the positive electrode, the formation of staged compounds was deduced; after deintercalation the single G-band (1600 cm^{-1}) was reversibly returned. Upon intercalation the D-band (1329 cm^{-1}) disappears; after deintercalation it reappears with greater intensity, indicating possible lattice damage during the process.

5.2.11 Surface Raman Spectroscopy

At first glance, Raman spectroscopy seems to be a rather unsuitable tool for interfacial investigations. The photon yield (scattered intensity) in the Raman process is notoriously low—values around 10^{-6} going down to 10^{-8} to 10^{-10} have been frequently quoted. Considering the rather limited numbers of potential Raman scatterers at an electrochemical interface, the measurement of a Raman spectrum of an adsorbed species seems to be a pointless undertaking. Surface enhanced Raman spectroscopy will be treated first because of both its overwhelming importance and its popularity. Subsequently other methods, including those wherein presumably no surface enhancement as claimed with the initially described approach is operative, will be discussed. Because a well-defined distinction is lacking and most likely even in the latter some enhancement (because of surface roughness and associated electromagnetic enhancement) is operative, this arrangement seems to be plausible. Conceivable overlap with normal Raman spectroscopy as described above should also be kept in mind.

5.2.12 Surface Enhanced Raman Spectroscopy

The discovery of a particular enhancement effect (up to 10^6) that affects only species in close contact with the metal electrode surface (i.e. adsorbed species) by Fleischmann et al. [352] and slightly later by Jeanmaire et al. [353] and a report on the utilization of resonance enhancement in surface Raman spectroscopy [354] demonstrated surprisingly the feasibility of vibrational studies of electrochemical interfaces with Raman spectroscopy (for a recent introductory overview, see [355]). Although the applicability was initially limited to electrodes prepared from coinage metals (i.e. copper, silver and gold) or species showing electronic transitions close to the wavelength of the illuminating light in resonance Raman studies, these initial reports opened a floodgate to the application of Raman spectroscopy in interfacial electrochemistry. The wide range of accessible wavelengths (100–4000 cm^{-1}), the possible resolution (1 cm^{-1} or better), the fast response of the detection setup (ns) and the lack of interference by water (the most frequently employed electrolyte solvent in electrochemistry) because of its weak Raman scattering have propelled this development. The number of review papers is considerable; however, they are mostly of a general type. Thus information pertaining to a specific adsorbate system is hard to find. Early overviews of investigated adsorbates are helpful [356, 357]; unfortunately they have not been updated. The terminology is also rather complex. Simple surface Raman spectroscopy of non-coinage metals like platinum, the use of illuminating wavelengths that are known not to support surface enhancement (e.g. $\lambda_0 = 514.5$ nm for a gold electrode) or the use of smooth surfaces might be called surface Raman spectroscopy (SRS) because of the absence of the surface enhancement features, particularly typical of the *d*-metals. This methodology has also been called surface unenhanced surface Raman spectroscopy (SUERS) [358]. Spectroscopy employing this particular enhancement feature is generally called SERS. Spectroscopy performed with surfaces where enhancement has been quenched by deposition/adsorption of foreign metal atoms [359, 360] has been designated deenhanced SERS (DESERS) [362] (see p. 104). Raman spectroscopy of molecules, adsorbates, films, etc. that is resonantly enhanced because of the optical absorption properties of the species on the electrode surface is called surface resonance Raman spectroscopy SRRS (see p. 125). When rough, surface enhancement active coinage metals are used as support, the method is termed SERRS. The possibility of using focused laser beams for illumination, particularly when combined with a microscope, provides considerable spatial resolution in surface studies. Since light from sites providing particularly effective surface enhancement will dominate, the scattered signal assignment of observed modes to specific surface features (kinks, edges, local adsorbates, etc.) is possible.

Fundamentals. A polyatomic molecule exposed to a monochromatic light scatters electromagnetic radiation. The radiation scattered exactly (elastically) at the illuminating frequency is called Rayleigh scattering and is of minor interest in surface studies; mostly it is a nuisance because of its large intensity, which extends up to a frequency (wavenumber) region where sometimes low-energy vibrational modes of adsorbate-surface interactions occur. In addition, scattered light at higher wavenumbers (anti-Stokes) and lower wavenumbers (Stokes) can be observed. The former results from species being initially in an excited vibrational state and relaxing into a vibrational ground state, the latter is caused by species starting in the vibrational ground state and ending in an excited vibrational state. In this model, assuming excitation into a virtual electronic state, resonance enhancement can be explained by invoking excitation into a higher real electronic state of the molecule.

Scattered light is observed in all cases only when the polarizability of the illuminated species changes during the involved vibrational mode (Raman selection rule) [165, 205].

In a quantitative study of some representative adsorbates, Weaver et al. [361] have determined surface enhancement factors for numerous internal modes of species that are considered to be specifically adsorbed ("inner-sphere", chemisorbed) or non-specifically adsorbed ("outer-sphere", physisorbed). Factors were found to always be between 1 and $10 \cdot 10^5$ provided that no resonance effect interfered. This implies that a particularly strong adsorptive interaction is no prerequisite for effective enhancement. Numerous reviews that cover fundamental aspects of enhancement mechanisms [289, 362, 363, 369] and various applications have been published [370–380]. Under near field conditions, enhancement factors up to 10^{10} can be rationalized [381]. Surface enhancement as observed with rhodamine 6G deposited on a self-affine silver surface is localized down to less than 200 nm portions of the surface [382]. These "hot spots" are not necessarily topographic elements like interstices or gaps between nanoparticles. This localized enhancement is assumed to be 10^3 larger than the enhancement observed with conventional far-field measurements. Despite the still ongoing discussion of the various enhancement mechanisms and their contribution to the observed overall enhancement, there seems to be general agreement that an electromagnetic effect (EM) associated with microscopic roughness features and the correspondingly increased local electric field strength of both the incident radiation and the inelastically scattered Raman radiation and a chemical effect (CT,[23] this contribution is sometimes also called chemical mechanism [383]) associated with adsorbate-metal interactions (but not necessarily with strong chemisorption) are operative. Their actual contributions depend on the type of metal, adsorbate, metal surface roughness and, conceivably, further influences. The EM contribution can be associated with surface plasmons, which can be excited with visible light in the case of the coinage or "free-electron" metals copper, silver and gold [384]. In an early overview, optical resonances were used as a more general term [385]. Using this concept, the dependency on the illuminating light wavelength, the variation between metals (in this case between Cu, Ag, Au and Li) and the distance between scattering molecule and surface as well as surface roughness effects could be explained. It is closely related to the particle size and shape ("lightning rod effect") [386] and it scales with the distance d between the involved scattering species and the radius r of a spherical particle according to $[r/(r+d)]^{12}$ [387]. The actual dependence on d at distances beyond 2–3 nm shows a slight decay only by a factor of $1/10$ [388, 389]. An additional image dipole enhancement effect has been proposed [389]. Support for this model (also called the image field model) has been gained from measurements of pyridine adsorbed on copper with various illuminating wavelengths [390]. Interactions between size features of illuminated particles and plasmon bands have been investigated with nanorods of silver

[23] CT refers to charge transfer because some explanations of the chemical enhancement effect assume a charge transfer between the metal and the adsorbate, resulting in the observed enhancement of scattered Raman radiation.

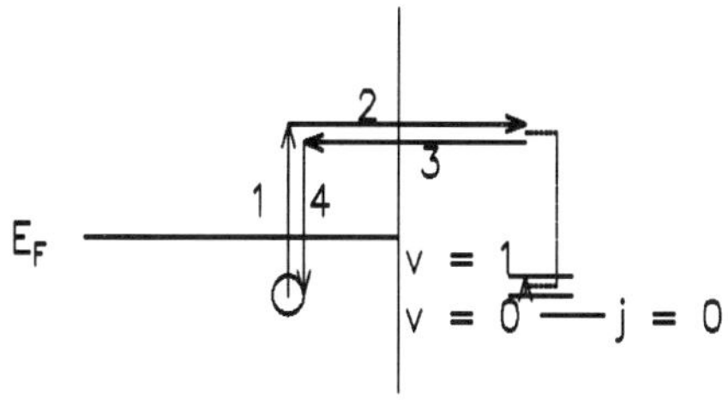

Fig. 5.63. Scheme of the four steps involved in the CT mechanism according to [398]

and gold [391]. Nanorods of suitable size (aspect ratio) that show an overlap of plasmon band and excitation source show an enhancement 10 to 10^2 greater than without this match. The relationship found between the size of nanostructures composed of different numbers of gold nanoparticles and the enhancement factors implies distinct excitation of various structural features by different wavelengths of the illuminating light [392]. Surface enhanced Raman spectra of pyridine adsorbed on gold nanorods have been reported [393]. Silver island films prepared by pulsed laser deposition showed increased enhancement only after annealing [394]. According to AFM images, the initially oblate islands coalesce into larger, less oblate and longer particles. The SERS activity of the silver islands depends on a complex interplay of several structural features beyond merely the nanometer scale. The dependency of surface enhancement on active sites has been illuminated in a comparative study of cyanide adsorption on gold with SERS and SFG [395].

The CT effect seems to be related to molecular resonance effects or, more precisely, to resonant electron transfer effects between the metal and the adsorbate that involve electronic states created by the adsorptive interaction [396, 397]. A schematic diagram depicting the conceivably participating electron transfers is shown in Fig. 5.63.

An incident photon induces an intraband transition. An electron is transferred from the metal to an excited state of the adsorbate. For physisorbed species, this may occur via tunneling; for chemisorbed species, hybridization has been proposed. Subsequently an electron is transferred back to the metal, leaving the adsorbate in an excited vibrational state. Finally this electron recombines with the hole in the metal below the Fermi level, resulting in photon emission (Stokes scattering). The electron transfer across the interface may occur under resonant conditions; for further details, see [401]. The energy of the Fermi level E_F is connected to the electrode potential when the metal is immersed in an electrolyte solution. Consequently the whole scheme as displayed might show some dependency on electrode potential. This might affect the scattered intensity in particular, because shifts in energies in this scheme will most likely affect transition probabilities. Correlations between SERS intensities, electron affinities of various organic adsorbates and the applied electrode potential have been studied [399] and the results support the described assumption and the model of a resonant CT mechanism in SERS. A detailed review of this model is available [400]. The contributions from both enhancement effects have been summed up in

$$\mu = \alpha E_{\text{local}} + \frac{1}{3}\beta \nabla E,$$

with the induced dipole μ, the molecular polarizability α (which is increased because of the suggested formation of a CT metal-molecule complex with a vibrational transition in resonance with both the incidence and the scattered light) and the local field E_{local} increased by surface roughness and excitation of SPPs by both incoming and scattered light. The second term describes the field gradient contribution to the induced molecular dipole.

The additional enhancement provided by coadsorbed halide ions on a colloidal silver surface has been pointed out [401]. As an explanation, morphological changes effected by the strongly adsorbed anions have been invoked. The photo-driven CT that is assumed to proceed from filled metal states near the Fermi level of the metal to the first and second excited state in the case of pyrazine adsorbed on polycrystalline gold has been invoked as the cause for the breakdown of Raman selection rules upon adsorption (i.e. the activation of originally Raman-forbidden vibrational modes) [314].

This effect is not inherently dependent on surface roughness but, as in most studies where rough surfaces are investigated, it is hard to separate the influence of roughness on the enhancement from CT enhancement. Nevertheless, there have been studies of smooth (up to low indexed single crystal) coinage metal surfaces reported [402–404]. They are of particular interest because the use of a well-defined smooth surface provides a fixed geometrical reference with respect to orientation of molecular adsorbates. The deduction of molecular orientation of an adsorbate is one of the most frequently investigated topics with *in situ* vibrational spectroscopies. Whereas in case of infrared spectroscopy the surface selection rules as discussed above provide a well-defined frame of reference with Raman spectroscopy, a less clear-cut situation is present here. Basically, selection rules that are considerably more complicated than those for an ideally reflecting metal surface in infrared spectroscopy can be derived [405–407]; for an overview, see also [408]. In practical applications the easy use of these rules is prevented by the mostly rough or otherwise imperfect surfaces and the rather complicated formalism of the rules. Instead, in numerous investigations (e.g. [409]) of adsorbates with well-defined adsorption geometries that have been verified with other methods (e.g. infrared spectroscopies), features of SER spectra that are useful for the identification of adsorbate orientation have been derived. In general, molecular modes perpendicular to the electrode surface will be particularly intense [410].

The conceivable effect of laser-induced local damage and the possible effect of carbonaceous deposits on surface enhancement have been reviewed by Cooney and Mernagh [411].

Beyond the enhancement provided by metals exhibiting surface plasmon resonances (i.e. copper, silver and gold in particular) with suitably rough surfaces, other means of enhancement have become available. The use of tip-enhancement is particularly promising. The tip of a scanning tunneling microscope that is in close proximity to the surface under investigation is illuminated. Its sharp apex provides the required enhancement of the electric field locally [412–414]. So far, adsorption of cyanide ions on a gold surface has been studied with a silver tip. With the tip in a distance enabling tunneling a Raman band of the adsorbed cyanide ion at the position previously observed with roughened gold electrodes was found. An additionally observed band at the position where cyanide had previously been adsorbed on a rough silver electrode was seen and taken as evidence that some cyanide ions are also adsorbed on the silver tip. The scattering of the adsorbed cyanide ions on the smooth gold surface is tip-enhanced and the scattering of the ions on the silver tip is surface enhanced. The use of fiber tips as localized probes wherein the fiber both guides laser illumination to the surface under investigation and guides the scattered light to the spectrometer has been reported [415]. An additional enhancement caused by the use of the fiber tip and depending on the tip preparation, in particular its coating, has been found. Use in electrochemical systems seems to be feasible, but has not yet been reported.

Ab initio calculations of vibrational frequencies of adsorbates on metal (surfaces) have been reported. In the case of pyridine adsorbed on silver, all calculations imply an edge-on interaction with an Ag^+ species as part of the active site [416]. Assuming this particular adsorbate-metal model, the mode $\nu_{Ag\text{–}N}$ could be predicted satisfactorily. The potential dependence of the position of this mode can thus be modeled assuming a chemical bond with a strength influenced by charge distribution and donation instead of assuming a vibrational Stark effect [417–420]. The use of other theoretical approaches, e.g. density functional theory (DFT), has been reviewed—in particular with respect to adsorbed CO; for details, see p. 85. The position of vibrational bands of adsorbates as observed with all vibrational spectroscopies depends on various factors. Besides the already mentioned electric field, which can show extremely large values inside the electrochemical double layer, the electrical charge on the electrode, the local crystallographic arrangement of those metal atoms interacting with the adsorbate and the presence of particular constituents in the electrolyte solution are only some of many conceivable influences. Because of the strong dependency of band positions of simple adsorbates like CO or CN^- on the electrode potentials, this relationship has attracted particular attention. An electric dipole, i.e. the infrared active part of an adsorbate, that is exposed to an external electric field will show a change of its observed vibrational frequency as a function of the effective electric field; this is called Stark tuning (see p. 109). The direction of the change of the vibrational frequency as a function of the electrode potential depends on the type of molecular orbitals involved; both positive and negative Stark tuning have been observed. Any adsorbate showing strong interaction with the electrode surface (indicated by free enthalpies of adsorption in excess of 20 $kJ\cdot mol^{-1}$) is bound implicitly with a considerable covalent contribution to the adsorptive interaction (in contrast with the purely electrostatic interaction during the weaker physisorption). This interaction implies electronic charge donation/back donation between molecular and atom orbitals of the adsorbate and suitable electronic states of the metal. Consequently a change of the charge density on the metal—

closely connected with a change of the electrode potential—will also affect this electronic interaction. In turn, bond orders and/or bond strength inside the adsorbate may be affected. Changes of the vibrational frequency will result, which are not an effect of Stark tuning.

Instrumentation. Because water is a weak Raman scatterer (for a review of vibrations of water molecules adsorbed on a SERS-active metal surface, see [421]) and most other electrolyte solution constituents are present only in small concentrations, standard cells with the working electrode surface close to a flat window are suitable; no thin layer arrangement is required. This results in good electrode potential control and current distribution. A typical design is depicted in Fig. 5.64; for further examples and details of both the electrochemical and the spectrometric setup, see also [372].

The cell is mounted in the sample chamber of the Raman spectrometer. The illuminating light is guided towards the electrode surface with an angle of incidence of about 60°; this angle has been identified as being suitable for maximum scattering intensity [422–425]. Further details of the cell setup are depicted in Fig. 5.65. A similar design has been described elsewhere [426]. In investigations of sensitive systems (e.g. systems that are prone to photochemical conversion or to photothermal degradation effects), a moving cell with continuously changing sections of the electrode surface exposed to the laser beam may be helpful [427].

Illumination is provided most frequently from laser systems. The light is filtered in order to remove plasma emission lines (in case of gas ion lasers) and preferably shaped to a rectangular cross section. Thus the illuminated surface can match the entrance slit of the spectrometer in an almost perfect way. Various types of spectrometers and detectors have been used and there are no special requirements with respect to their application in surface studies. With respect to the metal (or other

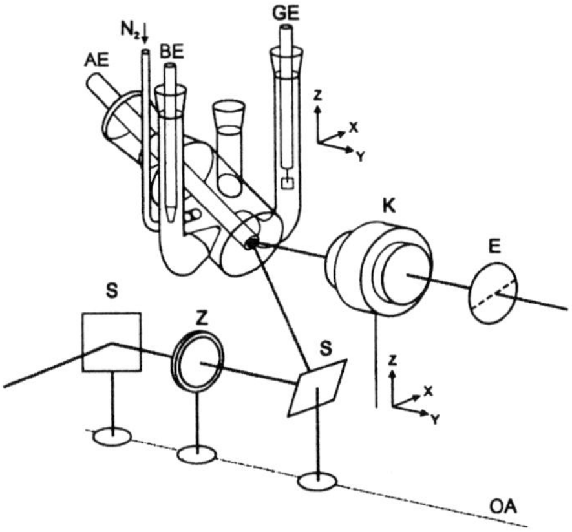

Fig. 5.64. Spectroelectrochemical setup and cell for *in situ* surface Raman studies; AE: working electrode; BE: reference electrode; GE: counter electrode; N_2: nitrogen purge inlet; S: mirror; Z: cylindrical lens; K: camera lens; E: entrance slit of spectrometer; OA: optical axis of spectrometer

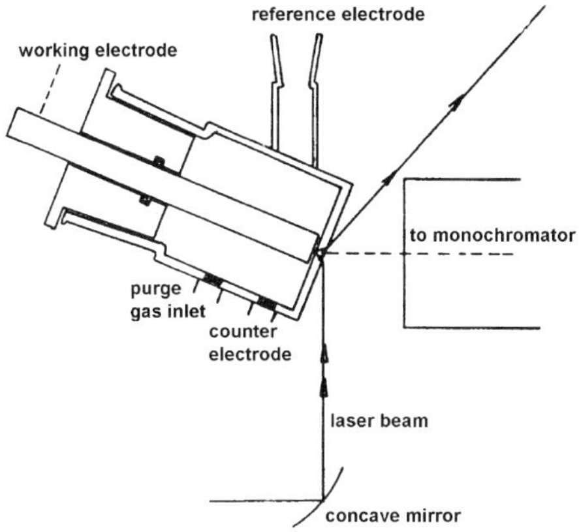

Fig. 5.65. Schematic vertical cross section of a spectroelectrochemical cell for *in situ* Raman measurements

substrate) to be investigated, the choice of illuminating light wavelength has to be made. Whereas for silver surfaces laser light of $\lambda_0 = 488$ nm, 514.5 nm or shorter as emitted from rather popular Ar^+-gas ion laser systems is suitable, d-metals like gold or copper require longer wavelengths above 600 nm as provided by Kr^+-gas ion laser systems because of the interband transitions of the latter metals. This in turn might have consequences with respect to the choice of detector, particularly when photomultipliers are used. The use of solid state lasers with emitted wavelengths in the near-infrared (e.g. 1064 nm emitted by Nd:YAG lasers) combined with Fourier transform spectrometers is possible in most cases. The advantage of excitation far away from any electronic absorption of the illuminated species can result in reduced fluorescence [404, 428]. The additional advantages of Fourier transform spectrometers for infrared spectroscopy, as discussed above (p. 73), are also helpful [429]. The dependency of Raman scattered light intensity on the illuminating wavelength (the intensity increases with the fourth power of the frequency of the illuminating light) might nevertheless result in poor spectra, which can in part be compensated by coaddition of spectra (employing the Connes advantage) [430]. With ultraviolet illumination ($\lambda_0 = 325$ nm), no surface enhancement was observed [431]; for details (see p. 121).

Preparation of the electrode surface is a central step in any SERS study. Although surface Raman spectra from smooth surfaces can be obtained with a considerably more demanding experimental setup, in most studies reported so far surfaces were roughened to provide the desired electromagnetic enhancement and to provide an artificially increased surface area with a correspondingly higher number of Raman scatterers illuminated by the exciting light. Roughening of the electrode surface is most conveniently done by applying oxidation-reduction electrode potential cycling (ORC) in a suitable electrolyte solution. In the case of coinage metals, chloride-containing solutions have frequently been used. Although chloride is known to adsorb strongly on these metal surfaces, numerous studies have demon-

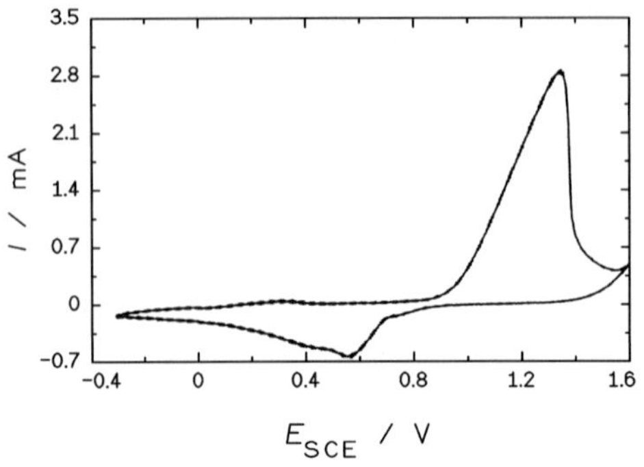

Fig. 5.66. Cyclic voltammogram of a gold electrode in contact with an aqueous solution of 0.1 **M** KCl, $dE/dt = 0.1\ \mathrm{V\,s^{-1}}$, nitrogen saturated

strated that adsorbed chloride can be washed off completely, leaving no spectroscopical evidence in subsequently recorded spectra. In the case of silver, only a few electrode potential excursions into the range of anodic silver dissolution are needed. Electrode potentials that are too negative should be avoided due to irreversible loss (quenching) of surface enhancement that presumably occurs because of surface reordering. In the case of gold, potential cycling beyond the first peak in the cyclic voltammogram as depicted in Fig. 5.66 is required.

Keeping the electrode potential at the upper limit for some time (a few seconds), as proposed by Gao et al. [432] and studied later in more detail [433], results in particularly stable surfaces. Nevertheless it could be shown elsewhere that even without this delay period highly active surfaces could be obtained [434]. The electrode potential program proposed by Gao et al. can nevertheless be obtained with an inexpensive trapezoidal voltage function generator [436]. Surface roughening should be done preferably in a solution that does not contain the species to be adsorbed later. As has been reported, adsorbate species already present during the roughening might be trapped in the modified surface, causing considerable spectral artifacts [435–437]. Trapping can also be observed when electrochemical processes (reduction, oxidation) occur that involve species that adhere to the surface, particularly in combination with electrode metals like copper or silver, which are easy to oxidize and subsequently to reduce [438]. A review of electrochemical surface treatment (most commonly called activation) procedures is available [372].

The roughness and thus the surface enhancement are stable even at very negative electrode potentials in the case of a gold electrode. With a silver electrode, the potential window extends only to less negative values. In Fig. 5.67, the scattered intensity of the silver–chloride stretching mode is shown as function of the slowly scanned electrode potential.

In the negative-going scan, a maximum of scattered intensity is reached, which can be assigned to an electrode potential dependent degree of coverage and to a change of the enhancement conditions for the CT contribution. These contributions have to be kept in mind generally when evaluating potential dependent SERS in-

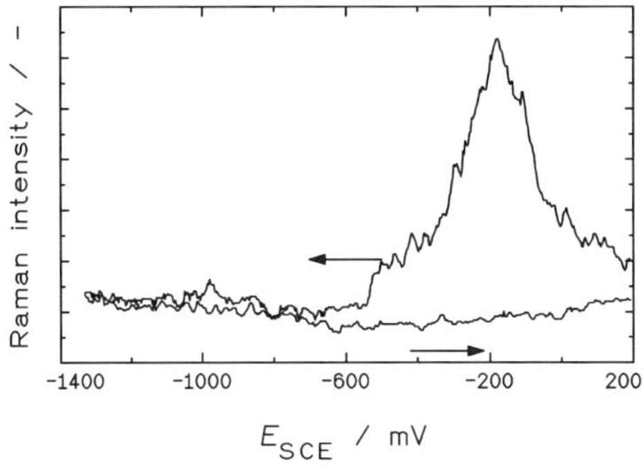

Fig. 5.67. Raman intensity of the silver–chloride stretching mode at 240 cm^{-1}; aqueous solution of 0.1 **M** KCl, $\lambda_0 = 488$ nm, $P = 400$ mW, resolution $\cong 10\ cm^{-1}$, $dE/dt = 2.5\ mV\,s^{-1}$, nitrogen saturated

tensities. Beyond this maximum the intensity decreases. The maximum is located at an electrode potential $E_{SCE} = 0.2$ V positive to the electrode potential of zero charge for a silver electrode in this solution [439]. Assuming a considerable contribution of the potential dependent coverage to the scattered intensity, this was to be expected because of the negative charge of the adsorbate. When the electrode potential is scanned back from the negative limit, this mode is not observed anymore and the surface enhancement property of the silver electrode is irreversibly lost. The influence of the electrode potential on the adsorbate orientation has been studied frequently; in a typical example, the orientation of interfacial methanol on a silver electrode has been reported [440]. Whereas the C–O bond is parallel to the metal surface at all investigated electrode potentials, at potentials positive to the E_{pzc} only one oxygen p-orbital seems to interact with the silver surface. Near the E_{pzc}, two methanol orientations appear to be present.

The surface roughness induced by electrochemical roughening as well as by other means (e.g. metal deposition, evaporation under vacuum, etc.) has been characterized with various techniques (for early reviews, see [441, 442]). Although simple double layer capacitance measurements did not show a significant increase and thus no considerable increase of electrochemically active surface area [437], investigations of scattered intensity as a function of size of nanometer-scale roughness features imply a relationship. In a study of the polycrystalline gold surface, features with a size of about 100 nm showed a maximum scattered intensity [407]. The availability of polystyrene and silica nanospheres of narrow and well-defined diameters has opened another access route to nanoscale surface features. A solution containing these nanospheres (typical diameter: 50 nm) is spin-coated on an inert substrate (e.g. gold-coated glass). After drying, a further layer of gold is deposited. These layers show large enhancement factors [443]. This procedure has been developed further [444] and applied to selected systems [445]. Use in surface enhanced resonance Raman spectroscopy (SERRS) has been reported [446].

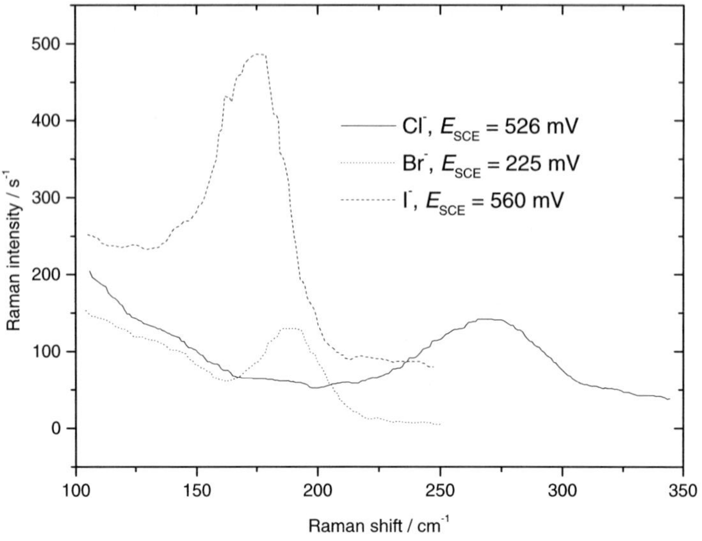

Fig. 5.68. SER spectra of a gold electrode in contact with various halide solutions (0.1 **M**), electrode potentials as indicated, $\lambda_0 = 647$ nm, $P_0 = 400$ mW, resolution 7 cm^{-1}

Even very simple adsorbates without internal vibrational modes, like halide ions, show well-defined spectra (Fig. 5.68).

The position of the vibrational band depends primarily on the mass of the adsorbed halide ion, which could be approximately explained based on a harmonic oscillator model [437]. Whereas the mass of the adsorbed ion used in this calculation poses no problem, the metal surface was assumed to be of infinite mass. This assumption has been questioned before [447]. Because of the notion of "SERS active sites", which are presumably adatoms or nanoscale metal clusters, various more complicated models were used. The adsorbed atom was assumed to interact with a single atom, which in turn was interacting with a very large mass (i.e. the metal surface). In a second model, the adsorbate molecules were assumed to be interacting with a metal cluster of four metal atoms in a square array. Correlations between band positions and masses of the assumed species were slightly better for the first model, thus a distinct identification of the actual mode of interaction is not reliable. The dependence of the band position on the electrode potential implies chemisorption with a considerable degree of covalence of the halide–gold interaction [448, 449]. An unusually high wavenumber of 865 cm^{-1} for terminal oxygen atoms in paramolybdate anions ($[Mo_7O_{24}]^{6-}$) that interact with a copper surface has been reported [450]. The assignment was based on the appearance of a band downshifted from the position at 940 cm^{-1}. This band has been assigned to the stretching mode of terminal oxygens of the Mo-O moiety.[24] The adsorption of both organic and inor-

[24] This assignment leaves open the possibility that the downshifted mode is actually not an O-Cu mode, but a shifted Mo-O mode. Such shifts of internal modes upon interaction with an electrode surface are conceivable and have been observed frequently (mostly with organic

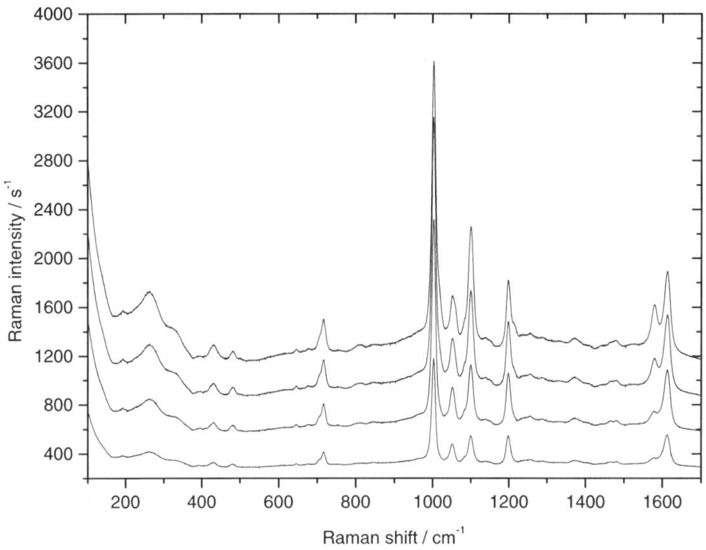

Fig. 5.69. SER spectra of a gold electrode with a self-assembled monolayer of 4-mercaptopyridine in a solution of 0.15 **M** KF at $E_{Ag/AgCl} = -700; -300; 100; 500$ mV (from bottom to top, traces offset by 300 s^{-1}), $\lambda_{Laser} = 647$ nm, $P_0 = 30$ mW, resolution 2 cm^{-1}

ganic molecules acting as corrosion inhibitors is of great practical importance. Accordingly, the understanding of their action requires considerable attention. Raman spectroscopy has been frequently employed, particularly with copper metal (widely used in boilers, heat exchangers, etc.). Because copper shows a surface enhancement typical of a coinage metal (i.e. both EM and CT enhancement, particularly when even only slightly roughened), most studies should be considered applications of SERS; a review is available [345].

Self-assembled monolayers (SAM) are another class of systems investigated with SERS [451]. Figure 5.69 shows a set of SER spectra obtained with a roughened gold electrode modified with a SAM of 4-mercaptopyridine:

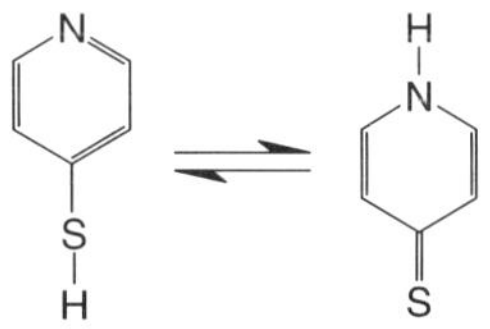

The observed bands can be assigned to modes of the adsorbate that is predominantly present in its thiole form. Particularly noteworthy is the band around

adsorbates, because inorganic adsorbates like small oxyanions show rather strong internal binding that is hardly affected by such adsorptive interaction). The observed direction of the shift supports this suggestion.

264 cm^{-1}. It is caused by the gold–sulfur stretching mode; this implies that the adsorbate is bound via the sulfur substituent, which has lost its attached hydrogen atom. Taking into consideration the free enthalpy of this bond, this mode of interaction is highly likely; it is supported by the fact that a complete assignment of observed bands supports a perpendicular adsorbate orientation [452]. Mapping of surface gradients of gold (metal film over nanosphere surfaces) covered with self-assembled various aromatic thiols using SERS has been described [446]. The effect of the electrode potential on the acid–base behavior of a SAM of ω-functionalized 2-aminoethanethiol has been studied with SERS [453]. The explanation invokes the effect of the electrochemical double layer on the surface pK_a. For a review of further studies of SERS applied to SAMs, see [454]; studies of surfactant adsorption have been reviewed by Matejka [455]. In a study of self-assembled monolayers of 4-mercaptobenzoic acid on a silver electrode, the feasibility of SERS at a silver electrode prepared by electrodeposition of silver from a room-temperature ionic liquid (RTIL) and studied in the presence of an RTIL was demonstrated [456].

The use of rough surfaces for SERS has been considered a drawback. The electrochemical roughening employed most frequently is suspected of introducing surface sites that have properties different from the rest of the surface. In the case of cold-deposited copper, the vibrational spectrum of adsorbed ethylene enabled the identification of Cu(110), Cu(111) and other surface defect sites of unknown configuration. When the latter disappear upon annealing, the resulting scattered intensity followed a simple electromagnetic enhancement model. Consequently these sites were called "SERS active" (this implies a somewhat more narrow sense of surface enhancement by limiting the term to the "charge transfer enhancement") [457]. The surface changes introduced by roughening are the formation of metal clusters, defects and other irregularities. They may show different adsorption properties and geometrical environments. In addition, the roughening excludes studies of single crystal surfaces, which are fairly popular with various other spectroelectrochemical and surface analytical techniques. Thus attempts have been reported to use both atomically smooth (polished, not roughened) and single crystal surfaces. With smooth polycrystalline gold, the scattered intensity of adsorbed pyrazine tracked the molecular coverage up to about $2/3$ of a monolayer; beyond this value the intensity decreased, presumably because of interactions between the induced dipoles of the adsorbed molecules [407]. With a Au(210) electrode, spectra of adsorbed pyrazine could be detected. No simple relationship between molecular coverage and scattered intensity was observed [406]. The relationship between the intensity of Raman bands was not observed with pyrazine itself (forbidden bands) but was seen for adsorbed pyrazine (indicative of changed, presumably lowered, symmetry); the orientation of the adsorbed molecule on a silver electrode and the two mechanisms of surface enhancement have been discussed [458].

The limitation of SERS studies to electrodes made from coinage metals has seriously hampered a widespread application to other metals and further substrates of interest. Deposition of the metal of interest as a thin (a few nm), pinhole-free layer on electrochemically roughened gold [459–461] or smooth [462] or again electro-

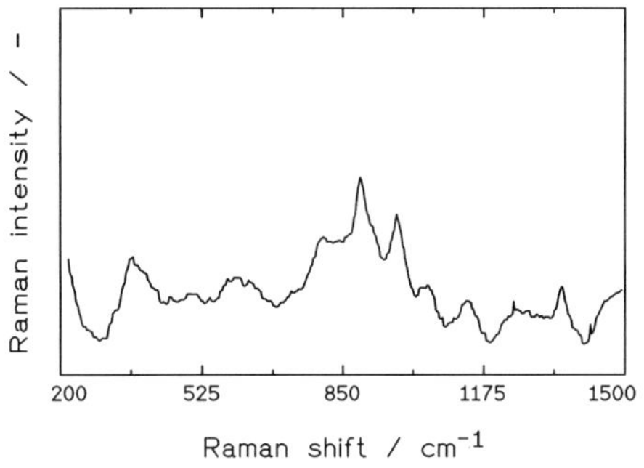

Fig. 5.70. SER spectra of a gold electrode coated with a rhodium overlayer in an aqueous solution of 0.5 **M** H_2SO_4 + ca. 10 m**M** benzene at $E_{RHE} = -0.2$ V (based on data in [465])

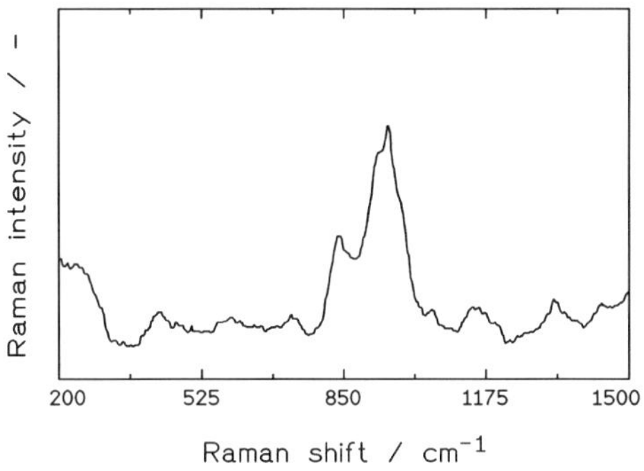

Fig. 5.71. SER spectra of a gold electrode coated with palladium overlayer in an aqueous solution of 0.5 **M** H_2SO_4 + ca. 10 m**M** benzene at $E_{RHE} = -0.2$ V (based on data in [465])

chemically roughened [463] silver electrode surfaces yields SER spectra showing features typical of the deposited metal instead of the supporting substrate. Surface enhanced Raman spectra of numerous adsorbates that show spectral features typical of the overlayer metal have been reported. The use of this method has met some criticism as the applied methods show some limits in generalization; in some cases, the deposited thin transition metal films may have crystallinities different from those encountered with bulk metal samples [464]. Typical examples are shown in Figs. 5.70–5.73.

The vibrational bands observed with both metal overlayers are clearly different from those observed with uncoated gold electrodes. The band positions show no appreciable dependence on electrode potential and their intensity decreases towards more positive electrode potentials. Evaluation of band position, band intensity and symmetry reveals a flat orientation of the adsorbed benzene molecule with both rhodium and palladium.

Toluene is also adsorbed on the palladium surface in a flat orientation, whereas benzonitrile is adsorbed pendant with the nitrile group interacting with the sur-

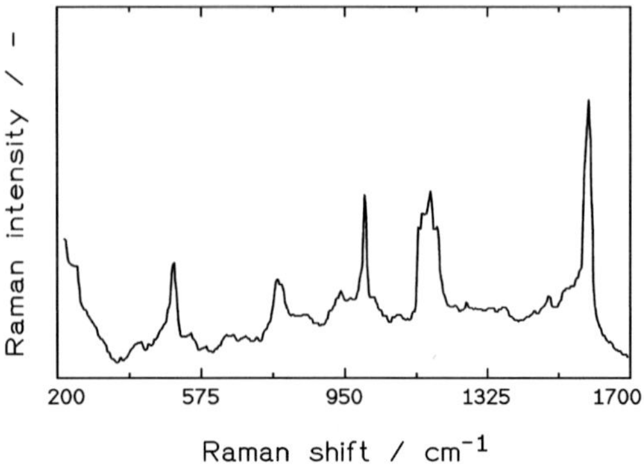

Fig. 5.72. SER spectra of a gold electrode coated with a palladium overlayer in an aqueous solution of 0.1 **M** $HClO_4$ + ca. 20 m**M** benzonitrile at $E_{RHE} = 0$ V (based on data in [465])

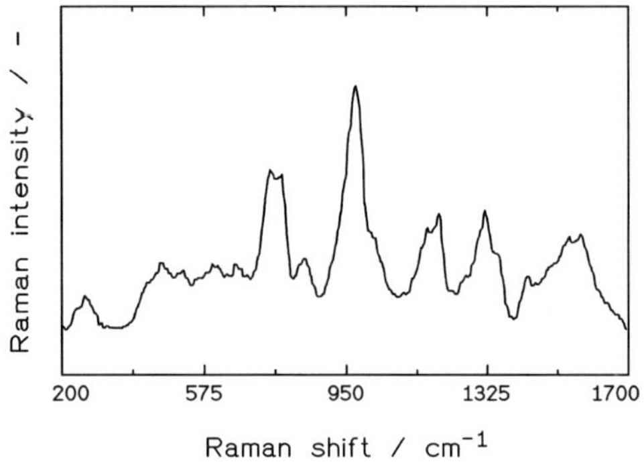

Fig. 5.73. SER spectra of a gold electrode coated with a palladium overlayer in an aqueous solution of 0.1 **M** $HClO_4$ + ca. 5 m**M** toluene at $E_{RHE} = -0.2$ V (based on data in [465])

face [465]; this is in agreement with similar studies with gold [466] and silver electrodes [467, 468].

In a study of the adsorption of ionic species on gold and upd-modified surfaces (not perfect, pinhole-free overlayers), the displacement of adsorbed sulfate by oxalate anions can be shown by comparing SER spectra obtained without and with oxalate anions present (Fig. 5.74) [469]. The gold-sulphate stretching mode at 177 cm^{-1} dominates the spectrum in the absence of oxalate; in the presence of the latter anion only one band assigned to the oxalate-gold interaction at 259 cm^{-1} is observed; this implies a complete displacement.

Upon upd-modification with tin (Fig. 5.75) and nickel (Fig. 5.76), the band assigned to the oxalate-metal stretching mode is shifted only very slightly to 256 cm^{-1} and to 257 cm^{-1}, respectively. The band position is almost independent of the nature of the upd-metal, implying rather unspecific and weak interactions (physisorption). Based on the evaluation of further internal modes of the adsorbed oxalato anion, small changes in band position indicative of the presence of the upd-metal could be identified.

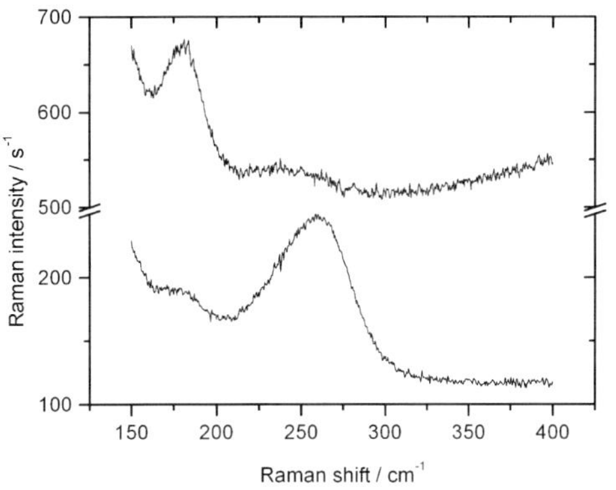

Fig. 5.74. SER spectra of a polycrystalline gold electrode in contact with an aqueous solution of 1 **M** Na_2SO_4 (*top*) and with added 0.01 **M** Fe(II)/Fe(III)oxalate (*bottom*), E_{HgSO_4} = 613 mV (based on data in [469])

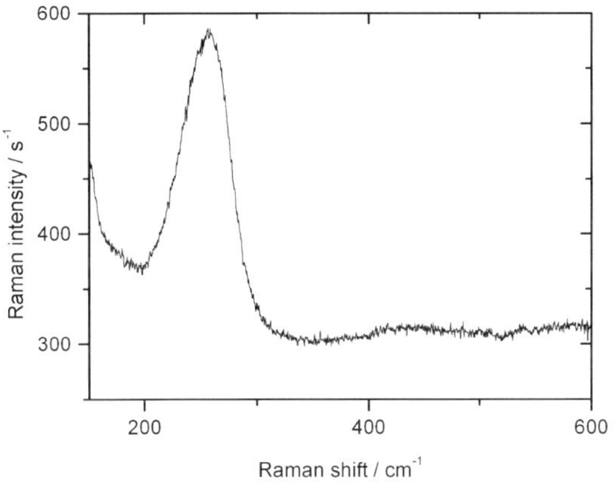

Fig. 5.75. SER spectra of the upd-tin modified gold electrode in an aqueous solution of 1 **M** Na_2SO_4 + **M** 0.01 Fe(II)/Fe(III)oxalate at E_{HgSO_4} = 613 mV (based on data in [469])

As an alternative approach, gold particles have been deposited on iron electrodes in studies of passive films on iron [470]. A layered structure was found. The inner layer is composed mostly of Fe_3O_4; the outer layer contains Fe(III) species and may also contain γ-Fe_2O_3.

Surface Raman spectra of non-coinage metals have been reported repeatedly [471]. In practically all cases the surface under investigation was roughened (preferably by applying electrode potential cycling) and thus the electromagnetic contribution was certainly effective. An alternative approach employs an aluminum foil etched with aqueous alkaline solution as a substrate [472]. When brought into contact with aqueous solutions of both coinage and non-coinage metals, the dissolved metal ions are reduced and deposited onto the aluminum foils, which are presumably somewhat roughened by the preceding etching; the foils nevertheless show no

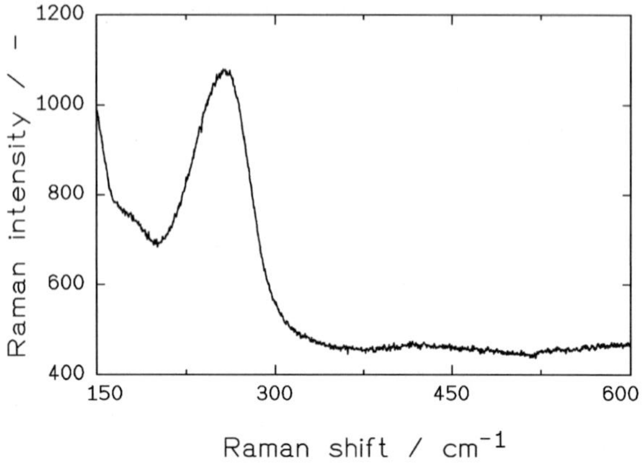

Fig. 5.76. SER spectra of the upd-nickel modified gold electrode in an aqueous solution of 1 **M** Na_2SO_4 + 0.01 **M** Fe(II)/Fe(III)oxalate at E_{HgSO_4} = 613 mV (based on data in [469])

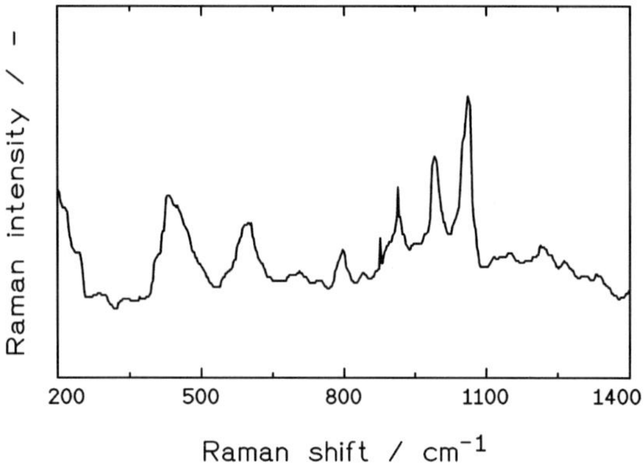

Fig. 5.77. SR spectra of an electrochemically roughened polycrystalline platinum electrode in contact with an aqueous solution of 0.5 **M** sulfuric acid and 0.244 m**M** $Sn(SO_4)_2$, E_{RHE} = 1.191 V, nitrogen saturated, λ_0 = 488 nm, P = 40 mW, resolution $\cong$ 10 cm^{-1} (based on data in [472])

surface enhanced Raman spectrum of adsorbed pyridine. The deposited metals adsorb pyridine. The vibrational spectra of the adsorbate show typical features already observed before implying the formation of α-pyridyl species and end-on adsorbed pyridine as well as side-on adsorbed α-pyridyl species. The metal deposits (e.g. deposits of nickel, cobalt) have also been prepared with this template technique in the form of nanorod arrays [473] showing significant surface enhancement with pyridine as a probe molecule.

In a typical application the oxidation state of upd-tin on a platinum surface electrochemically roughened by fast electrode potential cycling (roughness factor approx. 2.7) was studied using adsorbed hydrogensulfate or sulfate ions as a probe [474].

Based on a comparison with Raman spectra obtained with tin electrodes and with solutions of various tin salts, the observed bands and their potential dependence support an oxidation state of 2^+ or 4^+ for the tin that is underpotentially adsorbed on the platinum electrode [472]. In a study of the adsorption of thiourea TU on a roughened platinum electrode, interaction via the sulfur atom with tilted, potential dependent orientation was found [475]. Dissociation and formation of surface sulfide was observed at open circuit potential (i.e. in the absence of potential control). This suggests the need for proper potential control during immersion of the electrode in the case of particularly reactive adsorbate species. Further examples refer to pyridine adsorbed on platinum and nickel electrodes [476] or the adsorption of saccharin on a nickel electrode [477]. In a study of the adsorption of *p*-hydroxybenzoic acid on a roughened gold surface with ultraviolet laser light illumination ($\lambda_0 = 325$ nm), the obvious mismatch of the excitation light with the electronic transitions in the metal resulted in extremely poor spectra despite the fact that scattered Raman intensity basically scales with υ^4 [434]. Considerable instrumental improvements that enable acquisition of surface Raman spectra from non-coinage metals have been possible by applying confocal optics (see p. 128 for details).

Modification (i.e. decoration) of graphite surfaces also results in considerable enhancement of vibrational bands of the carbon substrate; in addition, modes of electrolyte solution constituents may be observed. In the case of a glassy carbon surface that is electrochemically activated by repeated electrode potential, cycling deposition of silver micro- and nanocrystals results in SER spectra, as shown in Fig. 5.78.

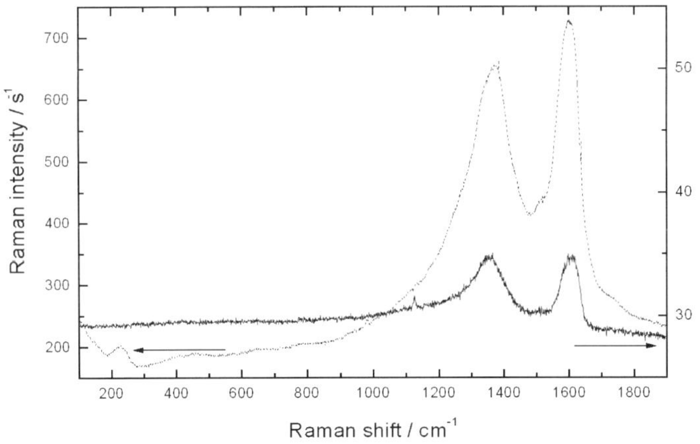

Fig. 5.78. SR spectra of a glassy carbon electrode activated electrochemically before (—) and after decoration with silver in contact with an aqueous solution of 0.1 M K_2SO_4, open circuit electrode potential, $\lambda_{Laser} = 514.5$ nm, $P_0 = 100$ mW, resolution 2 cm^{-1} (based on data in [354])

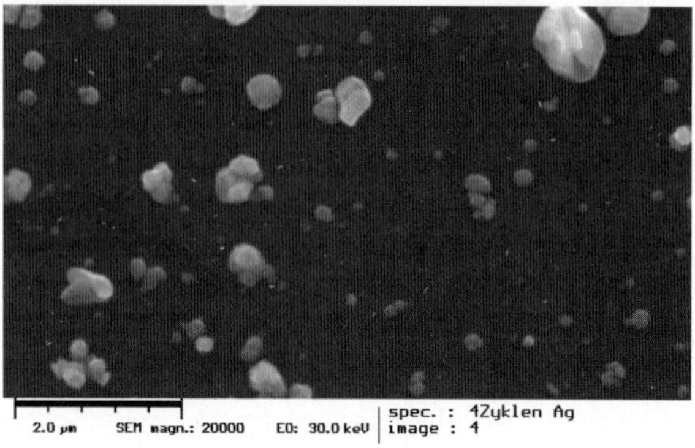

Fig. 5.79. Scanning electron microscope picture of the electrode surface as employed for the measurement illustrated in Fig. 5.78

The major bands of the carbon substrate around 1607 cm^{-1} (E_{2g} mode) and 1360 cm^{-1} (A_{1g} mode) are already visible with the activated surface. After decoration with silver particles, the scattered intensity of these bands is significantly enhanced. An additional band assigned to the silver–oxygen stretching mode caused by adsorbed sulfate ions is observed; the value of about 230 cm^{-1} is typical of this anion when it is adsorbed on a silver surface.

Although no average particle size of the silver deposits has been reported, the visible particles show dimensions in the range of about 100 nm; this size has been observed to be particularly effective for surface enhancement (see p. 106).

The E_{2g} mode has been monitored in a study of lithium intercalation into graphite [478]. A shift of about 7 cm^{-1} from the initial to the final state of intercalation was found. Initially intercalation occurs statistically and, upon formation of the phase LiC_{27}, two different types of graphite layers are observed, resulting in a splitting of the observed Raman band. The bands are assigned to the LiC_{27} structure and to a structure caused by statistical intercalation. Upon reaching the composition corresponding to LiC_{12}, no further bands are observed.

A spectroelectrochemical flow cell suitable for kinetic investigations has been described by Luo and Weaver [479]. A cell for time-resolved SERS that is used for the investigation of electron transfer dynamics based on a cylindrical rotating silver disc has been reported [480, 481]. Surface enhanced Raman spectroscopy measurements taken at a high temperature in molten salts have been reported for the identification of oxygen species that appear at gold electrodes during electroreduction of dioxygen [482]. The problem of black body radiation was solved by spectrally subtracting a spectrum obtained without laser illumination. The combination of Raman spectroscopy and optical microscopy that results in micro-Raman spectroscopy,[25] enabling surface studies with high local resolution, has been reviewed [483]; so far in numerous studies only localized measurements have been performed, whereas a

[25] The term "microsurface-enhanced Raman spectroscopy" is misleading and should not be used. Terms like "microprobe Raman spectroscopy" or "Raman microscopy" are more appropriate.

useful spatial resolution resulting in information about differences in surface properties has been employed only infrequently and, in most cases, in combination with other microscopic techniques (see Sect. 5.2.15 for confocal spectroscopy; see p. 128 for near field Raman spectroscopy). In a typical study wherein spatial resolution was of no particular importance, the oxidation of mercury at the mercury/aqueous solution of 0.1 **M** $KClO_4$ has been investigated [484]. Based on the appearance of the Hg–Hg band around 175 cm^{-1}, the solvated dimeric Hg(I) cation was the only oxidation product. At very anodic electrode potentials, a mercury-perchlorate precipitate was formed. In the presence of pyridine it yielded soluble Hg(II)-pyridine complexes. Azide anions were identified as participants in electrooxidation of ammonia in alkaline solution at nanostructured platinum electrodes with SRS[26] [485]. The high sensitivity of SERS has attracted analytical applications [486] and further interest from various fields, including medicine and biophysics [487–489] as well as the chemical industry [490].

Photochemical reactivity of adsorbed species is a topic that is easily overlooked. Although there have been only infrequent reports, photoinduced reduction of *p*-nitrothiophenol incorporated in SAMs [491] and photoinduced reduction of polyaniline [492] are illustrative examples.

5.2.13 Surface Enhanced Hyper-Raman Spectroscopy (SEHRS)

Fundamentals. In hyper-Raman scattering, two photons of the illuminating light at ω_0 and the scattered photon with ω participate. The scattered light corresponding to a vibrational mode of the scattering species is observed at a Raman shift of $2\omega_0-\omega$. Selection rules are relaxed in comparison to both Raman and infrared spectroscopy, which affects centrosymmetric molecules in particular. The theory of hyper-Raman scattering has been treated extensively elsewhere [493]. The scattering cross section is extremely weak—scattering intensities are about 10^{-7} to 10^{-5} for a laser field of 10^{10} W·cm^2. For an introductory overview of SEHRS in electrochemistry, see [494, 495]. Surface enhancements of 10^{13} have been reported [496].

Instrumentation. Instrumentation in the reported investigations is similar to that employed in SERS [497]. A powerful laser system is employed. In a study of pyrazine and pyridine adsorbed on a roughened silver electrode, it was observed, that spectra measured with adsorbed centrosymmetric pyrazine showed additional bands that were not observed with SERS [497]. With the non-centrosymmetric pyridine, the SEHR spectra were very similar to the respective SER spectra; no new bands were found. Surface enhanced hyper-Raman spectra were found to be more sensitive (than SERS) towards interaction between the adsorbate and the surface. This is illustrated in vibrational spectra of bis(4-pyridine)acetylene (BPA) adsorbed on a silver electrode [497].

---

26 Because of the floating boundaries between SRS and SERS (e.g. simply roughening an electrode may already yield the shift from the former to the latter), numerous authors simply claim to have performed SERS without providing any supporting arguments. In the present case, the nanostructuring most likely provided EM enhancement.

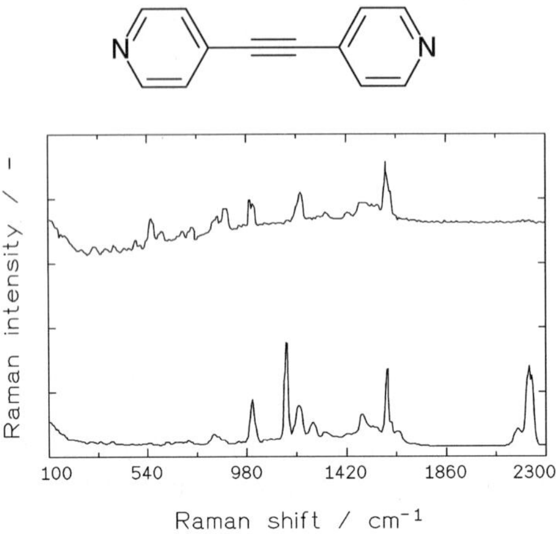

Fig. 5.80. Spectra of bis(4-pyridine)acetylene adsorbed on a silver electrode; *top*: SEHR spectrum, aqueous solution of 0.1 **M** KCl, $E_{Ag/AgCl} = -0.5$ V, $\lambda_0 = 1064$ nm, $P_0 = 2$ W; *bottom*: SER spectrum, aqueous solution of 0.1 **M** KCl, $E_{Ag/AgCl} = -0.6$ V, $\lambda_0 = 532$ nm, $P_0 = 10$ mW; data taken from [500]

In bulk state, BPA belonging to the symmetry group D_{2h} is centrosymmetric, thus it is particularly sensitive towards hyper-Raman scattering. The SEHR spectrum is considerably different from the corresponding SER spectrum (the small difference in electrode potential is of minor importance). A particularly striking feature is the absence of the acetylene stretching mode that is observed in SERS at 2226 cm^{-1} in the SEHR spectrum. In SEHR spectra obtained at less negative potentials, this mode appears weakly. These observations imply a significantly lowered adsorption symmetry at the latter electrode potential. A combination of infrared spectroscopy, normal Raman spectroscopy, SERS and SEHRS was used in a study of *trans*-1,2-bis(4-pyridyl)ethylene adsorbed on a silver electrode [498]. The centrosymmetric molecule should not have common Raman and hyper-Raman lines provided that its symmetry is not strongly disturbed upon adsorption. Besides a well-defined C=C stretch mode seen only in SERS, many other bands overlap in both spectroscopies. In addition, by using *ab initio* calculated spectra, observed bands could be assigned and good agreement between the results of all employed methods could be found. The results finally suggest that SEHRS spectra are consistent with the expected three-photon selection rules; they are not caused by surface second harmonic generation followed by surface enhanced Raman scattering that is excited at the second harmonic frequency, as proposed elsewhere.

In a study of phenazine adsorbed on a silver electrode that employed both SERS and SEHRS, the electroreduction product of phenazine and the reduction intermediates could be identified [499]. In a comparative study with SERRS and resonantly enhanced hyper-Raman spectroscopy SERHRS, several dyes adsorbed on a roughened silver electrode were investigated [500]. According to the results, the efficiency

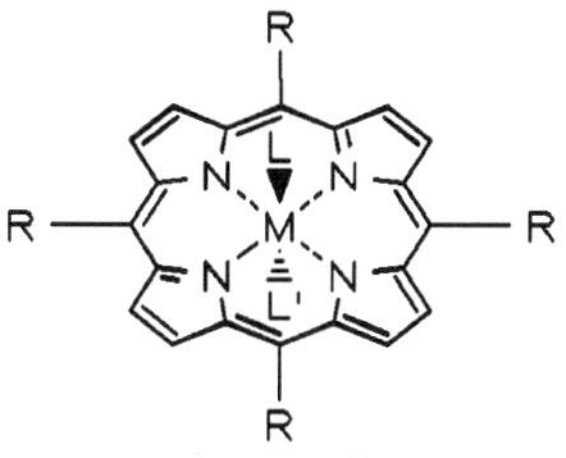

Fig. 5.81. Molecular formula of investigated transition metal complexes (R = phenyl, *p*-methoxyphenyl, 4-pyridyl, *p*-chlorophenyl; M = Fe, Co, Ni; L = Cl^-, Br^- for M = Fe)

of SEHRS depends not only on the type of surface employed (roughened silver as compared with silver colloid) but also on the type of dye.

5.2.14 Surface Resonance Raman Spectroscopy (SRRS)

Fundamentals. Illumination of a sample with light of a frequency matching the energy of an optical transition in the sample might result in vastly enhanced scattered Raman light intensity (10^4 to 10^6). Because of the closeness of the energy values involved, this is called resonance enhancement and the spectroscopy is termed resonance Raman spectroscopy (RRS). When applied to a surface it is called surface resonance Raman spectroscopy (SRRS). In cases where the species under investigation is present both as an adsorbate or adlayer and a dissolved species, resonance enhancement will be effective for all of them; thus Raman scattering from dissolved species might interfere.

Instrumentation. The experimental setup is basically the same as the one employed in SERS. For obvious reasons—the wavelength of the used light should match electronic transitions (i.e. optical absorptions) of the species at the investigated interface or in the interphase—sources with several tunable wavelengths are preferred. Dye lasers or gas ion lasers showing several laser lines are the source of choice.

Studied examples include dye molecules or other strongly colored molecules adsorbed on metal surfaces or other materials suitable to be used as electrodes (e.g. glassy carbon). In several studies transition metal complexes like porphyrines and tetraazaannulenes have been investigated [501–505]. An earlier, unfortunately rather incomplete and inconclusive report[27] is available [345].

Raman spectra recorded under resonant or pre-resonant (i.e. the laser line is somewhat away from the optical absorption maximum) conditions were obtained for the bulk complex cobalt tetra-*p*-methoxyphenylporphyrine Co(TMPP).

[27] Somewhat surprisingly, the authors fail to notice that resonance enhancement is operative despite the fact that they elaborate in detail on the UV-Vis spectroscopy of some transition metal complexes.

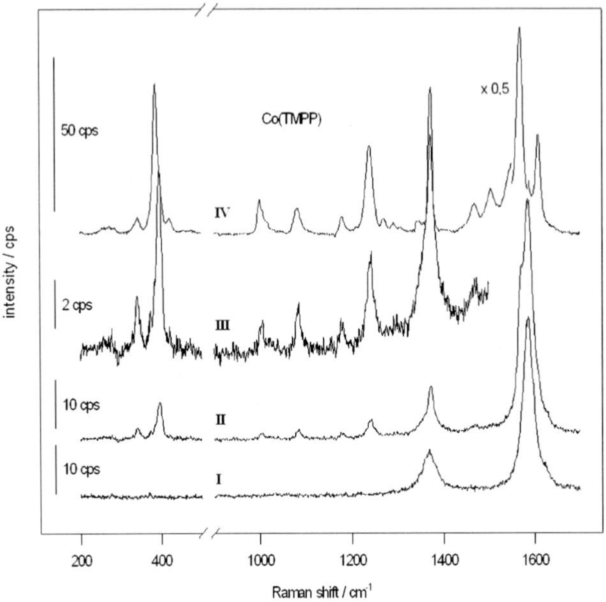

Fig. 5.82. Raman spectra of Co(TMPP) in KBr and on carbon supports: **I**: blank pyrolytic graphite (PG) surface; **II**: PG with Co(TMPP) film; **III**: identical to **II** but magnified by a factor 5; **IV**: Co(TMPP) in KBr. $\lambda_0 = 457.9$ nm, $P_0 = 200$ mW

Bands could be assigned to vibrations of the complex. In addition, two bands of the carbon surface were assigned. Changes upon exposure of the CoTMPP-coated PG-electrode to an acidic electrolyte solution could be explained by assuming solvent-complex interactions. When dioxygen is coordinated to the central metal ion in an axial position, depending on the electrode potential, a stretching mode of end-on coordinated dioxygen at 1204 cm^{-1} can be identified (see Fig. 5.83). Together with further results, this implies a reduction mechanism for dioxygen proceeding via the superoxide stage.

These spectra were obtained with a smooth, polished surface. Thus no surface enhancement was effective. The molecules under investigation can alternatively be deposited onto roughened surfaces and the obtained spectra show a combination of surface and resonance enhancement. The method is called surface enhanced resonance Raman spectroscopy (SERRS).

Because of the strong coloration, depending on their state of oxidation, intrinsically conducting polymers have frequently been studied with SRRS [507, 508]. Molecular vibrations could be assigned based on various approaches. Most frequently band positions of the monomers and of already known oligomers were compared with those of the polymers. Alternatively, band positions were calculated based on an effective conjugation coordinate [509–516]. In a typical example shown in Fig. 5.84, SRR spectra of polyaniline are displayed as a function of electrode potential.

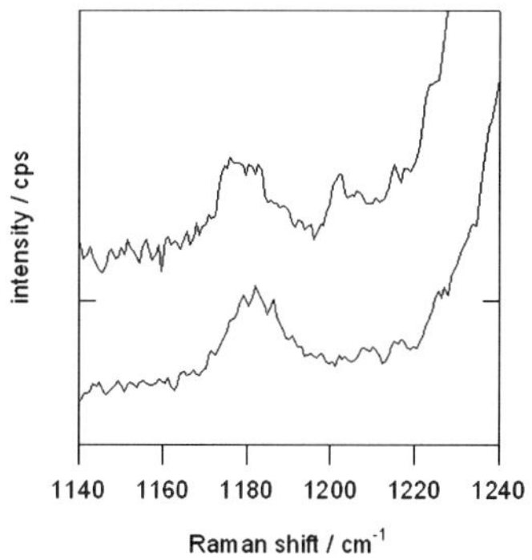

Fig. 5.83. Raman spectra of Co(TMPP) *in situ* Raman spectra of $(FeTMPP)_2O$ in dioxygen-saturated (*upper trace*) and dioxygen-free (*lower trace*) aqueous solution of 0.1 **M** $HClO_4$, $\lambda_0 = 457.9$ nm, $P_0 = 200$ mW, $E_{RHE} = 400$ mV (based on data in [506])

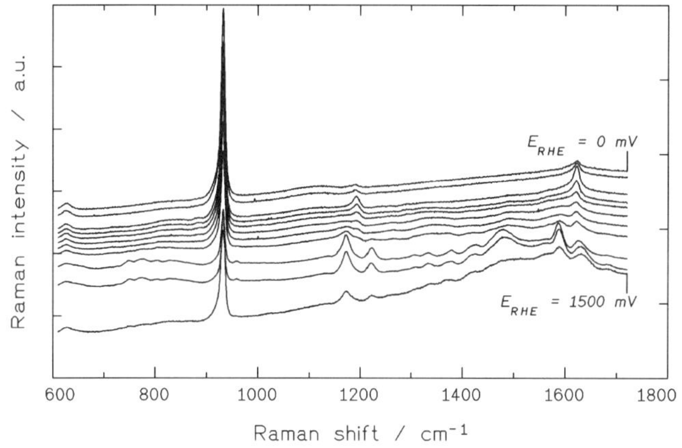

Fig. 5.84. SRR spectra of a polyaniline film deposited on a platinum electrode, in contact with an aqueous solution of 1 **M** $HClO_4$, electrode potentials as indicated, the band around 932 cm^{-1} is caused by the perchlorate ions

The observed bands can be assigned to internal modes of the aniline repeat units that are present in their benzoid form at lower electrode potentials and in their quinoid form at higher electrode potentials. In addition, modes assigned to molecular vibrations of the bonds connecting these repeat units predominantly in the para position can be identified. The resonant enhancement of a band may depend significantly on the illuminating wavelength. In the case of polyaniline, the pH-value of the electrolyte solution has a strong influence on the actual molecular structure (by protonation/deprotonation equilibria). This has been employed in an attempt to monitor the pH-value of a solution by registering the scattered light intensity of particularly sensitive bands [517].

The bands observed with a polypyrrole film under resonant and off-resonant conditions can again be explained in terms of inter- and intramolecular vibrational

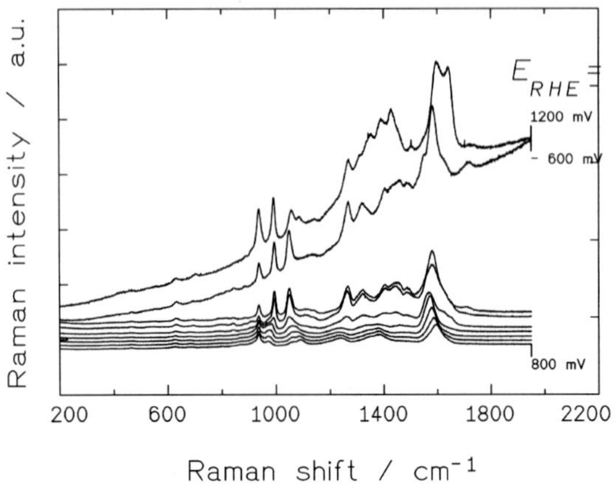

Fig. 5.85. SRR spectra of a polypyrrole film deposited onto a platinum electrode galvanostatically, in contact with an aqueous solution of 1 **M** $HClO_4$, electrode potentials as indicated

modes. The frequency dispersion of the band around 1580 cm^{-1} has been identified as an indicator of chain length. Alternatively, the effective conjugation coordinate approach has been employed.

5.2.15 Confocal Raman Spectroscopy

Fundamentals. In a confocal optical arrangement, a pinhole inserted in front of the optical detection system (photomultiplier, spectrometer, etc.) allows only light from the focal point to pass. Depending on the size of the pinhole and other properties of the optical setup, this can result in considerably high spatial resolution. In the case of confocal Raman spectroscopy, the sample under investigation is illuminated with laser light using a microscope objective. The scattered light is collected and guided using the same objective, i.e. a backscattering geometry is employed. For further details, see [518–521]. Single molecule surface enhanced Raman spectroscopy (sm-SERS or SMSERS) becomes possible when surface enhancement conditions are met (suitable metal, finely dispersed) [522, 523]. The structural basis of this giant enhancement has been discussed [524].

Instrumentation. The optical setup is depicted schematically in Fig. 5.86. By optimizing the optical geometry, a high collection efficiency can be achieved [467]. In the z-direction, a spatial resolution of about 20 μm is feasible. This might be of interest in attempts to separate scattered light from the interface from contributions originating from solution phase species. The laser beam is focused to a spot size of 1 to 2 μm, which allows measurement with corresponding lateral resolution. In the case of substrates showing no particular surface enhancement effect (e.g. in the case of non-coinage metals), suitable roughening procedures can be employed to enhance scattered light intensity. Details of procedures employed with various transition metals have been reviewed [467]. Despite the seductive gains achieved by

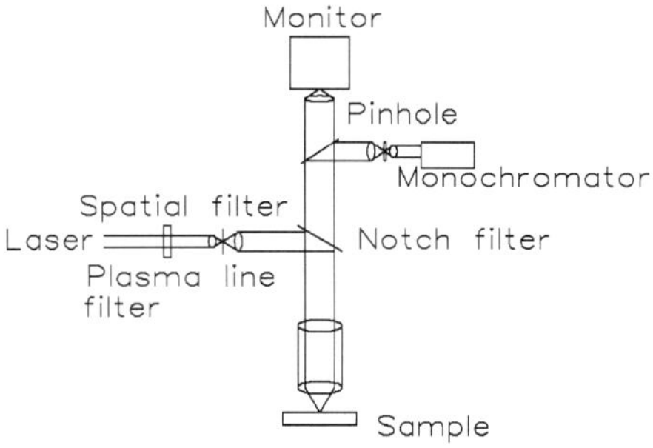

Fig. 5.86. Schematic setup for confocal Raman microscopy

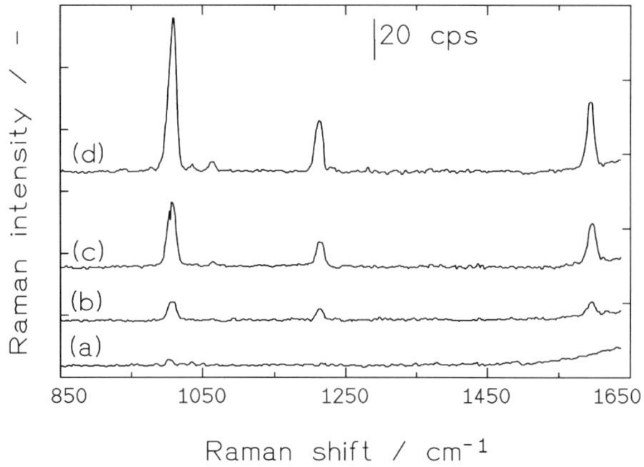

Fig. 5.87. Surface Raman spectra of an iron electrode in contact with an aqueous solution of 0.1 **M** KCl + 0.01 **M** pyridine with various surface treatments applied: (*a*) mechanically polished; (*b*) chemically etched in 2 **M** H_2SO_4 solution; (*c*) *ex situ* activation; (*d*) *in situ* activation (based on data in [467])

activating the metal surface under investigation in the presence of the species to be adsorbed [467], this procedure should be avoided because of the danger of experimental artifacts caused by trapping the adsorbate in the roughened surface [438, 439]. Reported investigations deal mainly with adsorption of various molecules on transition metal surfaces. The influence of the roughening procedure on the scattered light intensity is obvious in Fig. 5.87. The *in situ* roughening procedure based on a double step oxidation reduction cycling (for details, see [467]) yields a considerably stronger signal without any evidence of "trapping". Oxide films on nickel exposed to strongly alkaline electrolyte solutions that were composed of various nickel oxides were identified [525].

Because the use of colloidal silver or gold as a substrate for single molecule SERS, as observed with confocal microscopy, has met with some criticism (the chemically prepared colloids are exposed to the molecules to be adsorbed and are subsequently deposited onto a substrate, e.g. an ITO-coated glass slide; during this procedure, agglomerization may occur resulting in poor definition of size and shape of the particles that are finally studied), an electrochemical procedure for deposition of silver clusters of variable size and density has been developed [525]. The lower detection limits for selected dyes are still not as low as those observed with silver colloids. Silver and gold nanoparticles of 10 to 500 nm size that are prepared with this electrochemical double pulse procedure have been characterized with confocal and surface enhanced Raman spectroscopy [526]. The maximum enhancement factors observed were 10^{10} for silver and 10^8 for gold particles. Surface enhanced Raman spectroscopy was described as a localized phenomenon. Only a few particles were "Raman active". The strongest enhancement was observed with agglomerates; irregular structures like necks in agglomerates are preferential scattering sites.

5.2.16 Near Field Raman Microscopy (Micro-Spectroscopy)

Fundamentals. In near field microscopy (see also Sect. 7.3.1), close range interactions between sample and probe are utilized as with scanning probe microscopies (see Sect. 5.3: AFM, STM, etc.). Because of the miniaturization of the probe, spatial resolution beyond the Abbé limit (see Chap. 7) is possible. Since the probe is scanned across the investigated surface, the method is termed scanning near field optical microscopy (SNOM). A scanning near field optical probe for Raman spectroscopy has been described [527]. Localized evanescent fields associated with electromagnetic waves as also encountered with attenuated total reflectance (ATR) are of central importance. Probes are metal-coated sharpened glass fiber tips with a hole of the coating at the tip (about 50 to 100 nm in size, i.e. below the wavelength of visible light). At a short range (e.g. 10 nm) from the sampled surface, the optical near field is distorted by the sample surface. Light thus emitted by the surface can be detected in reflection or transmission.[28] Interaction between the light emitted from the fiber and the scanned surface may result in inelastic (Raman) light scattering. Optical setups with and without aperture (aSNOM) are possible. In the former case, light is guided to the surface as described above. The resolution is controlled by the size of the aperture. Limitations of this size are mostly given by the optical properties of the coating material (for an introduction and further details, see [528]). Calculations have indicated a resolution limit of 12 nm [529]. Because of the extremely low intensity of the normal Raman scattering effect, apertures of 50 to 100 nm are used. In addition, enhancement of scattered light intensity by resonance of the applied light and the species observed on the sample surface or surface enhancement by suitably prepared (roughened coinage metals surfaces, see SERS) [530] are utilized. Without an aperture, the surface under investigation is illuminated. The required near field interaction is caused by a tiny tip smaller than the wavelength of the employed light. Nanometer-sized metal particles (of metals employed in SERS) or the tip of

[28] In conventional microscopy, the evanescent field is not utilized; instead, far field effects are employed, which results in the diffraction limit of resolution.

a scanning probe microscope may be used. In the latter case, the method is termed tip-enhanced Raman spectroscopy (TERS) (see p. 79 ff.).

Instrumentation. In both cases, a near field probe is employed—either a metal-coated fiber (aperture-based) or a metal tip (apertureless). Distance regulation, as used with scanning probe methods (see Sect. 7.2), controls the probe–surface gap; it may also be used to obtain a topographic mapping of the studied surface. Scattered light is collected and guided to a Raman spectrometer. In a (non-electrochemical) study, dye-labeled DNA that had adsorbed onto evaporated silver layers on PTFE nanospheres was observed [531]. Special surface sites with particularly high enhancement could be identified.

5.3 Spectroscopy in the X-ray Range

Absorption of electromagnetic radiation in the X-ray range (gamma ray range) of the electromagnetic spectrum can occur in various ways that involve different parts of an atom, a molecule or an interface and can result in various effects. Photoelectron spectroscopy can be done using X-rays to generate photoelectrons. Absorption of gamma radiation close to the K-edge of a given atom can result in an absorption spectrum showing a fine structure (EXAFS, XANES). Recoilless emission and absorption of gamma radiation can also be used in Mössbauer spectroscopy. The various spectroscopies and their *in situ* applications in electrochemistry will be described here. An X-ray light source frequently used is a synchrotron; radiation provided by this source has been treated in Chap. 4 and elsewhere [532–534].

5.3.1 Mössbauer Spectroscopy[29]

Fundamentals. The energy of a gamma quantum emitted or absorbed by an atomic nucleus usually differs somewhat from the actual energy difference between the nuclear states because of recoil and Doppler-broadening induced by thermal movement. At room temperature, the value differs by about 10^{-2} to 10 eV from the true energy value $h\nu_0$ of the nuclear transition. The recoil energy E_R and the corresponding momentum $\boldsymbol{p}$ are

$$E_\mathrm{R} = \boldsymbol{p}^2/(2M) = (h\nu_0)^2/(2Mc^2),$$

with M as the nuclear mass. An emitted nuclear quantum has the energy

$$h\nu = h\nu_0 - E_\mathrm{R},$$

whereas for absorption the quantum needs to have the energy

$$h\nu = h\nu_0 + E_\mathrm{R}.$$

[29] Although the proper name is Mößbauer, the preferred spelling in the Anglo-American literature is Mössbauer, as employed here.

Emission and absorption lines consequently differ by 2 E_R ($\approx$ 1 eV as a typical value). Since nuclear transition energies are very narrow (10^{-10} to 10^{-4} eV), resonant absorption is only possible if the lacking energy is supplied somehow. With freely moving species, this can be achieved by the Doppler effect, e.g. in an ultracentrifuge. When the involved atom is part of a rigid crystalline structure, the recoil is transferred to the crystal instead of to the single atom. Because the mass of the crystal is much larger than the mass of the atom, the recoil energy and the Doppler shift practically disappear to about 10 meV, which is in the same order of magnitude as the phonons in the crystal lattice (Mössbauer effect). The corresponding gamma line (sometimes also called the Mössbauer line) shows only the natural linewidth (5×10^{-13} eV for ^{57}Fe; 5×10^{-16} eV for ^{67}Zn), which is reduced by about five orders of magnitude. Because of the extremely narrow linewidth resonance, the conditions might already be lost if the source or absorber crystal are moving even at low speed (fractions of mm/s). This provides a simple possibility to measure shifts of the line induced by other influences. Measuring such a shift requires simply to move the source or the absorber at a well-defined acceleration and look for resonant absorption. Consequently, the resulting Mössbauer spectra are usually displayed with the gamma transmission of the sample vs. the velocity of the source. In the emission mode, the intensity is displayed as a function of the speed.

Shifts of the absorption line can be induced by changes in the electron density around the atomic nucleus. These changes can be effected by filled inner-core orbitals and partially filled valence orbitals; they are called isomer shifts. The change influences both the ground state and the excited state; in sum, a change of the absorption energy is observed as depicted in Fig. 5.88.

This shift is indicative of the nature of the chemical bond between the investigated atom and its neighbors or the state of oxidation. Figure 5.89 provides a rough correlation between the state of oxidation of iron and the isomer shift.

The deviation of the nuclear charge distribution from an ideal spherical one is described with the nuclear quadrupole moment. This moment can interact with the gradient of the electric field generated by charged species (ions and electrons) around this nucleus and thus the degeneracy of the nuclear states may be lifted, resulting in a splitting of the single absorption line into several lines. This may oc-

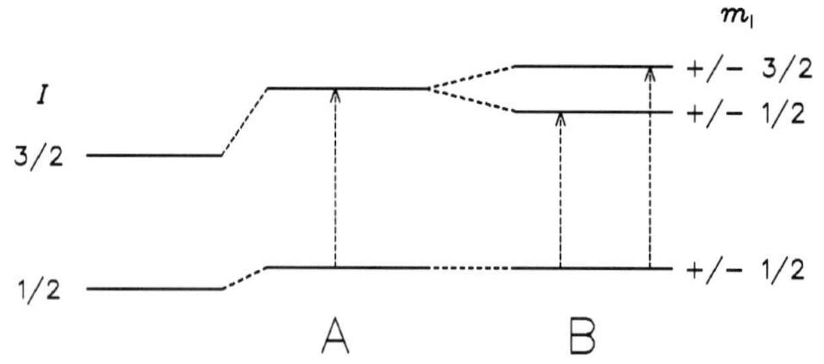

Fig. 5.88. Effects of isomer shift (*A*) and quadrupole splitting (*B*) upon allowed nuclear transitions

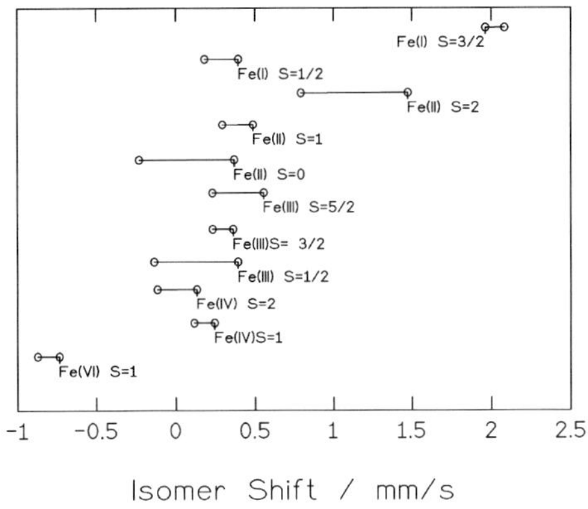

Fig. 5.89. Correlation between state of oxidation of iron ions and the observed isomer shift (based on literature data [535])

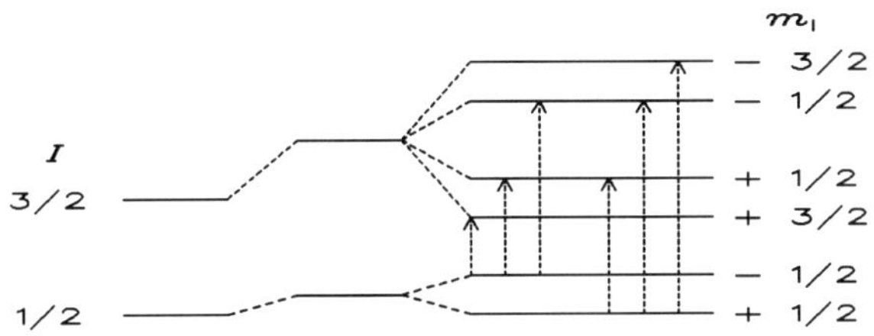

Fig. 5.90. Effects of magnetic splitting upon allowed nuclear transitions (Dependencies of energies of states upon m_{I} are slightly complicated, resulting in arrangements contrary to simple expectations)

cur only when the nuclear quadrupole moment is different from zero and when the nuclear state spin quantum number is larger than 1/2. This effect is called quadrupole splitting, where Δ describes the actual line splitting. A similar splitting can be caused by an external magnetic field that results from magnetic species present in the sample (magnetic hyperfine interaction). It causes splitting of the nuclear state with spin quantum number I into $2I + 1$ equally spaced states, as shown in Fig. 5.90.

Further details and tabulated values can be found elsewhere [536, 537]. The various parameters, observables and properties are collected in the table below.

Mössbauer parameter	Observed quantity	Obtained information
Isomer shift	Charge density at nucleus	Valence/oxidation state
Quadrupole splitting	Electric field gradient at nucleus	Symmetry of charge environment
Magnetic splitting	Magnetic field at nucleus	Magnetic ordering, particle size

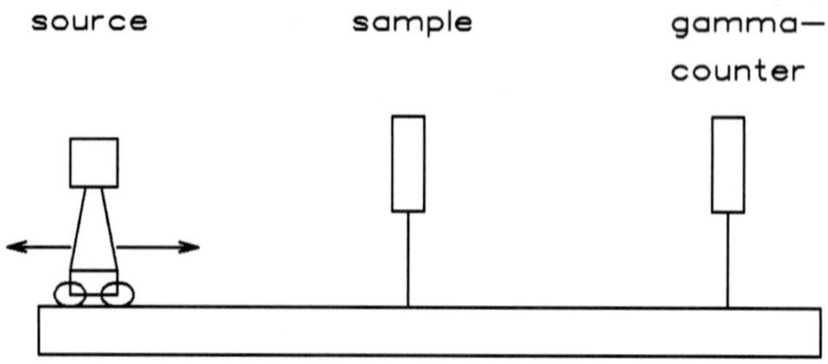

Fig. 5.91. Schematic experimental setup for Mössbauer spectroscopy in the transmission mode

Instrumentation. In order to record a Mössbauer spectrum, an emitter producing the requested monochromatic gamma radiation is needed. The minute modulation of the energy of the emitted radiation is accomplished by moving the emitter at a low speed in the range of mm/s. To probe the whole range of resonance conditions, the speed is varied slowly from slow approaching to slow departing. The absorbing sample is kept in a fixed position. Absorption is detected by positioning a gamma counter behind the sample. Upon resonance, the amount of passing radiation arriving at the detector will be diminished. A simple setup is shown schematically in Fig. 5.91.

Besides this transmission mode, an emission mode is possible. In this mode the gamma emission of the sample containing suitable isotopes of the atom to be investigated is passed through an absorber of known properties to detect the desired transition energies. Finally measurements of signals formed during the deexcitation of nuclei after absorption of gamma rays is possible. Gamma rays, X-rays or conversion electrons can be formed during this process. This technique is particularly useful with samples that are too thick for transmission measurements. In addition, the technique can be made particularly surface selective when the emitted electrons (conversion electrons) are measured, because these electrons have only a very short mean free path length in solids.

The source most commonly employed with ^{57}Fe Mössbauer spectroscopy is elemental ^{57}Co, which is incorporated into rhodium or copper metal. During the radioactive decay of the cobalt isotope into ^{57}Fe, the needed gamma radiation is emitted. For measurements with tin (^{119}Sn), sources of $CaSnO_3$ or $BaSnO_3$ enriched with ^{119m}Sn are used, which again release the proper radiation during their radioactive decay. The source is moved at constant positive and subsequently negative accelerations (i.e. linearly varying speed) to probe the resonant absorption.

The sample, i.e. the electrode material to be investigated, is mounted in an electrochemical cell particularly suited for the employed mode of Mössbauer spectroscopy. A cell designed for measurements in the transmission mode is shown in Fig. 5.92. In order to achieve a maximum surface sensitivity, the electrode should be made as thin as feasible because Mössbauer-active species deep inside the electrode that are not affected by the electrochemical processes occurring at the solution/solid

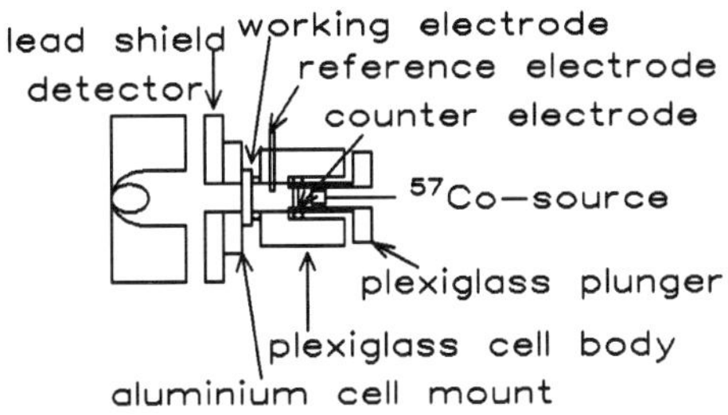

Fig. 5.92. Electrochemical cell for *in situ* Mössbauer spectroscopy in the transmission mode [538]

interface will also absorb gamma radiation. Besides making thin porous electrodes wherein a maximum amount of the electrochemically active surface of the electrode material is in contact with solution, thin film electrodes deposited onto insulating carriers or electrochemically inert materials are used.

In this cell the iron is deposited as a thin film onto a gold substrate. Gold deposited onto a thin plastic film serves as a counter electrode.

Because of the very limited escape depths of conversion electrons (about 1.8 μm in water, 0.25 μm in metallic iron), their detection is somewhat difficult. This seeming drawback provides a unique surface sensitivity. In a rotating disc electrode arrangement Kordesch et al. [539] have used a disc-shaped electrode that slowly rotates with part of the disc immersed in the electrolyte solution. As a thin electrolyte film thin enough to permit escape of conversion electrons adheres to the metal surface, potential control is always maintained. Conversion electrons were detected using a suitable gas-filled detector mounted close to the upper emersed part of the disc. In a study of passive oxide films on iron, the advantage of this approach was demonstrated; beyond an unmatched surface sensitivity, the measurement time was reduced to a small fraction of that needed for transmission measurements [543]. An inherent drawback of the setup is the poor current distribution inside the very thin electrolyte film (its thickness is around 4 nm as reported by Gordon [540]).

Mössbauer spectra are mostly displayed as gamma transmission/absorption vs. velocity of the source. Investigations deal preferentially with metal surfaces containing suitable Mössbauer-active species. They include studies of corroding iron surfaces, electrode surfaces modified with iron-containing complexes (e.g. hexacyanoferrates, transition metal macrocyclic complexes), tin electrodes and cobalt-electrodes. In a typical experiment, Vertes et al. [541, 542] have recorded Mössbauer spectra of a ^{119}Sn enriched tin electrode deposited on an aluminum substrate in contact with a borate buffer solution (pH = 8.4) at various electrode potentials in order to elucidate the passive film (Fig. 5.93). At $E_{\mathrm{SCE}} = -0.9$ V β-Sn ($\delta = 2.5$ mm s^{-1}) and SnO_2 or $Sn(OH)_4$ ($\delta = 0.03$ mm s^{-1}) can be identified. The weak absorption shoulder at $\delta = 4.3$ mm s^{-1} was attributed to Sn(II) species being present in amorphous form. This feature could not be observed at more positive electrode potentials. Thus the authors concluded that at electrode potentials negative to $E_{\mathrm{SCE}} = -0.78$ V, the passive films contain highly amorphous $Sn(OH)_2$ besides SnO_2 and/or $Sn(OH)_4$.

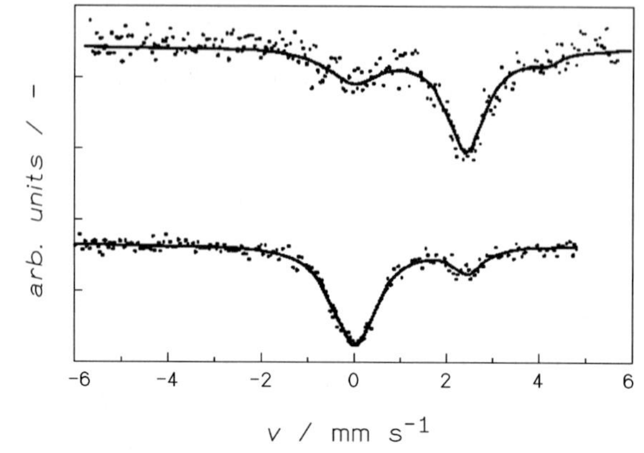

Fig. 5.93. *In situ* Mössbauer spectra of a tin electrode at various electrode potentials in a borate buffer solution (pH = 8.4) (based on data in [546])

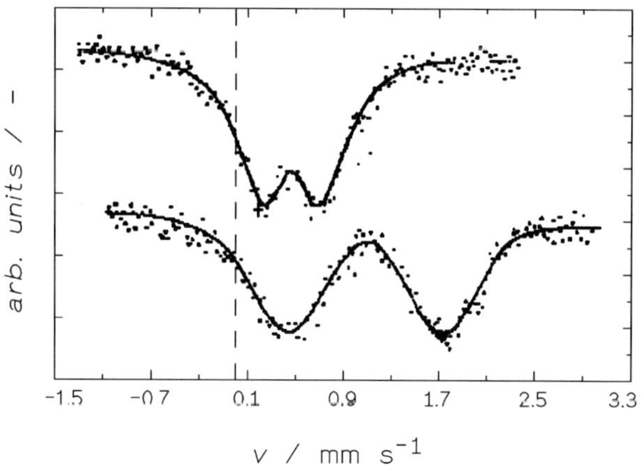

Fig. 5.94. *In situ* Mössbauer spectra of a film of Prussian blue deposited on a glassy carbon electrode in an aqueous solution of 1 **M** KCl (pH = 4) at $E_{SCE} = 0.6$ V (*top*) and $E_{SCE} = -0.2$ V (*bottom*) (based on data by [549])

Prussian blue that is formed as a highly colored colloidal precipitate by adding Fe(III) ions to a solution of Fe(II) hexacyanide [543] has been identified as being electrochromic [544]. Redox processes occurring during the reversible change of colour have been studied with Mössbauer spectroscopy [545]. At $E_{SCE} = 0.6$ V, the Prussian blue film formed from Fe(II) hexacyanide and highly enriched ^{57}Fe(III)Cl_3 shows two absorptions with $\delta = 0.37\ \mathrm{mm\,s^{-1}}$ and $\Delta = 0.41\ \mathrm{mm\,s^{-1}}$ typical of an Fe(III) in its high spin state (Fig. 5.94). The Mössbauer spectrum at $E_{SCE} = -0.2$ V shows two lines typical of high spin Fe(II) ions ($\delta = 1.14\ \mathrm{mm\,s^{-1}}$ and $\Delta = 1.31\ \mathrm{mm\,s^{-1}}$). The electrochromic redox reaction obviously involves the uncoordinated iron species; no iron exchange seems to take place.

Mössbauer spectroscopy has been used frequently in *in situ* studies of electrodeposits [546, 547], passivation phenomena [548–551], battery materials [552] and transition metal complexes [553–555]. Co_2SnO_4 has been studied with respect to its use in lithium ion batteries *ex situ* with ^{119}Sn Mössbauer spectroscopy [556]; results indicate a consumption of Sn(IV) species accompanied by formation of a cobalt–tin intermetallic compound. Finally, a tin-rich lithium phase is found. Upon charging the initial state is finally restored. $CoSn_2$ as an electrode material has been investigated with *in situ* ^{119}Sn Mössbauer spectroscopy [557]. Upon discharge a lithium-rich phase of approximate composition $Li_{3.5}Sn$ and cobalt nanoparticles are formed. Upon charging, the more complex processes include conversion of the lithium-rich phase into a modified $CoSn_2$ nanocompound (i.e. a matrix of $Li_zCo_ySn_2$), which is the active fraction in subsequent cycling.

5.3.2 X-Ray Absorption Spectroscopy[30] (XAS)

Fundamentals. Electromagnetic radiation in the X-ray range of the spectrum may be scattered (as employed in diffraction methods, see below) or absorbed. The relationship between the incident radiation intensity, I_0, and the transmitted one, I, is given by Lambert's law

$$I = I_0\,\mathrm{e}(-\mu_\mathrm{m} m),$$

with the mass absorption coefficient μ_m and the mass per unit area m. The value of μ_m increases with wavelength and there is no simple relationship. When the energy of the incident radiation is sufficient to eject a core level electron, the μ_m shows a sharp increase (edge) and the respective energy is typical of the absorbing element. The actual value depends slightly on the state of oxidation of the element, e.g. with increasing positive valency the energy shifts to higher values. For an introduction, see [558]. A general overview of recent technological advances is available [559]. The element specificity of the absorption is employed in microradiology (microradiography)—see Sect. 7.6.

Instrumentation. An overview of experimental approaches and results has been provided elsewhere [560]. A cell suitable for time-resolved studies with XAS has been described [561]. *Ex situ* studies of battery materials have been reported [562, 563].

5.3.3 X-Ray Absorption Fine Structure Spectroscopy

Fundamentals. The absorption of X-rays at high resolution close to the absorption edge of atoms under investigation can be measured in order to obtain information about the structural and chemical environment of these atoms [564–566]. Measurements at energies (wavelengths) slightly below the absorption edge are called XANES[31] (X-ray absorption near edge structure/spectroscopy) or NEXAFS (Near

---

30 Somewhat confusingly studies employing investigations of the absorption fine structure (see p. 137) are sometimes also labeled simply XAS studies.

31 The final letter S in XANES, EXAFS, etc. refers both to structure and spectroscopy, thus causing confusion in proper designation of results.

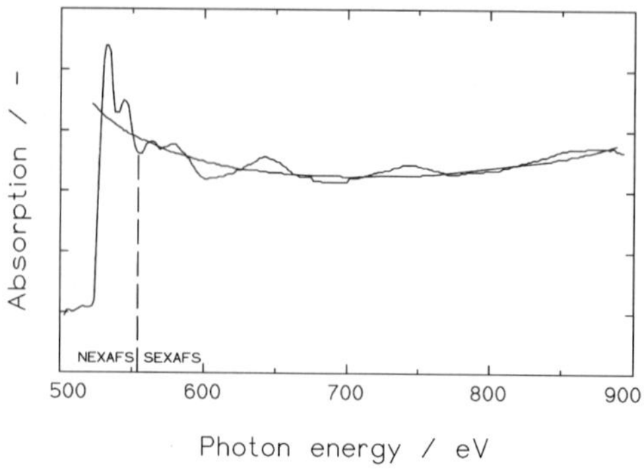

Fig. 5.95. Characteristic SEXAFS spectra of atomic oxygen adsorbed on Cu(110) in a Cu(110)-(2 × 1)-O arrangement, smooth line shows absorption of free atoms (based on data in [571])

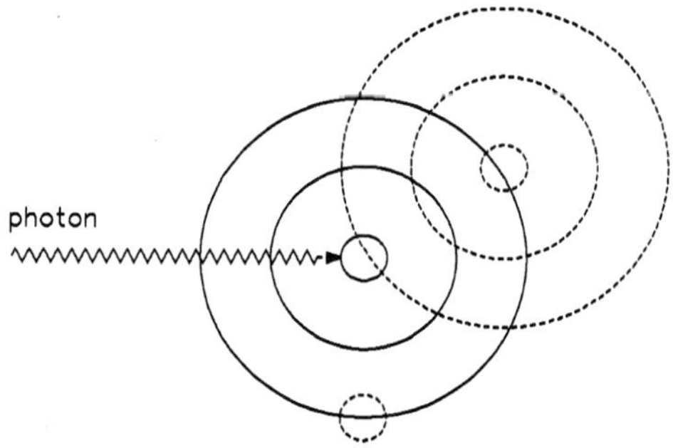

Fig. 5.96. Principle of photon interference in X-ray absorption

edge X-ray absorption fine structure/spectroscopy), whereas measurements around the absorption edge and slightly above it are called EXAFS[32] (extended X-ray absorption fine structure/spectroscopy). Because of the penetration depth of X-rays, these methods are not intrinsically surface sensitive. Depending on the experimental setup and/or the selection of the X-ray energy, and in turn the atoms to be probed, species at the interface can be studied specifically.

The absorption coefficient μ of an atom in a molecule or a solid has a fine structure[33] amounting to about 15% of the absorption jump at the edge and extending up to several hundred electron volts above the absorption edge, whereas a free atom only shows a smooth absorption as seen in Fig. 5.95 [567].

The oscillations of the observed absorption are caused by interference between the photon emitted from the absorbing atom and photon waves emitted from neighbor atoms, as depicted in Fig. 5.96.

---

[32] In its surface sensitive adaptation, the method is called SEXAFS (surface extended X-ray absorption fine structure spectroscopy).

[33] This absorption is sometimes called AXAFS (atomic X-ray absorption fine structure).

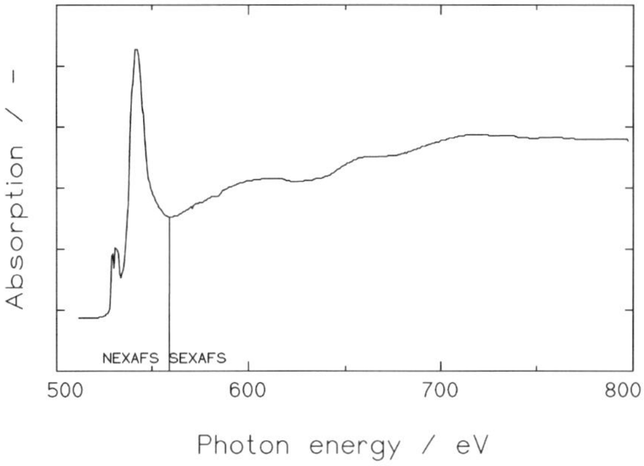

Fig. 5.97. Characteristic SEXAFS spectra of O adsorbed on Ni(111) in a Ni(111)-c(4 × 1)-NO arrangement (based on data by [571])

When the neighbor atoms are surface atoms the distance between the atoms of interest and the surface atoms r_{AS} can be determined. This distance determines the frequency of the absorption oscillation, whereas the amplitude is influenced by the number and identity of the neighbors and their distance. The polarization of the incident X-rays strongly influences the observed oscillations. Bonds with interatomic vectors in the plane of polarization contribute significantly, whereas bonds perpendicular to this plane do not (for further details, see [568]).

With atomic species, the near edge structure is of no further interest. With molecular adsorbates, the near edge X-ray absorption is dominated by intramolecular transitions (μ- and σ-resonances). Their dependency on the polarization provides information on the orientation of the molecule. From the energy of the σ-resonance intramolecular bond lengths, r_{AB} can be estimated for simple molecules. Figure 5.97 shows an X-ray absorption spectrum of NO adsorbed on a nickel surface.

These arguments are of course also valid for species on top of a surface. Figure 5.98 shows the structural information available from both spectroscopies. Distances can be determined with a precision of about ±1 pm and the number of atoms can be determined with a precision of about 15%. Since no diffraction is involved, samples without any long range ordering can be studied.

As already indicated, X-ray absorption spectroscopy is inherently not surface sensitive. By adjusting the photon energy to a value matching the absorption edge of species sitting on a surface, this sensitivity can be easily obtained. With photons impinging at a grazing angle below the angle of total reflection, contributions from the substrate are suppressed to a large extent [569] and the depth of information is reduced to only a few nanometers. The method is called grazing incidence EXAFS (GIXAFS).

Instrumentation. X-ray absorption spectroscopy is done mostly with synchrotron radiation (see Chap. 4); for an introductory overview, see [570, 571]. The linearly

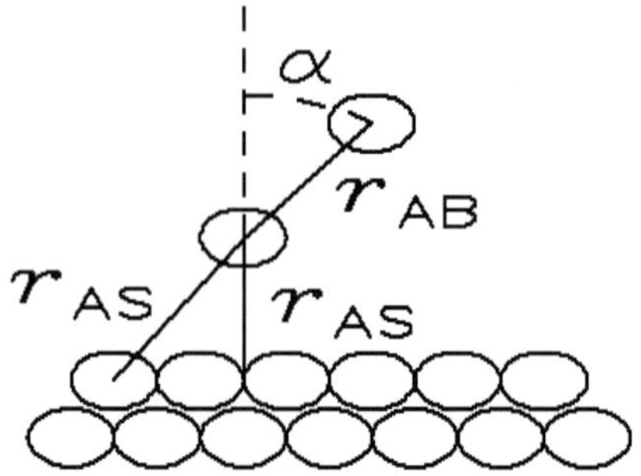

Fig. 5.98. Structural features determined with X-ray absorption spectroscopy

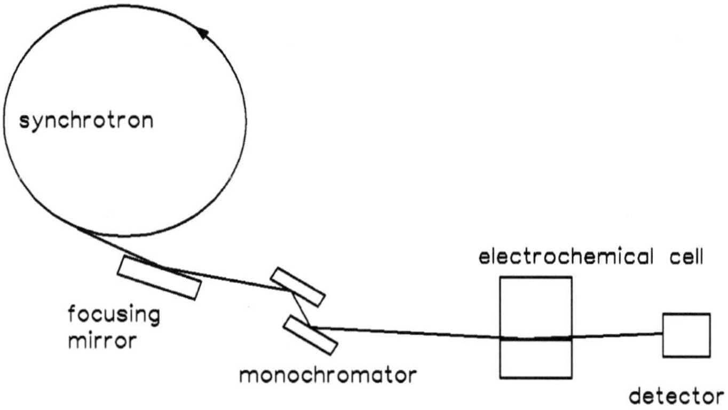

Fig. 5.99. Setup for an X-ray absorption experiment in the fluorescence mode

polarized light is monochromatized (resolution down to 100 meV or better) and guided to the sample as outlined in Fig. 5.99.

The electrochemical cell is mounted on a moveable holder, allowing measurements at different angles and polarizations of the incident light. Unfortunately, the number of atoms absorbing radiation is very small (a monolayer of atoms or even less), thus simple absorption measurements like the ones that can be achieved with bulk samples are impossible. Instead, the measurement of Auger electrons (possible only under UHV conditions) or of fluorescence intensity is possible.

Alternatively, measurement in the transmission mode is possible. A simple cell design has been described [572]; see Fig. 5.100.

The solution level is kept at the upper limit during electrochemical experiments, providing complete participation of the working electrode surface in any electrochemical process. In order to minimize solution absorption of the X-rays, the level is lowered before X-ray absorption measurements. As the lower edge of the working electrode is still immersed in the electrolyte solution, electrode potential control is maintained.

Evaluation of the measured absorption starts with subtraction of the smooth underground and normalization. The oscillating absorption in SEXAFS contains infor-

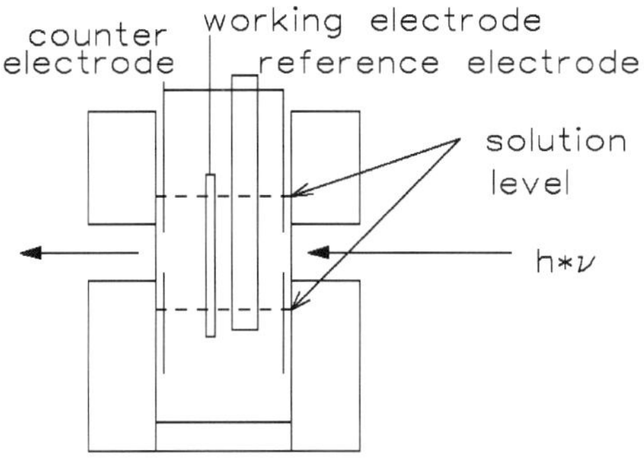

Fig. 5.100. Electrochemical cell for XANES/SEXAFS measurements in the transmission mode (based on a design in [576])

mation on distance, number and identity of neighbor atoms. Same atoms are summarized in shells. Evaluation of polarization dependencies and further spectroscopic data in XANES provide information about intramolecular bond length and the relative arrangement of adsorbate species with respect to the surface (for further details, see [571]).

Depending on the type of measurement (absorption or emission), different types of cells and samples have been employed. A slightly modified working fuel cell using an ion exchange polymer membrane as solid electrolyte coated with various catalyst layers has been described [573] that enables investigations of the platinum and ruthenium atoms incorporated in the catalyst layers. In order to avoid convolution of signals of both electrode layers (which are penetrated by the X-ray beam), one catalyst layer (the cathode) was removed (cut out) in the area of the beam path.

Time-resolved measurements can be made at storage rings with high flux insertion devices that use a quick-scanning mode of operation of the monochromator [574]. In a reported study, products of Mo corrosion in KOH solution could be identified and quantified [578]. Application of time-resolved dispersive high-energy X-ray absorption fine structure (DXAFS) measurements on platinum nanoparticles in a fuel electrode have been described [575]. Results indicate severe surface reconstruction of the nanoparticle surface, showing at least three types of Pt–O bonds (adsorbed OH, adsorbed atomic O and amorphous PtO_x) under oxidative conditions.

Numerous studies pertaining to upd-adsorbates have been reported that contain data on the bond distance between substrate atoms and deposits and the relative positions of the involved atoms [572–579]. Beyond these data the valence state of the adsorbed atoms and modifications of the electronic structure of the support upon adsorption of the upd-layer could be determined. In an approach where only the absorption modulation intensity is evaluated, changed electronic properties of platinum particles deposited on a carbon support could be explained by invoking changing charge on the metal as a function of the electrode potential [580]. In a comparison of carbon-supported platinum modified with upd-Sn and a carbon-supported

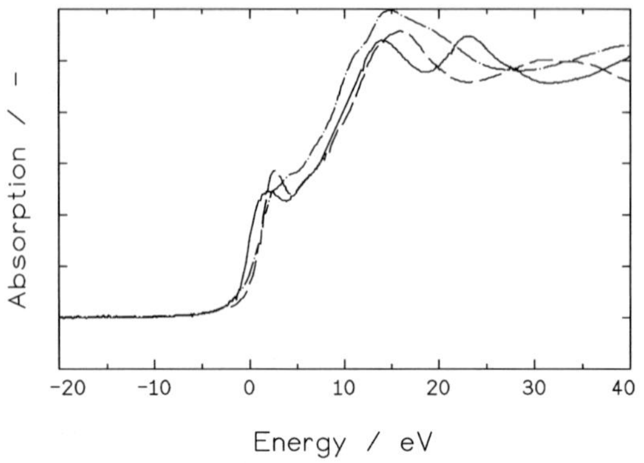

Fig. 5.101. Normalized XANES of copper foil (——), Cu_2O (- - -) and upd-Cu on Pt/C (· · · · ·) (based on data in [576])

platinum–tin alloy, the significantly higher electrocatalytic activity of the former system was attributed to electronic effects. In the alloy, tin causes partial filling of the d-band vacancies of platinum and an increase in Pt–Pt bond length. Upd-Sn does not perturb platinum, either electronically or structurally. In both systems, tin is associated with oxygen containing species at all electrode potentials [581]. In a similar investigation of upd-Cu on platinum deposited on a carbon support, a result as depicted in Fig. 5.101 was obtained.

A copper coverage of 0.57 was deduced. The position of the absorption edge of upd-Cu is close to that of Cu^+. Further differences between the traces for upd-Cu and Cu_2O were explained by invoking crystal symmetry; in this case upd-Cu is present in a tetrahedral environment [576]. Stimulated by the considerable and still growing interest in binary and ternary electrocatalysts, XANES and SEXAFS (mostly combined) have frequently been used. The coverage of a single crystal platinum surface with iodine was monitored [575] and results were found to be in good agreement with those of AES at an emersed electrode at electrode potentials below the estimated potential of zero charge E_{pzc} [582]. The composition of a Pt–Ru alloy with 25% of Ru in the actual alloy and an excess of 90% of Ru not alloyed at all was reported [583]. Further details of the electronic band structure and changes in Pt–Pt bond distances could be elucidated. The structure of small two-dimensional clusters and linear nanostructures of copper and cadmium deposits on Pt(533) at characteristic electrode potentials in the upd-range has been studied with GIXAFS combined with GIXD [584]. Catalytic mechanisms at binary platinum–metal (metal = Co, Cr, Ni, Fe) alloys deposited on carbon for water activation and dioxygen reduction in perchloric acid have been elucidated with XANES and EXAFS [585]. A cell suitable for *in situ* investigations of catalysts in operating PEM[34] fuel cells has been described and applied for determination of coverage of PtRu catalysts with CO and other species (O[H]) involved in the electrocatalytic processes as a function of cur-

[34] PEM = polymer electrolyte membrane.

rent [586]. Both a bifunctional mechanism and an electronic ligand effect are operative in CO oxidation at PtRu surfaces. Changes of the electronic properties of a cobalt porphyrin as a function of electrode potential and of interaction with solution phase species has been monitored with XANES [587]. An overview of applications in fuel cell studies is available [588].

Studies of passivation layers and corrosion phenomena, particularly with iron and magnesium using EXAFS, NEXAFS and related X-ray absorption spectroscopies have been reviewed elsewhere [589]. Using GIXAFS, depth profiling has become possible and spatially resolved studies of silver dissolution and lithium intercalation have been reported.

The formation and dissociation of S–S bonds in poly(tricyanuric acid), which is proposed as electrode material for lithium batteries, has been studied [590, 591]. The reversibility of the process essential for the use of this material in a secondary battery could be established. Further studies of battery materials have been reported [592, 593]. X-ray absorption near edge structure spectroscopy has been successfully employed in studies of inhibiting species in passive films and the adjacent electrolyte solutions.

5.4 Magnetic Resonance Spectroscopy

Magnetic resonance spectroscopies are methods capable of detecting transitions of spin orientations of electrons or atomic nuclei between states separated energetically under the influence of an external magnetic field. Transitions involving the spin of the nucleus of an atom with a nonzero magnetic moment are studied with nuclear magnetic resonance spectroscopy (NMR), whereas transitions involving the spin of unpaired electrons in paramagnetic samples are investigated with electron spin resonance spectroscopy (ESR).[35] Both methods are widely employed in analytical chemistry. Nuclear magnetic resonance spectroscopy, preferably of protons, ^{13}C-atoms and further selected atoms, is presumably the most important method in analytical organic chemistry. Electron spin resonance spectroscopy is used less frequently in studies of free radicals (chemical species with unpaired electrons) observed typically in organic reactions and in investigations of transition metal ions and paramagnetic substances.

Because in many electrochemical reactions, particularly in electroorganic ones, radicals are formed as reactive intermediates, ESR has been applied frequently to studies of the mechanism and the kinetics of these reactions [594–596]. Although possible, NMR spectroscopy has been used infrequently and only in very recent experiments, mainly because of the considerably larger experimental effort [597]. With NMR spectroscopy, information about surface structure, surface diffusion and electron spillover from the metal electrode onto an adsorbate can be obtained. So

[35] This method has frequently been called electron paramagnetic resonance spectroscopy (EPR) because the presence of one or several unpaired electrons, being a precondition for this spectroscopy, is also closely related to the phenomenon of paramagnetism.

far, an application as broad as that of ESR has not materialized. Both methods have been employed infrequently to monitor the concentration of species involved in electrochemical reactions in order to established reaction kinetic parameters, activation energies, etc. These applications do not fall within the scope of this book, so they are not considered here.

Fundamentals. Atoms are composed of nucleons: electrons, protons and neutrons (except the hydrogen atom, which has no neutron). These elementary particles have a spin—a property described in classical physics as a rotational movement of the particle around its own axis. It is a quantum property showing only certain quantisized values; for the electron, the spin is $\boldsymbol{s}$ with the corresponding spin quantum number $s = 1/2$. The spin of an atomic nucleus, $\boldsymbol{I}$, is composed of contributions of the nucleons. The corresponding angular momentum is $|\boldsymbol{s}| = \hbar/\sqrt{s(s+1)}$. This momentum is a vector, and its component parallel to a magnetic field H oriented in the z-direction can be $+\hbar/2$ or $-\hbar/2$. Because the proton and the electron are charged particles, their rotation corresponds to a flow of electricity. This results in a magnetic momentum for the electron:

$$\boldsymbol{m}_{\mathrm{s}} = g_{\mathrm{e}}\gamma\boldsymbol{s} = \gamma_{\mathrm{e}}\boldsymbol{s},$$

with g_{e} = electron g-factor, γ = magnetogyric ratio and γ_{e} = electron magnetogyric ratio.

When considering an atom containing these nucleons in numbers characteristic for a given element, the resulting properties of the atom are slightly more complex. The resulting spin of the atomic nucleus depends upon the number of protons and neutrons and the relationship between both as listed in Table 5.1.

The value of the magnetic momentum $\boldsymbol{m}_{\mathrm{I}}$ for the atomic nucleus is

$$\boldsymbol{m}_{\mathrm{I}} = g_{\mathrm{N}}\gamma\boldsymbol{I} = \gamma_{\mathrm{N}}\boldsymbol{I},$$

with g_{N} = nuclear g-factor and γ_{N} = nuclear magnetogyric ratio.

In atoms or molecules containing several electrons, only unpaired electrons present in singly occupied atomic or molecular orbitals show an effective magnetic momentum.

The vectors representing the nuclear or the electron angular momentum and the associated magnetic momentum of particles in a given sample are randomly orientated. This is changed drastically when the sample is exposed to an external

Table 5.1. Resulting nuclear spins as a function of protons and neutrons

n_{p}	n_{n}	Resulting nuclear spin $\boldsymbol{I}$	Example
odd	odd	integer number	^{2}H, ^{14}N, ^{10}B
odd	even	half-number value	H, ^{13}C, ^{15}N, ^{19}F, ^{31}P, ^{11}B, ^{29}Si
even	odd	half-number value	see above
even	even	zero	^{12}C,^{16}O, ^{28}Si, ^{30}Si

magnetic field. Because of the interaction between the magnetic field and the magnetic momentum of the electron or the atomic nucleus, the momentum is oriented in one of two possible directions relative to the external field vector. The momentum is not oriented exactly parallel to the field vector; consequently the vector shows a precession movement with a characteristic Larmor frequency. A calculation of the potential energy of the species in the magnetic field is based on the size of the vectorial component of the magnetic moment in the z-direction. This is given for an electron by

$$|\boldsymbol{m}_{\mathrm{sz}}| = |-g_{\mathrm{e}}\mu_{\mathrm{B}}m_{\mathrm{s}}|,$$

with μ_{B} = Bohr magneton.

The corresponding value for a nucleus is

$$|\boldsymbol{m}_{\mathrm{lz}}| = |g_{\mathrm{N}}\mu_{\mathrm{N}}m_{\mathrm{N}}|,$$

with μ_{N} = nuclear Bohr magneton.

The potential magnetic energy E for a dipole $\boldsymbol{m}_{\mathrm{z}}$ in a magnetic field is

$$E = -\boldsymbol{m}_{\mathrm{z}}\boldsymbol{B}.$$

With the electron, the result is

$$E = g_{\mathrm{e}}\mu_{\mathrm{B}}m_{\mathrm{s}}\boldsymbol{B},$$

for the nucleus it is

$$E = -g_{\mathrm{N}}\mu_{\mathrm{N}}m_{\mathrm{l}}\boldsymbol{B}.$$

For a proton with $m_{\mathrm{N}} = 1/2$ only two orientations corresponding to $m_{\mathrm{l}} = 1/2$ and $m_{\mathrm{l}} = -1/2$ are possible; this case will subsequently be considered as an example for NMR. Of course, other nuclei with $m_{\mathrm{N}} > 1/2$ possess more than only two nondegenerate energy levels. They are spaced equally, which implies that only one resonant transition is observed.

For both the electron and the nuclear momentum the calculated energy depends upon the actual orientation of the momentum vector. The values that are energetically lower are those corresponding to $m_{\mathrm{s}} = -1/2$ and $m_{\mathrm{l}} = +1/2$. Energies for the electron are larger by three orders of magnitude. The actual energies and the differences between various orientations depend upon the magnitude of the magnetic field. A calculation of the difference is possible by assuming a temperature of 300 K and a magnetic flux density of $B = 0.3$ T for investigations of the electron spin and $B = 1.409$ T for those of the nuclear (in this example, the proton) spin. Results are:

$$\Delta E_{\mathrm{s}} = g_{\mathrm{e}}\mu_{\mathrm{B}}B = 5.57 \cdot 10^{-24}\ \mathrm{J}$$

and

$$\Delta E_{\mathrm{N}} = g_{\mathrm{N}}\mu_{\mathrm{N}}B = 3.97 \cdot 10^{-26}\ \mathrm{J}.$$

The differences in population between the corresponding states can be calculated by assuming a Boltzmann distribution. For the electron spin, the ratio $n_{\mathrm{s,high}}/$

$n_{s,low} = 0.99866$; for the proton, $n_{N,high}/n_{N,low} = 0.99999039$. A transition of the orientational state of the spins from the lower into the higher state is possible only if the necessary energy is supplied under resonant conditions to the system. Continuous absorption of this specific energy occurs only when there is a smaller population of the higher energy level. This in turn depends upon relaxation of the spins from the higher state into the lower one, which is possible only when effective ways of dissipating the energy exist. Spontaneous emission, which provides an effective path of depopulating higher levels in optical spectroscopies, is extremely low and can be neglected. Consequently, stimulated emission initiated by the incoming radiation and interactions with other particles (spin–spin relaxation) and the surrounding matter (spin–lattice relaxation) provide the required energy dissipation. Since the difference in energies between both states of orientation depends upon the strength of the magnetic field B, the resonance condition is described properly by stating both B and the frequency of the electromagnetic radiation. For a proton in a magnetic field with a flux density of $B = 1.409$ T, the frequency is $60 \cdot 10^9\ s^{-1}$; at a stronger magnetic flux density of $B = 11.74$ T, which is employed in advanced spectrometers, resonance occurs at $500 \cdot 10^6\ s^{-1}$. For an electron in a field of $B = 0.34$ T the value is $9.5 \cdot 10^9\ s^{-1}$.

The previous discussion assumes that both the unpaired electron(s) and the atomic nuclei interact solely with the magnetic field. Correspondingly, the resonance condition should be fixed as described. In reality, both are surrounded by other species that influence the effective magnetic field at the location of the electron or nucleus by means of their own electric and magnetic properties. Based on the type of chemical bond and the electronegativity of the participating elements, the electron density at the atom under investigation can change considerably. In the case of the nucleus, the resulting effect is particularly pronounced. The actual resonance condition is changed accordingly. Appropriately, this is called a chemical shift; because of its high specificity, it can be used as a tool in identification of the molecular structure. The influence upon the unpaired electrons is much smaller and of only very limited analytical value. Besides this effect, which is related to the local electron density, the magnetic spins of neighboring atoms can influence the local field. Depending upon the relative orientation, the field is increased or decreased and the resonance condition is changed again. Because of the number of combinations of orientations of magnetic spins in a multi-atom system, several additional resonant transitions may occur. The effect is termed hyperfine splitting with ESR; in NMR, it is called spin–spin splitting because it is the result of an interaction of nuclear spins. In solid samples anisotropic effects of electrostatic or magnetic fields caused by the atomic or molecular constituents of the sample and their particular arrangement can also change the effective magnetic field, resulting in both shifts and splitting. This effect is not present in liquid and gaseous samples. With both spectroscopies the number of additional lines, their relative intensities and energetic difference provide valuable information for the elucidation of the molecular structure. In the electrochemical applications discussed below in more detail, ESR is used more often, thus some example will illustrate the brief outline of magnetic resonance spectroscopies.

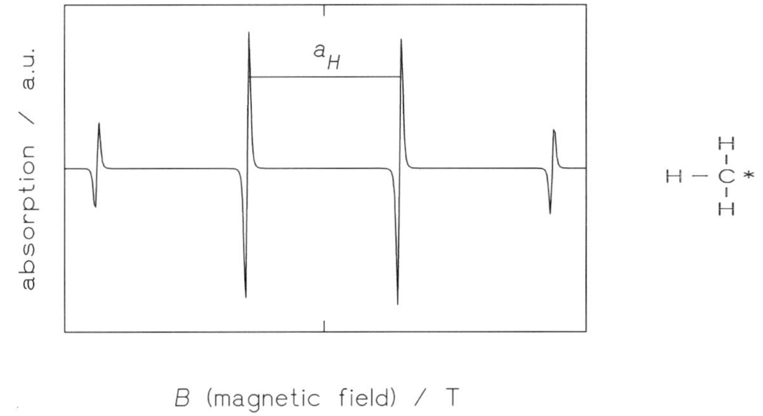

Fig. 5.102. Calculated ESR absorption spectrum of the methyl radical

The number N of lines created by n nuclei labelled k of spin $\boldsymbol{I}_k$ is

$$N = \prod_k (2n_k \boldsymbol{I}_k + 1).$$

The distance of the lines corresponding to the energetic difference is stated by giving a splitting constant a (sometimes also called the coupling constant). The relative intensity of the lines depends upon the probability of transitions between the states involved; this in turn depends on the probability of the various combinations of orientations. In Fig. 5.102, the calculated ESR spectrum of the methyl radical is depicted.

As in ESR spectra, because of the mode of detection of the absorption signal (see p. 149) the first derivative is always measured and the simulated spectra shown here are plotted accordingly. With a spin $\boldsymbol{I}_\mathrm{H} = 1/2$ and $n = 3$, the number N of lines is four. A surprisingly simple spectrum (as shown in Fig. 5.103) results with a considerably more complex molecule: di-tert-butyl nitroxide in its radical form. We can understood it easily when taking into account that the unpaired electron is located at the nitrogen atom ($\boldsymbol{I} = 1$), whereas both oxygen and carbon have $\boldsymbol{I} = 0$ (see p. 144).

A check for the correctness of a simulated spectrum is provided by comparing calculated and measured spectra. In Fig. 5.104, the measured and the calculated spectra of the electrochemically generated nitropropane radical are shown.

Upon the rather wide splitting caused by the nitrogen atom, a further splitting caused by the two methylene hydrogen atoms at the carbon atom next to the nitrogen is superimposed. The hydrogen atoms at the vicinal methyl group are too far away and do not cause any further splitting. Further details are discussed in textbooks of magnetic resonance spectroscopy (see suggested reading).

Instrumentation. From the description of the fundamentals of magnetic resonance spectroscopy, basic building blocks and functional elements of the experiment can

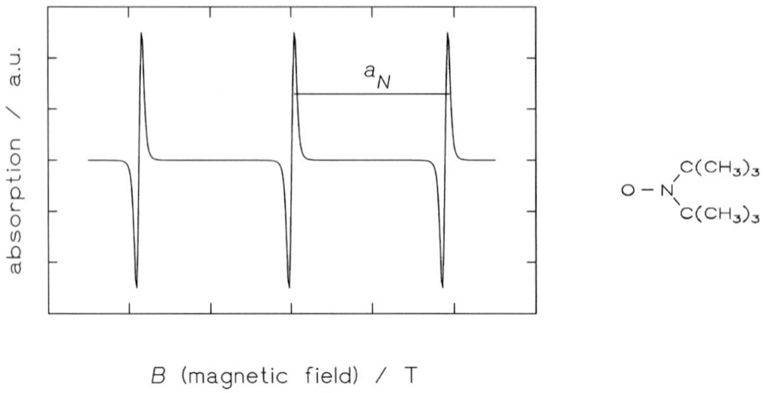

Fig. 5.103. Calculated ESR absorption spectrum of the di-tert-butyl nitroxide radical

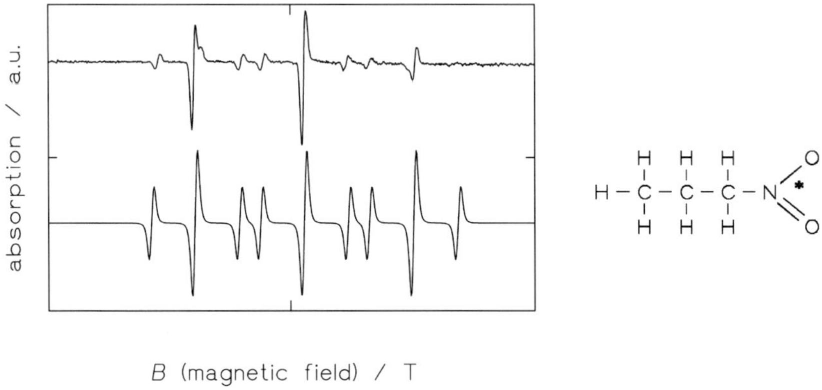

Fig. 5.104. ESR absorption spectrum of the electrochemically generated nitropropane radical anion; $E_{\mathrm{NHE}} = -1740$ mV; *top*: measured spectrum, *bottom*: calculated spectrum; hyperfine splitting constants $a_{\mathrm{N}} = 2.48$ mT, $a_{\mathrm{H}} = 0.998$ mT

be derived easily: The sample has to be brought into a magnetic field of appropriate strength. The energy corresponding to the difference between the two states of the nuclear or electron spin is supplied as electromagnetic radiation of suitable frequency—generally in the radio frequency range or, more precisely, in the micro-wave range for ESR and in the UHF range for NMR.[36] Because of the modes of propagation of radiation in the micro-wave range waveguides have to be used instead of cables. The sample is inserted into a cavity at the end of a waveguide in between the poles of the magnet. Since the actual resonance conditions are shifted from the reference values stated above, either the strength of the magnetic field (i.e. the magnetic flux) or the radiation frequency has to be tuned in order to detect the actual resonance. A modulation of the frequency of the radiation supplied at con-

[36] UHF: ultra high frequency.

stant intensity (cw: continuous wave) with the necessary precision and stability is somewhat difficult to obtain for reasons related to high frequency electronics. A direct modulation of the magnetic field by simply changing the current flowing across the coils of the electromagnets, which are employed frequently in NMR and practically exclusively in ESR, is cumbersome because of the need to control pretty large electric currents. This is impossible in the case of very strong magnets that use superconducting coils, anyway. A more effective approach is the addition of small coils attached to the poles of the main magnet. The additional magnetic flux provided by these modulation coils can be easily controlled. Consequently, the resonance conditions are probed by slowly changing the magnetic field. Absorption of electromagnetic radiation can be detected by various means. The actual absorption, particularly with ESR, is rather small because of the small difference in occupation of both states. An increase of sensitivity can be obtained by applying sophisticated amplification and detection methods (phase sensitive detection with lock-in amplifiers). In the case of ESR, this results in spectra that are equivalent to the first derivative of the actual absorption spectrum. More recently, the Fourier transform (FT) technique already described in Sect. 5.2 as applied to vibrational spectroscopy has been adapted for ESR and NMR. The electromagnetic radiation is supplied as a pulse. Detection and data manipulation is more complex, the advantage is a greatly enhanced sensitivity. This is described in more detail in textbooks of spectroscopy.

Basically, an NMR or ESR spectrometer is composed of a magnet; a radio frequency generator; an additional generator driving the field modulation coils; and the necessary detector, data manipulation, and storage electronics. The sample is inserted into the magnet between its poles. Some preparation of the sample is generally necessary; materials containing ESR- or NMR-active substances have to be avoided as sample holders. Various types of cuvets for solid, liquid and gaseous samples are in use. In NMR spectroscopy, quartz tubes of various diameters that are free of paramagnetic impurities are used almost exclusively. In ESR spectroscopy, different shapes (tubes, flat cells) of cuvets are used, depending upon the type of cavity (cylindrical ones for the former, rectangular ones for the latter cuvets).

For electrochemical applications, the experimental arrangement is rather simple. Because of the broad application of ESR, this method is treated first. Some additional information on NMR in electrochemistry can be found at the end of this section. In ESR experiments the spectrum can be recorded when the species under investigation is created either just inside the spectrometer (intra muros generation, subsequently treated as the *in situ* method) or outside the spectrometer (extra muros). In the latter case the sample has to be transferred by means of a flow apparatus or by removal of a small sample from the electrochemical cell, which is put into a standard ESR cuvet. For reasons already outlined in Chap. 4, the latter procedure, which is similar to an *ex situ* experiment, carries some inherent sources of error because of the limited lifetime or subsequent chemical reactions of the species initially created by the electrochemical reaction. Since no particular design of the cuvet is necessary with respect to the ESR spectrometer, the latter procedure will not be discussed in detail.

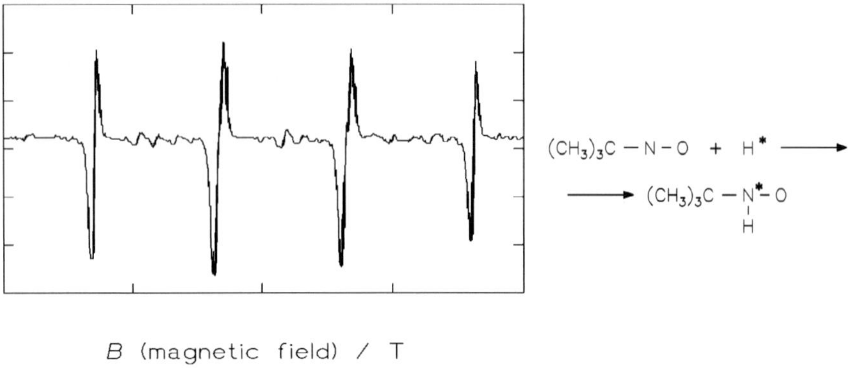

Fig. 5.105. ESR spectrum of the hydrogen adduct of NtB generated electrochemically at $E_{SCE} = 0.2$ V in an aqueous solution of 0.5 **M** $LiClO_4$ [598]

The experimental requirements of the spectrometer and an appropriate electrochemical cell design need to be considered in the generation of species and their detection inside the cavity of the ESR spectrometer. Since most electrochemical reactions proceed at metal electrodes, a very fundamental problem is encountered in any attempt to obtain ESR spectra of radicals that are still adsorbed, i.e. interacting strongly, on the electrode surface. This interaction between the unpaired electron of the radical and the electrons in the metallic conductor will quench the free spin and no ESR spectrum will be observed. This is different with semiconductor or insulator electrode surfaces. The quenching can be suppressed by coating the electrode with a layer of nonmetallic material (chemically modified electrode), but obviously this may change the interesting properties of the electrode considerably. Nevertheless ESR-active species can be detected as soon as they leave the electrode surface and stay in the electrolyte solution at a sufficiently large concentration for a time long enough to allow the measurement of a spectrum. The first requirement seems somewhat odd at first glance because ESR is a rather sensitive spectroscopy. Paramagnetic species at a concentration as low as 10^{-10} mol dm^{-3} can be detected easily. Unfortunately the high reactivity of organic radicals tends to keep the stationary concentration low. In addition, the components of the electrochemical cell (electrolyte, solvent, electrodes) reduce the sensitivity of the spectrometer considerably—particularly by increasing dielectric losses. Detection of radicals that are present in only very small concentrations or have a short lifetime can be facilitated by using "spin traps". These are mostly organic compounds that form stable radicals by reaction with the radicals formed during the investigated process. In many cases, these spin traps contain a nitroso group. The observed ESR spectra are more or less complicated depending on the type of spin trap and the trapped radical. A fairly simple spectrum is obtained by using t-nitroso-butane (NtB) as a spin trap (for further details, see pp. 150, 158). The adduct formed with hydrogen radicals causes the spectrum depicted in Fig. 5.105; for further details, see pp. 150, 158.

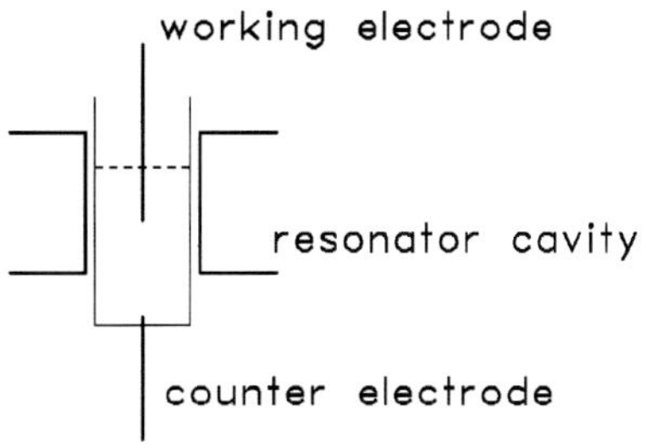

Fig. 5.106. Cell design for ECESR spectroscopy according to Maki and Geske [599]

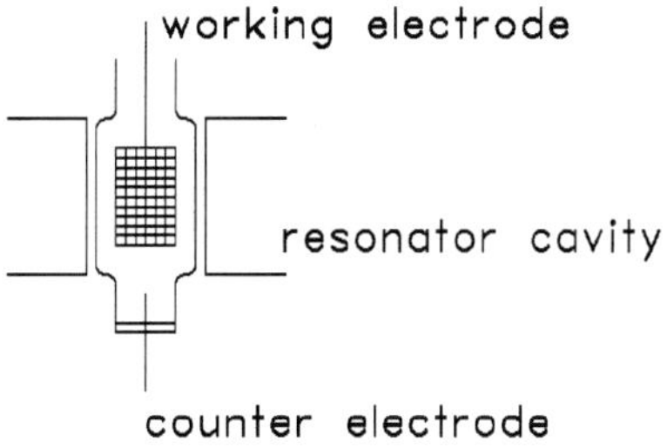

Fig. 5.107. Cell design for ECESR spectroscopy according to Piette et al. [606]

A first cell design with two electrodes was reported by Maki and Geske [599–603] (Fig. 5.106). A platinum wire used as a working electrode was mounted in the center of a quartz tube that served as a cell vessel, which is placed in the middle of the ESR cavity at the position of highest sensitivity. A platinum wire as the counter electrode was mounted in the tube at a position outside the cavity of the spectrometer. Any species created at this counter electrode were not detected; in addition, the distance between both electrodes reduced the risk of unwanted electrochemical reactions at the working electrode of species created at the counter electrode. The small actual surface area of the working electrode limited the rate of formation of species to be studied; very precise positioning was required. Despite its simplicity, this cell design has nevertheless been used continuously even with its obvious drawbacks and limitations [604]. Using a vanadium wire as working electrode, VO^{2+} ions could be identified as electrooxidation products [605].

An increase of the surface area was realized by Piette et al. [606] (Fig. 5.107). A platinum gauze electrode was put inside a flat cuvet used as the electrochemical cell. The counter electrode was mounted inside a glass tube attached to the flat cell below the working electrode and outside the ESR cavity.

A cell of this design was employed in studies of microcrystal solids attached to a platinum flag electrode [608]. The measured ESR spectrum of electrochemically reduced 7,7′,8,8′-tetracyanoquinodimethane was in perfect agreement with the respective spectrum of the chemically prepared compound and the simulated spectrum.

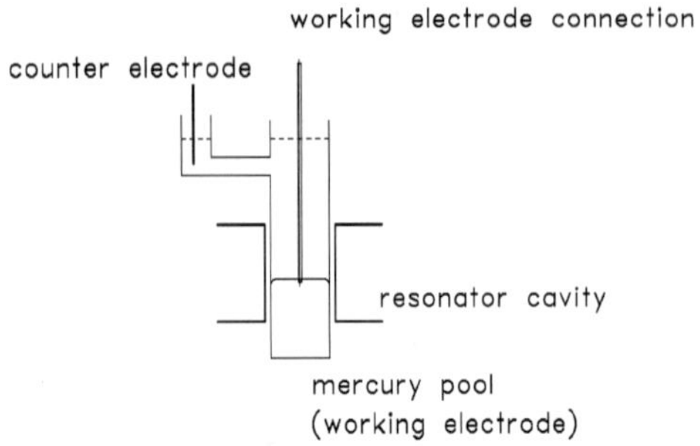

Fig. 5.108. Cell design for ECESR spectroscopy according to Möbius [609]

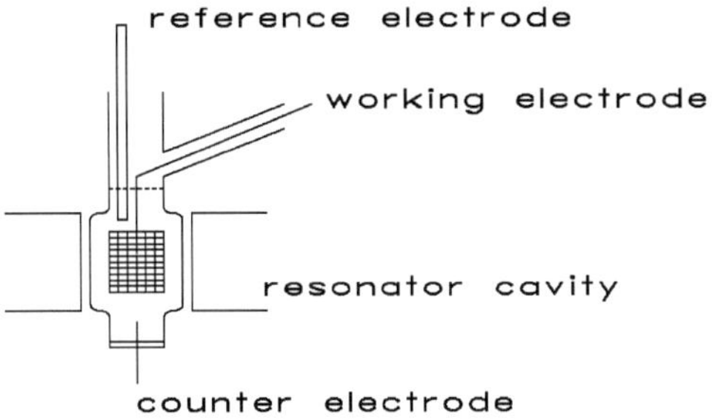

Fig. 5.109. Cell design for ECESR spectroscopy according to Piette et al. [611]

A cell design suitable for investigations with a mercury electrode was described by Möbius [609, 610] (Fig. 5.108).

Unfortunately these two-electrode cell designs did not allow for proper control of the electrode potential of the working electrode. In a modified design, Piette et al. [611, 612] added a reference electrode connected to the electrolyte volume via the glass tube on top of the flat cell (Fig. 5.109).

This design has been employed to study the electrochemistry of microdroplets attached to an electrode in contact with an electrolyte solution [613]. A microdroplet of N,N,N',N'-tetrahexylphenylene diamine deposited onto a gold electrode immersed into an aqueous electrolyte solution showed only a symmetric single line ESR signal similar to those observed with intrinsically conducting polymers. A well-resolved spectrum showing the expected hyperfine structure was observed in a dilute solution of the same molecule in a suitable organic solvent. Using a commercially available cell of this design, the electrochemical properties of some naturally occurring α-hydroxyquinones have been studied [614].

A design similar to that of Piette et al. was reported by Koopman and Gerischer [615] (see Fig. 5.110).

A significantly simpler cell design with a smaller working electrode surface and consequently smaller currents flowing through the cell in the case of potentiostatically controlled measurements has been reported by Bagchi et al. (Fig. 5.111) [616].

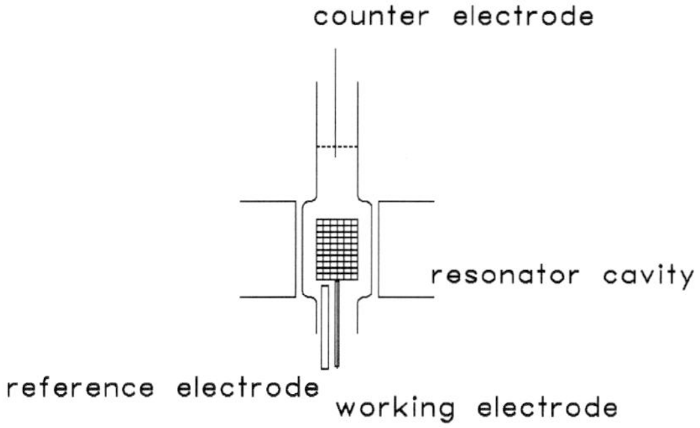

Fig. 5.110. Cell design for ECESR spectroscopy according to Koopman and Gerischer [615]

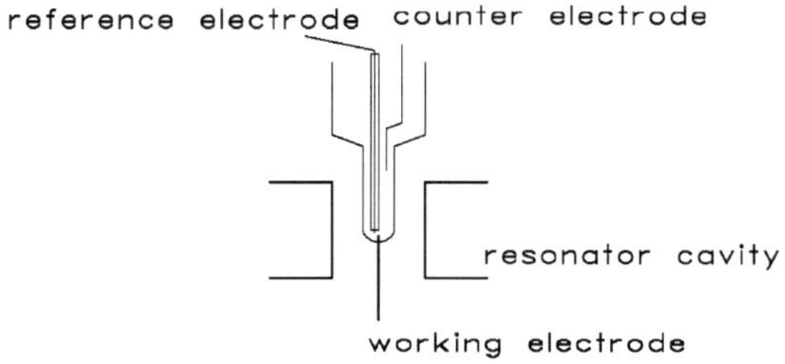

Fig. 5.111. Design of an ECESR cell proposed by Bagchi et al. [616]

A capillary tube housing a platinum wire that acts as a reference electrode fits into a modified cuvet that is placed in the microwave cavity of the ESR spectrometer, thus displacing most of the electrolyte solution in the lower part of the cell that is exposed to the microwave. Accordingly solvents with high dielectric constants can be used with a satisfactory ESR response. The platinum wire sealed into the bottom of the Pyrex cuvet that acts as the working electrode can be coated easily with a drop of mercury for experiments where this type of electrode is required. The Pyrex glass may cause unwanted ESR signals; unfortunately, platinum cannot be sealed directly to quartz glass.

Mu and Kadish have described an ECESR cell with a thin layer design suitable for measurements at both ambient and low temperatures [617]. The thin layer of electrolyte solution is enclosed by the quartz tube inserted into the microwave cavity of the ESR spectrometer and a solid quartz rod fixed in the center of the tube. An expanded platinum mesh in the gap is used as the working electrode. At low electrode potential scan rates, the cell shows an acceptable electrochemical response.

A mechanically robust ECESR cell that is suitable for measurements even at very low temperatures with all kinds of electrolyte solvents and that employs a platinum wire loop as the working electrode located at the bottom of a 4 mm ESR cuvet has been developed by Fiedler et al. [618]. The reference and counter electrodes are placed above the working electrode outside of the sensitive region of the ESR cavity.

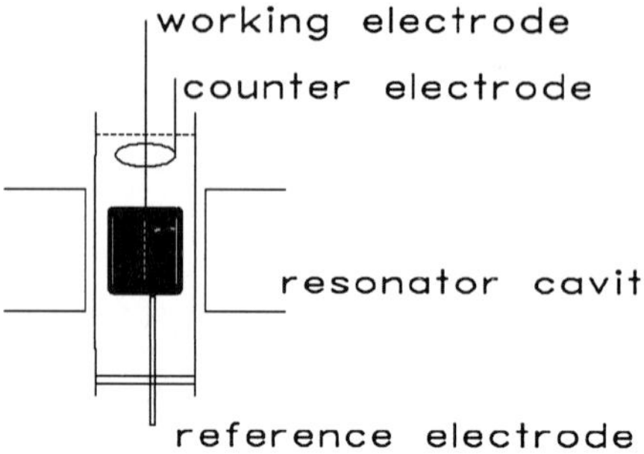

Fig. 5.112. Cell design for ECESR spectroscopy according to Dohrmann and Vetter [619–622]

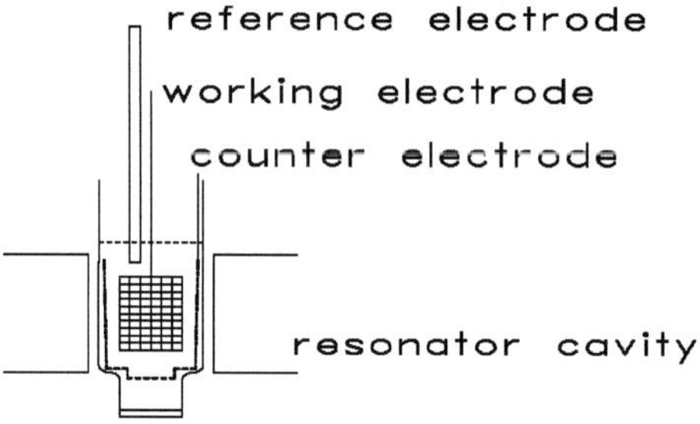

Fig. 5.113. Cell design for ECESR spectroscopy according to Goldberg and Bard [623]

The very small electrolyte volume present inside the flat cell poses a serious problem, because the concentration of the reacting species is rapidly diminished upon continuous flow of a Faradaic current. In the latter design, this was corrected by a slow upward flow of electrolyte solution through the cell. A very similar setup was described by Dohrmann and Vetter (Fig. 5.112) [619–622].

The fairly long solution pathways in the three-electrode cell designs caused a poor dynamic response of the cell under instationary conditions; the electrode potential control was imperfect. A flat cell design by Goldberg and Bard [623] provided considerable improvement (Fig. 5.113). The limited amount of electrolyte solution resulted in fast depletion of reactand.

A considerable improvement in terms of electrode potential distribution inside the ECESR cell, available electrolyte solution volume and ease of manufacturing was provided with the cell design of Allendoerfer et al. [624, 625], which was subsequently improved by Heinzel et al. [606, 626]. A metal wire coil of the working electrode material is inserted into a quartz glass tube. Because of the limited penetration depth of the microwaves, only the solution volume enclosed by the wire surface and the inner glass wall was probed. The insertion of the counter and the reference electrode centrally inside the working electrode coil can be made without major constraints caused by cell or working electrode design. Because of the size

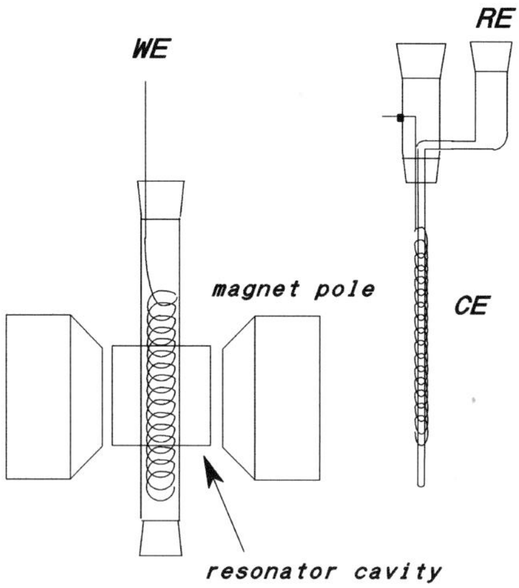

Fig. 5.114. The components of an ECESR cell according to Heinzel et al. [606]

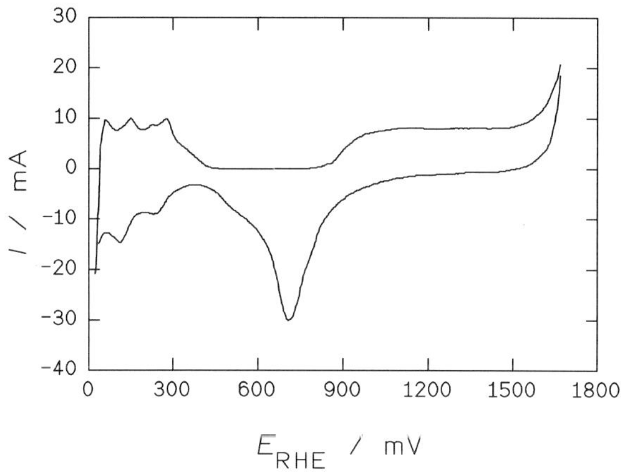

Fig. 5.115. Cyclic voltammogram of a platinum wire working electrode inside an ECESR cell according to Heinzel et al.; electrolyte solution 0.5 **M** H_2SO_4, $dE/dt = 50\ \mathrm{mV\,s^{-1}}$; nitrogen purged [627]

of the cell, a less common large cylindrical microwave resonator cavity is required. The cell components and their placement inside the resonator between the magnet poles are shown in detail in Fig. 5.114.

The superior electrochemical performance of this design is demonstrated with a cyclic voltammogram, as shown Fig. 5.115.

The cell has been employed in studies of the electrooxidation of organic fuels [606, 608–628] and of the nitrogen-containing monomers for the generation of intrinsically conducting polymers [629–631]. In numerous studies, films of intrin-

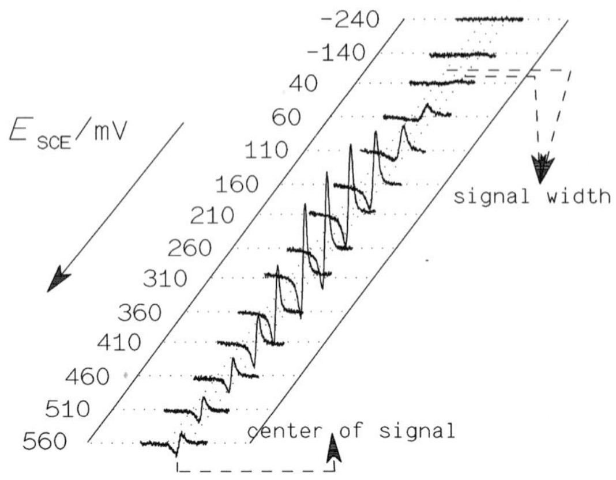

Fig. 5.116. ESR spectra of a film of polyaniline in a solution of 0.1 **M** $LiClO_4$ in acetonitrile

sically conducting polymers deposited onto the working electrode have been investigated [632–634]. Because of the high degree of delocalization, the radical cations created by the oxidation (*p*-doping) of the film show only a single line. The amplitude of the signal corresponds to the concentration of free spins and thus presumably to the number of mobile charge carriers. In a plot of ESR spectra obtained as a function of electrode potential, this can be illustrated (Fig. 5.116). The striking similarity, which in some cases even extends to small asymmetries in line shape, has caused early erroneous assignments of the line to mobile electrons as observed with ESR in case of metals by Dyson (Dysonian line, [635]).

The major drawbacks of cells with stationary solutions or very low solution flow, as described above, are the limited supply of reactand and consequent limited maximum concentration of species to be studied. Various designs of flow-through cells have been proposed. A review by Bagchi et al. [636] describes selected examples. Since it has been shown that the stationary concentration of the species to be detected decreases with an increasing flow rate [637], the actual operating conditions have to be optimized individually. A cell design as proposed by Bagchi et al. [623] is shown in Fig. 5.117.

An optimized design employing a tubular electrode in a cylindrical cavity has been described [638]. The mechanism and kinetics of the electrooxidation of several para-haloanilines and the follow-up reactions in acetonitrile have been investigated with this cell [639]. A similar design that is suitable for low temperature measurements (233 K) has been reported [640]. It has been employed in a study of the temperature dependence of the reduction of bromonitrobenzene in acetonitrile solution. The electroreduction of perinaphthenone in a single electron process has been investigated with this cell [641]. The lifetime of the neutral radical formed by deprotonation of the radical anion has been estimated to be around 1 min. A similar electrochemical behavior of benzanthrone was observed.

Free radicals formed by the electroreduction of several different nitroso phenyl 1,4-dihydropyridines at a mercury electrode were studied with a flat ECESR cell

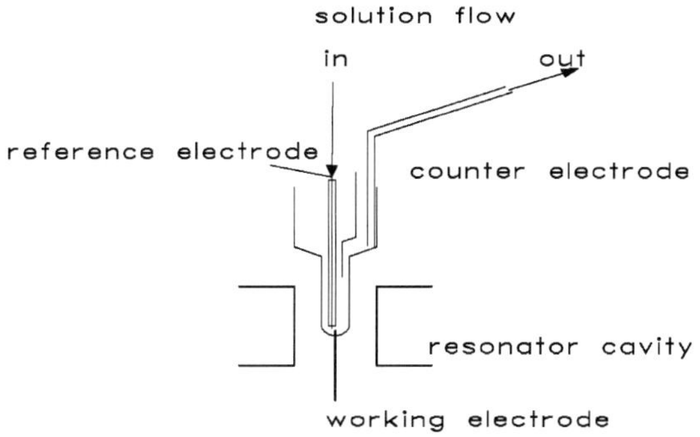

Fig. 5.117. Design of a flow-through ECESR cell proposed by Bagchi et al. [623]

(not further specified) in an aprotic medium [642]. Increasing stabilization of the free radical with increasing bulkiness of the alkyl substituents from methyl to isopropyl in 3- and 5-position was found.

A combination of ECESR and *in situ* UV-Vis spectroscopy has been proposed by Petr et al. [643]. In the case of a cell that was designed to be similar to the ECESR cell proposed by Piette [613], a UV-Vis spectrometer is coupled with the cell via fiber optics. The working electrode is of a minigrid type. A cell design with an electrochemical cell directly coupled with a cuvet fitting into the ESR spectrometer has been described by Friedrich and Baumgarten [644].

Although the detection limit for ESR spectroscopy per se is extremely low, the use of electrochemical cells filled with solvents that have high dielectric constants results in considerable losses in the cavity of the ESR spectrometer. This in turn increases the limit of detection. In the case of electrode reactions that have only very small stationary concentrations of radicalic intermediates, detection may be impossible. The use of spin traps may help. These compounds are rather simple organic molecules that react easily with radicals forming adducts (see Fig. 5.118). The molecular structure of the intermediate may be deduced from the known structure of the spin trap and the observed ECESR spectrum. Unfortunately, this technique doesn't necessarily trap the major reaction intermediate; rather, it only traps those which react easily with the spin trap. Consequently, misinterpretations are possible.

Cell designs for *in situ* NMR spectroscopy with electrochemical cells are scant. Because of the low sensitivity, designs with working electrodes that have large active surface areas (powder electrodes, metal-coated inert supports like silica or alumina) have been described [645, 646]. This is a result of the fact that, in a typical NMR experiment, about 10^{19} spins are required. Assuming 10^{15} surface atoms on 1 cm^2 metal surface area at complete coverage of all adsorption sites, a 1 m^2 working electrode surface area is required. A typical design is shown in a schematic drawing in Fig. 5.119. If continuous electrode potential control during the acquisition of the NMR spectrum is desired, particular attention has to be paid to proper shielding and decoupling of the electrical wiring between the potentiostat and NMR probe head.

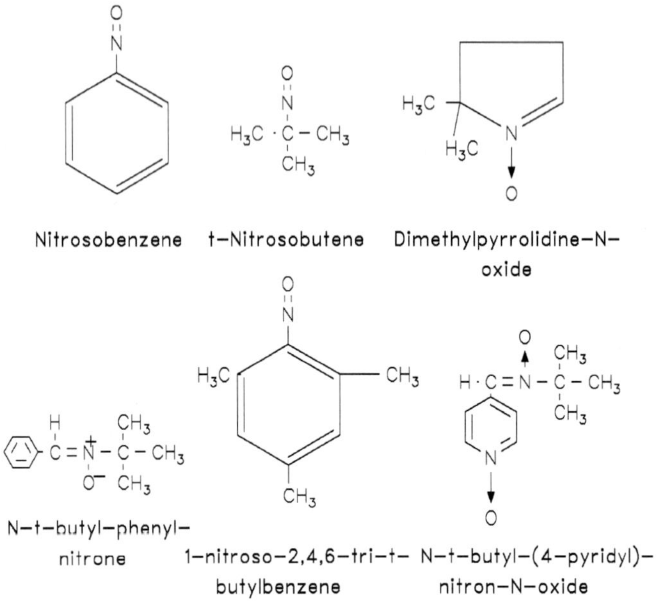

Fig. 5.118. Examples of organic compounds used as spin traps in ECESR spectroscopy

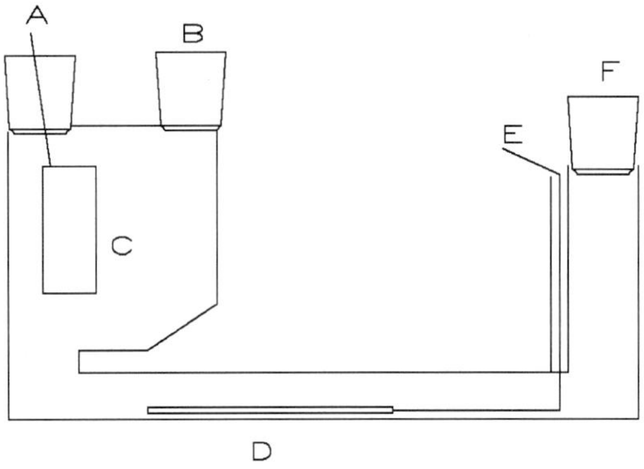

Fig. 5.119. Schematic drawing of an electrochemical cell for *in situ* NMR spectroscopy [649]. *A*: counter electrode connection; *B*: reference electrode joint; *C*: counter electrode; *D*: platinum black working electrode; *E*: working electrode connection; *F*: purge connection

Results reported so far pertain in particular to ^{13}C NMR spectroscopy of adsorbed CO and CN^- ions on platinum [649, 647, 648]. Besides information about surface diffusion and electronic adsorbate–surface interaction, data on the effect of the strong electric field at the electrochemical interface (typically 10^7 V cm^{-1}) have been reported. With both adsorbates the ^{13}C resonance becomes more shielded at more positive electrode potentials as expected when assuming an adsorbate attachment via the carbon atom. These results were supported with data from ^{195}Pt

NMR spectroscopy [649]. Differences in the electrooxidation of methanol and CO on carbon-supported platinum have been associated with different linewidths of the ^{13}C signal [650].

The use of NMR spectroscopies with various species, e.g. H, F, Na, Al or Li, in studies of solid electrolytes has been reviewed elsewhere [651].

Further Reading

On ESR spectroscopy:

R. Kirmse, J. Stach, *ESR-Spektroskopie* (Akademie-Verlag, Berlin, 1985)

J.E. Wertz, J.R. Bolton, *Electron Spin Resonance* (Chapman & Hall, New York, 1986)

N.M. Atherton, *Principles of Electron Spin Resonance* (Ellis Horwood/Prentice Hall, Chichester, 1993)

K. Scheffler, H.B. Stegmann, *Elektronenspinresonanz* (Springer, Berlin, 1970)

C.P. Poole, Jr., *Electron Spin Resonance* (Dover, Mineola, 1996)

C.P. Poole, H.A. Farach, *Handbook of Electron Spin Resonance*, vol. 1 (AIP, 1997)

C.P. Poole, H.A. Farach, *Handbook of Electron Spin Resonance*, vol. 2 (Springer, 1999)

J.A. Weil, J.R. Bolton, *Electron Paramagnetic Resonance* (Wiley, Hoboken, 2007)

On NMR spectroscopy:

J.W. Akitt, *NMR and Chemistry* (Chapman & Hall, New York, 1983)

E.D. Becker, *High Resolution NMR* (Academic Press, New York, 1980)

F.A. Bovey, *Nuclear Magnetic Resonance Spectroscopy* (Academic Press, New York, 1988)

R.K. Harris, *Nuclear Magnetic Resonance Spectroscopy* (Longman Scientific & Technical, Essex, 1986)

5.5 Magnetooptic and Magnetic Methods

5.5.1 Magnetic Circular Dichroism

Fundamentals. Linearly polarized light passing an optically transparent material in a magnetic field will show a rotation of the plane of polarization (Faraday effect). This phenomenon is also called magnetooptical rotation (magnetooptical rotatory dispersion) and is closely related to the Zeeman effect. Electronic transitions are affected by the external magnetic field in various ways. The ground or the excited state may split or a mixing of states may occur. Further details and accompanying changes in the observed spectra are described in detail elsewhere [652]. As a result, information pertaining to the electronic structure of the investigated molecules may be obtained. Spectroelectrochemical studies reported so far have only employed light in the UV-Vis range; basically, extension into the NIR range seems to be possible [653].

Instrumentation. The spectroelectrochemical cell described by Kobayashi and Nishiyama includes a minigrid working electrode in a glass cuvet (approx. 3 mm optical pathlength) attached to the bottom of a cylindrical cell body with counter electrode and reference electrode mounted therein. The cuvet is exposed to the magnetic

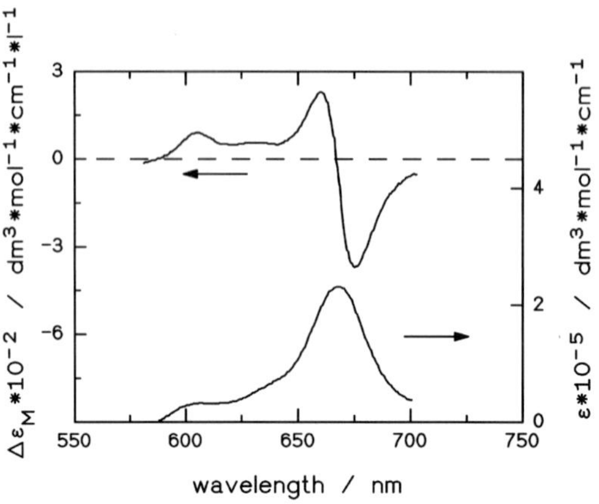

Fig. 5.120. MCD (*top*) and conventional UV-Vis spectra of an aqueous solution of 10^{-4} **M** tetrasulfonated cobalt(III)phthalocyanine in 0.05 **M** H_2SO_4, $H = 7.7$ T, based on data in [660]

field [654]. An experimental setup suitable for measurements with solid electrodes (e.g. polycrystalline gold) has been described by Zhao et al. [655, 656]. The loss in active surface area as compared to the previously described design is compensated in part by a modified beam path arrangement. The light of a tungsten halogen lamp passes a monochromator, a fixed polarizer and a photoelastic modulator. It is collimated and directed into the bore of a superconducting magnet. The cell mounted inside the magnet is of a thin layer design. The beam is reflected off the working electrode at near normal incidence. The reflected beam is guided with an additional mirror towards the photomultiplier detector. A conventional UV-Vis spectrum and an MCD spectrum obtained with this experimental setup are shown in Fig. 5.120.

The UV-Vis spectrum shows a prominent Q-band around 668 nm. The MCD spectrum implies an axially coordinated complex with at least a threefold symmetry axis. The degenerate excited state thus possible is non-degenerate in the magnetic field, resulting in further optical transitions at different energies selectively excited by right and left circularly polarized light. Further systems that have been investigated include additional transition metal complexes (tetracarboxylated phthalocyanines) [658].

5.5.2 Magneto-Optical Kerr Effect (MOKE)[37]

Fundamentals. The plane of polarization and the ellipticity of light of a fixed wavelength are changed upon reflection at the surface of a magnetic material ex-

[37] When applied in surface investigations, this effect is also called surface magneto-optical Kerr effect (SMOKE) [661].

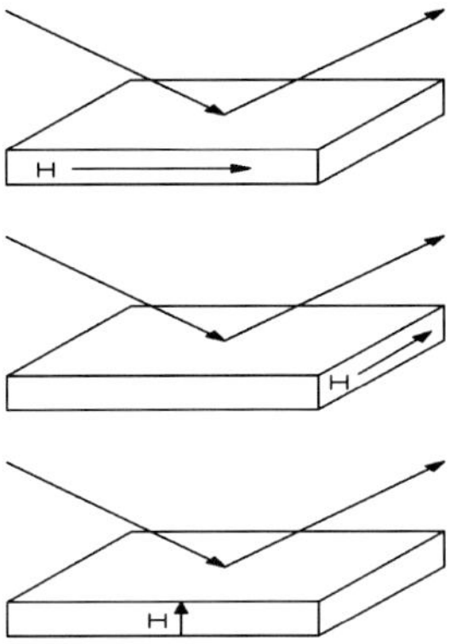

Fig. 5.121. Possible geometries in SMOKE experiments: (*top*) longitudinal; (*middle*) transverse; (*bottom*) polar

posed to an external magnetic field[38] (magneto-optical Kerr effect[39]) [657]. When the light is linearly polarized, it becomes elliptically polarized in the reflected beam. When only p- or s-polarized light is employed, the state of polarization is maintained and the plane of polarization is only rotated. In the polar Kerr effect, the plane of reflection of the employed beam of light and the magnetization are perpendicular with respect to the surface; in the longitudinal effect, the beam impinges non-perpendicularly and the magnetization is oriented in the plane of reflection. In the transverse (equatorial) Kerr effect, the magnetization is perpendicular to the plane of reflection. These different geometries (see also Fig. 5.121) enable detection of the orientation of the magnetization of the sample under investigation.

The amount of change is directly proportional to the magnetization of the reflecting surface layer. The probing depth is about 10–20 nm, thus the method is not particularly surface sensitive [658–660]. If surface specificity is required, magnetization induced second harmonic generation might be useful; applications of this method under *in situ* conditions in electrochemical studies have not been reported so far.

Instrumentation. The surface under investigation is exposed to an external magnetic field with adjustable strength. Illumination of the surface is accomplished by a laser beam. Although laser light from most sources is already polarized, a polarizer is employed between source and cell and an additional polarizer adjusted to maximum extinction at 90° rotation is mounted between cell and detector. Any change in the position of the plane of polarization causes a rise in the detector signal. A mechanical chopper at the source is added for better signal processing. A typical cell

[38] This effect is related to the Cotton–Mouton effect of magnetic birefringence.

[39] The Kerr effect is observed upon reflection, whereas the Faraday effect is observed in transmission [as employed in optical components for ellipsometry (Sect. 5.9.5)]. The Kerr effect itself relates to the change of the refractive index of a medium exposed to an electric field.

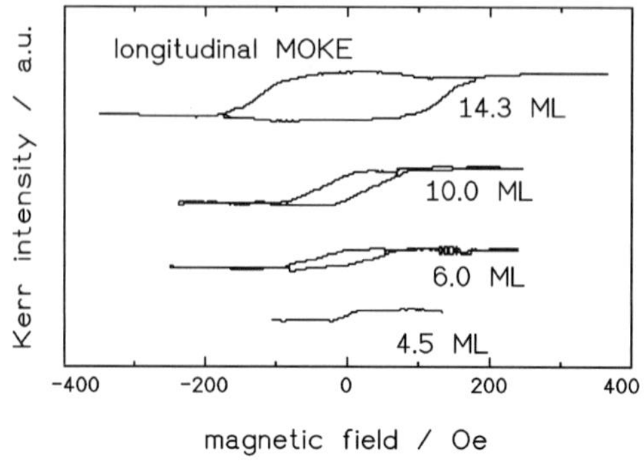

Fig. 5.122. Longitudinal MOKE results of a nickel film (thickness as indicated) on a Ag(111) supporting electrode; film deposited from aqueous solution of 1 m**M** $NiSO_4$ + 10 m**M** H_3BO_3, corrected for diamagnetic contributions from cell window and solution, based on data in [662]

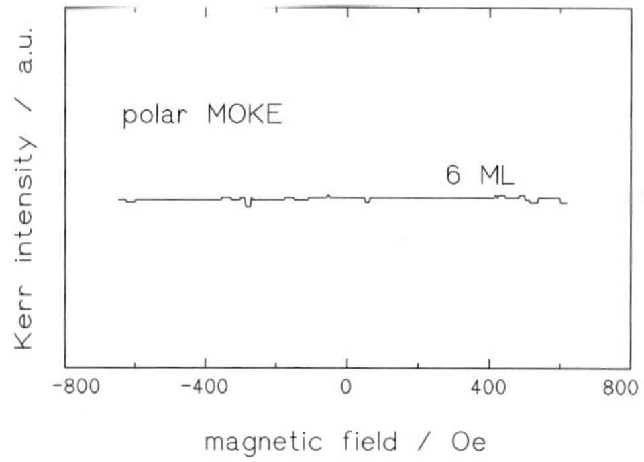

Fig. 5.123. Polar MOKE results of a nickel film (thickness as indicated) on a Ag(111) supporting electrode; film deposited from aqueous solution of 1 m**M** $NiSO_4$ + 10 m**M** H_3BO_3, corrected for diamagnetic contributions from cell window and solution, based on data in [662]

and optical setup have been described [662, 664]. A representative set of results is displayed in Figs. 5.122 and 5.123.

Applications in studies of magnetic layers and monolayers have been reported frequently [663, 664]. Ultrathin cobalt film that is deposited under particularly clean conditions has been studied [665]. In *ex situ* studies of electrochemically deposited mono- and multilayers, SMOKE has been employed [666, 664]. The capability of MOKE to study buried interfaces has been demonstrated [667].

5.5.3 Alternating Gradient Field Magnetometry (AGFM)

Fundamentals. A ferromagnetic sample at the end of a glass rod is suspended in an alternating magnetic field gradient ∇H generated by an AC current into the two

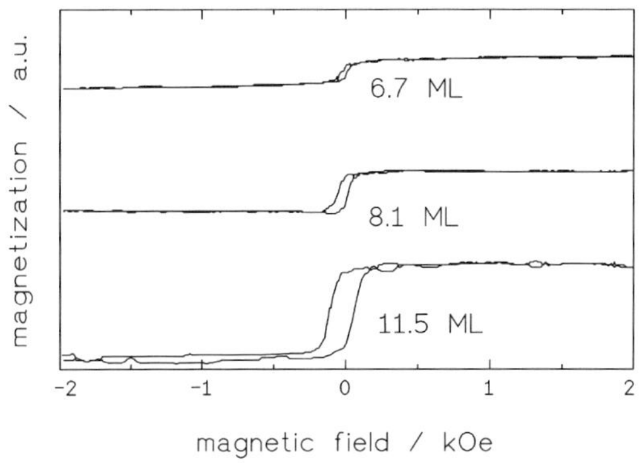

Fig. 5.124. AGFM results (parallel configuration) of a nickel film (thickness as indicated) on a Au(111) supporting electrode; film deposited from aqueous solution of 1 m**M** $NiSO_4$ + 10 m**M** K_2SO_4 + 0.1 m**M** KCl + 1 m**M** H_2SO_4, based on data in [668]

coils on both sides of the sample. The frequency of the AC current is tuned to the resonance frequency of the sample assembly. The force thus generated at the sample is measured by registering the electrical voltage across a bimorph piezo element at the lower end of the sample assembly. It is of particular interest when the plane of the sample is either parallel or perpendicular to the external magnetic field lines.

Instrumentation. The sample is mounted inside an electrochemical cell between the poles of a magnet. Details of the cell and the additional experimental setup have been described [668, 669]. In a typical result (see Fig. 5.124) obtained with nickel films, the similarity with MOKE results (see above) is obvious.

5.5.4 SQUID Magnetometry

Fundamentals. Movement of charged particles (e.g. an electronic current) causes a magnetic field. Very small fields can be detected with a SQUID.[40] The small size of the sensitive elements in a SQUID allows spatially resolved measurements of magnetic fields down to a resolution of about 1 mm.

Instrumentation. The sample under investigation is placed in an electrochemical cell mounted close to the pickup coils of the SQUID, generally inside a magnetic shield providing low magnetic gradients. Data processing provides images of local magnetic flux density, which in turn is related to local electric current density. Reported examples deal mostly with corroding metals like aluminum alloys and other metallic materials [670–684]. In particular, hidden corrosion in aircraft aluminum alloys has been studied.

---

[40] Supraconducting quantum interference device.

5.6 Photoelectrochemical Methods

In photoelectrochemical methods, the electrochemical response as a function of illumination either with white light or with light of a selected wavelength is studied by measuring photocurrents, photovoltages, etc. In most applications, this is done without spatial resolution. Attempts to provide photoelectrochemical images by recording these responses as a function of localization with a focused light spot have been described [685]; for an exemplary study of passivating oxide films on iron, see [686]. The function of a random array of gold disc microelectrodes has been evaluated using the photothermal effect at a local resolution of 1 μm [687]. The approach has been termed photoelectrochemical microscopy [688]. When using a scanning laser light source, the name is modified to scanning laser photoelectrochemical microscopy (SLAPEM); its application in a study of hexacyanoferrate(II) oxidation at a gold electrode has been reported [692].

5.6.1 Photoemission Spectroscopy

Fundamentals. A photon of sufficiently high energy (i.e. short wavelength) can be absorbed by a metal or a semiconductor, overcome the binding energy of an electron and thus generate an excited electron ("hot electron") with enough energy to penetrate the potential barrier at the metal/solution interface. This electron is emitted into solution. Because a free electron is not stable in aqueous media, it passes various relaxation steps to end in a capturing reaction with a suitable solution phase species (scavenger), which is left in its reduced form. The whole process corresponds to a photon induced electron transfer across the interface. Experimentally it can be watched by measuring the photoemission current. The yield of photon induced generation of emitted electrons is low because of the fast thermal recombination of photogenerated charge carriers in the metal and the high probability of the electrode itself recapturing the emitted electron without generating a measurable current. The photoemission current depends on the photon energy according to the following equation,[41] which is slightly different from the relationship for photoemission into the vacuum:

$$I_{\mathrm{PE}} = A \cdot (h \cdot \nu - h \cdot \nu_0 - eE)^{5/2},$$

with the photocurrent I_{PE}; a constant A; the energy of the photon $h \cdot \nu$; the threshold energy $h \cdot \nu_0$ for photoemission at zero charge electrode potential E_{pzc}; e as the electron charge and E as the electrode potential [689–696]. The low yield has limited studies to metals like gold or mercury, which show limited electrocatalytic activity and, in turn, may not cause additional Faradaic processes that add electrical current to the photocurrent. If the photocurrent itself is studied as a function of illuminating wavelength, suitable scavengers are nitrous oxide, nitrate ions or protons in acidic solutions. The solvated electron reacts according to

[41] This equation is also known as "five-halves law"; other authors have observed relationships similar to "Fowler's square law".

$$N_2O + e^-_{aq} \rightarrow N_2O^-$$
$$N_2O^- + H_2O \rightarrow N_2 + OH^- + OH^*.$$

The formed hydroxyl radical can be reduced at the electrode

$$OH^* + e^- \rightarrow OH^-,$$

resulting in a measurable electrical current. The factors influencing the photocurrents produced by UV irradiation of a mercury electrode have been discussed extensively [697]. Fundamentals and theoretical aspects of processes occurring at solution/semiconductor interfaces have been reviewed [693].

Instrumentation. The experimental setup is composed basically of an electrochemical cell suitable for *in situ* studies under illumination of the electrode surface with visible light. Because of the low yield of the photoemission process (10^{-2} to 10^{-5} electrons per incoming photon), careful design of the cell with respect to light detection, powerful sources of light (laser light or high-pressure xenon or mercury lamps) and sophisticated detection (lock-in amplification combined with chopped or otherwise modulated light intensity) are required. Because some common substances are powerful scavengers, the cell has to be perfectly air-tight (oxygen and carbon dioxide are electron scavengers). Electrolyte solutions have to be purified carefully.

Reported studies deal with measurements of the electrode potential of zero charge E_{pzc} [698, 699], double layer investigations [700–702] and studies of electrode reaction mechanisms [703, 704]; for an overview, see also [693, 697]. Numerous studies (beyond many conducted *ex situ*) deal with intrinsically conducting polymers, particularly photogenerated mobile charge carriers [705–707].

5.6.2 Photocurrent Spectroscopy (PCS)

Fundamentals. Illumination of a surface, especially of semiconducting materials, can result in the generation of electron-hole pairs, which in turn may be the cause of electrochemical charge transfers across the electrode/solution interface [708–714]. Both cathodic (i.e. electron emission) and anodic (hole recombination with oxidizable solution species) currents can be observed. Measurements with a single wavelength (monochrome) or with white light (polychrome) are termed photocurrent measurements, whereas studies of the photocurrent as a function of the illuminating wavelength are called photocurrent spectroscopy (PCS).[42] The dependence of the observed photocurrents on wavelength, electrode potential and solution composition may provide information about the nature of the ongoing photoprocess and its energetics and kinetics (e.g. the behavior of oxalic acid in an aqueous solution in contact with a polycrystalline anatase electrode and the associated kinetics) [715].

[42] Although the difference between both methods is substantial, in most published reports measurements of both types are frequently employed in a single study, thus no distinction is made here between these methods.

In addition, the characteristics of various materials of the semiconductor can be derived from the observed relationships. Photocurrents can also be observed because of photochemical (essentially photolytic or photothermal) processes in the solution that are close to the surface under investigation.

Instrumentation. The electrode under investigation is mounted in an electrochemical cell and illuminated with light of a single wavelength (tunable laser, light from a lamp passed through a monochromator) or with light from a white light source (tungsten or xenon lamp). A combination of a chopper and a lock-in amplifier allows measurements of the photocurrent–voltage relationship of the electrode, which has to be corrected for the spectral intensity of the used lamp. This way currents caused by competing Faradaic processes can be separated [718]. Conversion efficiencies can be calculated, indicating the yield of electrons per incident photon. Typical examples include measurements at inorganic and organic semiconductors and other semiconducting materials, like heavy or transition metal halides, chalcogenides and pnictides [716]. Reviews of the photoelectrochemistry of diamond have been provided [717–719]; for a broader overview, see [720]. The thickness of very thin semiconductor films that are otherwise hard to measure could be determined; in the case of PbO, layers down to 100 nm thickness were studied [718]. Changes in polarity of the measured photocurrent as a function of applied electrode potential at fixed wavelength provide information about semiconductor properties of the investigated films.

The determination of threshold energies as shown in the inset is based on a modified display of the obtained photocurrent spectrum (see Fig. 5.125). The two linear sections yield two threshold energies, for a detailed treatment, see [722]. Measurements of photocurrents at p-type synthetic diamond electrodes that have wavelengths ranging from $\lambda_0 = 193$ to 351 nm yielded results implying that only illumination with the short wavelength (supra-bandgap illumination) was able to excite electrons into the conduction band [723].

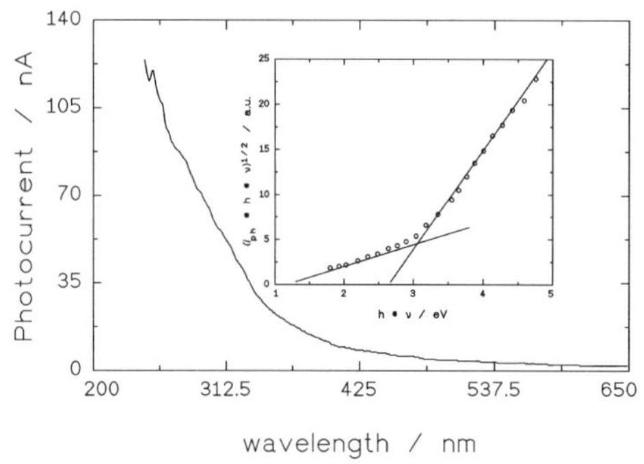

Fig. 5.125. Photocurrent spectrum of a polycrystalline diamond electrode in contact with an aqueous solution of 0.5 **M** H_2SO_4 at an electrode potential of $E_{NHE} = 0.05$ V (based on data in [721])

Beyond photocurrent measurements at constant light intensity, experiments with modulated light intensity have been proposed. A generalized theory for the use of intensity modulated photocurrent spectroscopy (IMPS) to determine the rate of charge transfer and recombination at the semiconductor/solution interface has been described by Ponomarev and Peters [724] and extended to the treatment of multi-step electron transfer mechanisms involving surface-bound intermediates [725]; for an overview, see also [726]. In this method, an AC perturbation (modulation) of the impinging light causes periodic changes of the photocurrent. This response is measured as a function of the perturbation frequency. Because of the competition between transfer of interfacial minority carriers and electron–hole recombination in a complex plane plot of the real and the imaginary part of the conversion efficiency (or an equivalent experimental signal), a semicircle in the first quadrant is observed. Using IMPS, transient photocurrents at a passivated iron electrode that is in contact with an alkaline solution have been studied [727]. Upon addition of $[Fe(CN)_6]^{4-}$, a change of the frequency response is observed and interpreted in terms of competitive multistep and single step reactions of photoexcited carriers generated in the passive film. Figure 5.126 shows the complex plane plot of the real and imaginary parts of the AC component of the electrode's photocurrent response.

A comparison of this method with photoelectrochemical impedance spectroscopy[43] as applied to photoelectrochemical hydrogen evolution at *p*-InP has been reported [728]. Intensity modulated photocurrent spectroscopy has been compared with intensity-modulated photovoltage spectroscopy (IMVS) [729]. Intensity modulated photocurrent spectroscopy has been applied in a study of the silicon dissolution in aqueous solutions of NH_4F [730] and n-GaAs electrodes [731]; a general review is available [732]. Photocorrosion of CdS has been investigated with IMPS and photoelectrochemical impedance spectroscopy (PEIS) [733]; a mechanism could be de-

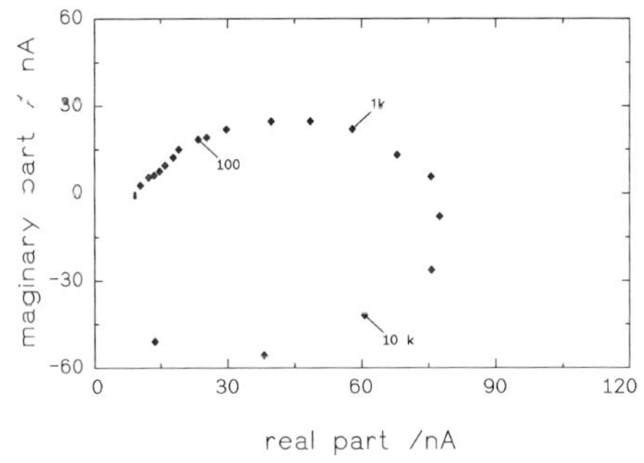

Fig. 5.126. IMPS plot for an iron electrode in an aqueous solution of 0.1 **M** KOH with addition of 1.7 m**M** $K_4Fe(CN)_6$, illumination with laser light of $\lambda_0 = 442$ nm (based on data in [731])

[43] The use of spectroscopy in connection with impedance measurements is not reasonable, thus it is frequently discouraged.

rived. Polythiophene has been studied as a material for photovoltaics devices with IMPS [734].

Semiconductors formed on various metals have been studied with PCS. In an investigation of passivated iron electrodes [735], the potential dependence of the photocurrent observed at a fixed wavelength is displayed in Fig. 5.127.

The oxidation of lead in various electrolyte solutions has been studied [736–738]. In a tetraborate solution, the photocurrent spectrum measured with white light showed a marked dependence on the positive (anodic) potential limit of the potential scans. With the most positive limit, the photocurrent–potential relationship displayed in Fig. 5.128 was obtained. The formation of tetragonal PbO by reduction of α-PbO_2 was monitored. The onset of the photocurrent indicates the appearance of the formation of PbO and the electrode potential coincides with the sudden onset of the photocurrent in the anodic scan. In the positive-going scan, the photocur-

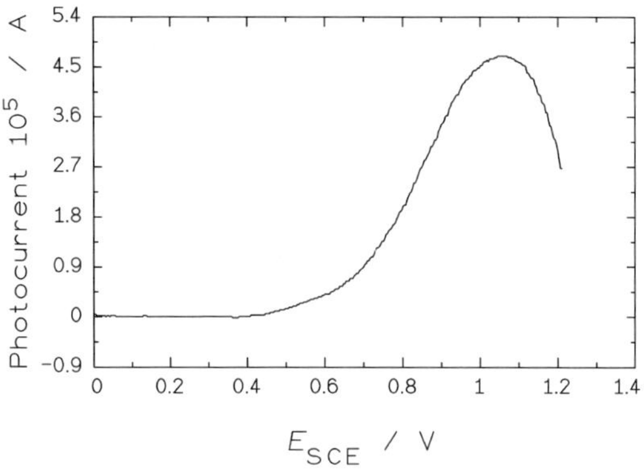

Fig. 5.127. Potential dependence of the peak photocurrent of an iron electrode in an aqueous buffer solution (pH = 4.8) (based on data in [735])

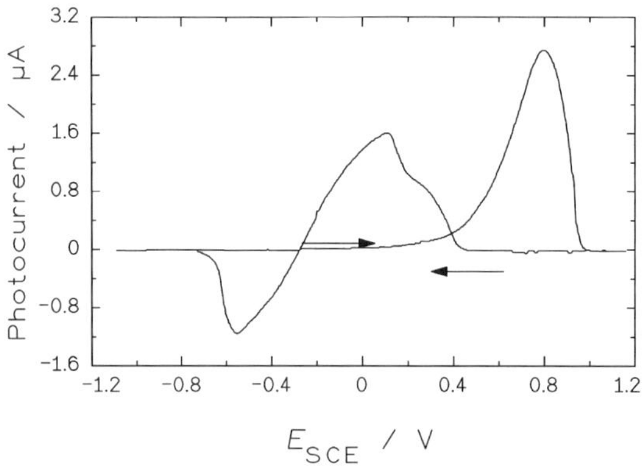

Fig. 5.128. Potential dependence of the photocurrent of a lead electrode in an aqueous solution of 0.1 **M** $Na_2B_4O_7$, $\lambda = 340$ nm (based on data in [736])

rent collapses rapidly when formation of α-PbO_2 proceeds. The disappearance of the cathodic photocurrent coincides with the complete reduction of PbO to lead as observed in cyclic voltammograms. The changes between anodic and cathodic photocurrents have been associated with intrinsic semiconductor properties of the oxide film.

In a sulfuric acid solution, the formation of PbO during the corrosion of lead without any other phase of PbO_x $(1 < x < 2)$ has been observed; the sensitivity of the method approached the monolayer level [742]. Passive films on nickel and iron surfaces have been studied both with polychromatic and monochromatic light [739]. Characteristic data of the semiconducting surface layers (flatband potentials, charge carrier densities, bandgap energies) could be obtained. The limitations of the traditional band model that is used in solid state physics for ideal crystalline solids with practically unlimited periodicity have been pointed out and the additional difficulties caused by the polycrystalline or even amorphous nature of these films were stressed.

In a study of anodic oxide films grown on titanium [740], photocurrent spectra were converted, yielding photocurrent conversion efficiencies Φ as a function of electrode potential (see Fig. 5.129).

The smooth increase of Φ with photon energy and the implicit absence of any reduction in efficiency at short wavelengths suggest negligible surface recombination under the employed experimental conditions. Composite oxide films on a nickel surface exposed to an alkaline electrolyte solution have been studied with PCS [741].

Measurements of the photocurrent as a function of the electrode potential at various fixed wavelengths of the illuminating light have been employed in a study of rhodium iodide photoconductors [742]. The quantum yield of a photosensitive film of Bi_2S_3 in contact with an aqueous solution of Na_2S measured as a function of the wavelength of incident monochromatic light has been found to approach unity

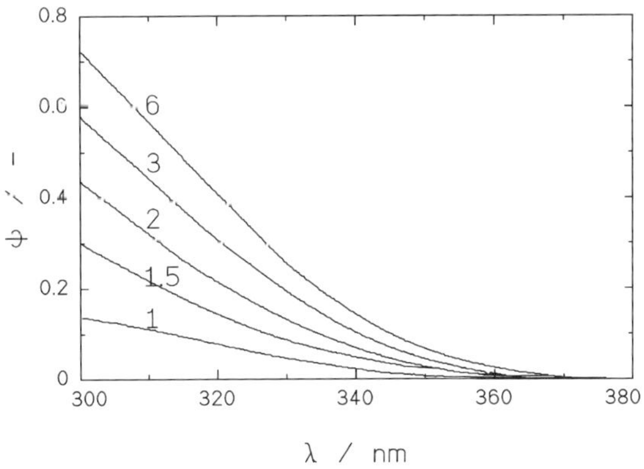

Fig. 5.129. Potential dependence of the photocurrent conversion efficiencies Φ at various electrode potentials (vs. E_{SCE}, as indicated) for a titanium electrode in contact with an aqueous electrolyte solution of 1 **M** H_2SO_4 (based on data in [740])

at light energies above 2 eV [743]. The absorption spectrum of the semiconductor was constructed from the results. It shows an indirect transition at 1.25 eV and possibly a direct transition around 1.7 eV. Films of CdS that were electrochemically grown from aqueous solutions of Na_2S have been characterized with PCS [744]. The corrosion of CdS films has been studied with intensity-modulated PCS [745].

A multilayer film assembled from highly sulfonated polyaniline and diazoresin has been studied with PCS [746]. The PCS response agrees with the optical absorption spectrum, implying that the multilayer film is responsible for photocurrent generation.

Photocurrent spectroscopy and a time-resolved variant of this method have been used to characterize hybrid thin films of crystalline ZnO modified by 5,10,15,20-tetrakis(4-sulfonatophenyl)porphyrinato zinc (TSTPPZn) and 2,9,16,23-tetrasulfophthalocyaninatozinc(II) (TSPcZn), which were prepared by electrochemical deposition [747]. In films containing both sensitizers, panchromatic sensitization was observed. Details of the modification procedure had considerable influence on the photoelectrochemical efficiency of the films.

The mechanism of photoreduction of CO_2 and water at the p-type GaP in contact with aqueous solutions has been studied with PCS under chopped light illumination [748]. Results show a photo-anodic sub-bandgap response and, as a consequence, a photo-cathodic current.

The bandgap of an intrinsically conducting polymer (polypyrrole) has been determined with photocurrent spectra [749].

5.6.3 Photovoltage Spectroscopy[44] (PVS)

Fundamentals. Illumination of a semiconductor with monochromatic light might result in a transfer of charge carriers between the surface and the bulk [750, 751]. Separation of the charge is caused by a surface barrier. The magnitude of the surface photovoltage dU_s depends on the net change of the electronic charge dQ_{SS} on the surface, which is caused by surface states lying within the energy gap of the semiconductor. Thus dU_s is proportional to dQ_{SS}. The type of photoionization transition determines the sign of the photovoltage. Photoionization of an electron from a surface state to the conduction band in a p-type semiconducor decreases the amount of negative charge and results in an increase of the surface barrier heights. An electron promoted from the valence band into a surface state increases the negative charge and consequently the height of the surface barrier. In this case, a negative photovoltage signal is observed. Measurements of electrode potentials under illumination have been called photopotential measurements; for an example, see [752]. Intensity-modulated photovoltage spectroscopy (IMVS) has been described [733].

Instrumentation. The electrode is illuminated with monochromatic light. The electrode potential is registered as a function of the wavelength of the incident light versus a reference electrode. Investigated systems include p-type indium phosphide in

[44] Because the central process occurs at the solution interface, the method is also named surface photovoltage spectroscopy (SPS).

contact with alkaline electrolyte solutions [754] modified by etching, foreign metal deposition or extended cathodic hydrogen evolution. Obtained spectra enabled the identification of surface states at the semiconductor electrode. In measurements at *p*-type InP in contact with aqueous solutions of sulphuric acid band-bending under illumination was found [753–755]. In a study of CVD-generated diamond electrodes with nanosecond laser flashes ($\lambda_0 = 337.7$ nm), photopotential transients were recorded. The decay behavior could be explained by invoking an equivalent circuit deduced from electrochemical transient measurements.

Heterojunctions formed by electrodepositing polypyrrole or polyaniline on *n*-type silicon were studied with PVS [756]. Experimental observations were explained invoking photosensitization by electrochemically generated polarons in the polymer film. The effects of nano-sized particles of Fe_2O_3 that were incorporated into films of polypyrrole were investigated with PVS [757].

In IMVS, the light intensity applied to the electrode under investigation is modulated by adding light from an additional source, which can be modulated easily (e.g. with the use of a light emitting diode). The observed photovoltage U_{photo} is composed of a steady state value U_0 and a varying component according to

$$U_{\text{photo}} = U_0 + \Delta U\, \mathrm{e}^{\mathrm{i}(\omega t - \varphi)},$$

with frequency of modulation ω and phase shift φ; for further details, see [758]. Results of electrochemical impedance measurements obtained with dye solar cells could be explained based on IMVS data [762]. Intensity-modulated photovoltage spectroscopy has been applied in a study of dye-sensitized photo electrodes [733]; for a description of the experimental setup, see [762].

5.6.4 Photoluminescence (PL)

Fundamentals. Luminescence, i.e. the emission of light from a substance not based on its (high) temperature can be caused in numerous ways. In the case of photoluminescence (PL), the emission of light is caused by illumination with light. The wavelengths of the illuminating and emitted light may differ. Photoluminescence of the semiconductor/solution interface is very sensitive to the generation of new phases as caused by metal deposition and potential distribution across the interface and to the surface charge carrier recombination velocity. Changes in the PL signal allow detection of the formation of a fraction of a metal monolayer [759].

Instrumentation. The electrode (e.g. a GaAs disc) is mounted in an electrochemical cell behind a suitable window (quartz glass) in a three-electrode arrangement. Illumination can be provided by a laser operating at a rather low power (a few mW). In order to facilitate detection of the PL light, the laser light is interrupted periodically with a mechanical chopper. Photoluminescence light is measured with collection optics, a monochromator, a photomultiplier and a lock-in amplifier connected with the mechanical chopper. Photoluminescence intensities are plotted as a function of the electrode potential. Typical results as observed during deposition on a GaAs surface are shown in Fig. 5.130.

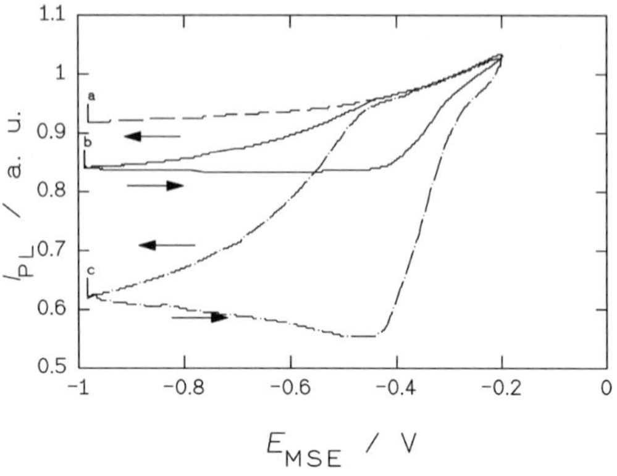

Fig. 5.130. Photoluminescence intensity of a *p*-doped (100)-oriented GaAs electrode as a function of electrode potential and concentration of $CuSO_4$: *a*: 10^{-4} M; *b*: 10^{-3} M; *c*: 3×10^{-3} M; supporting aqueous electrolyte solution 1 **N** H_2SO_4; $dE/dt = 20$ mV·s^{-1}, arrows indicating scan direction; based on data in [763]

In an optical arrangement used for surface plasmon resonance spectroscopy (see Sect. 5.9.6), PL light locally enhanced by the surface plasmon field can be detected as is done in surface plasmon field-enhanced light scattering (surface plasmon field-enhanced photoluminescence (SPPL) spectroscopy). A photomultiplier is mounted on the bottom side of the electrochemical cell opposite the prism arranged in the Kretschmann setup [760]. In a study of poly(3,4-ethylenedioxythiophene) (PEDOT), a change in SPPL intensity was observed. In the dedoped state the signal increased considerably under open circuit conditions. Obviously, in the absence of an applied electrode potential, ionic charge carriers migrate into the polymer layer, thus quenching the photoluminescence.

5.6.5 Micro-Optical Ring Electrode (MORE)

Fundamentals. The insulating material inside a microring electrode can be manufactured from an optically transparent material as employed in fiber optics. Light piped into the electrolyte solution inside an electrochemical cell close to the ring electrode may be used to cause photochemical processes, which in turn can result in species that can be detected at the ring electrode [761, 762]. Compared with the photoelectrochemical microscope (PECM) only low levels of light intensity are employed, resulting in a significantly reduced risk of thermal perturbation or other undesirable side effects.

Instrumentation. An electrode design has been reported [765, 766] together with applications pertaining to the quenching of photoexcited $Ru(bipy)_3]^{2+}$ by Fe^{3+} and the photoelectrochemistry of methylene blue [765].

5.7 Nonlinear Optical Methods

Illumination of a surface or interface with poly- or monochromatic light of very high intensity can result in non-linear optical phenomena like the generation of light with the second (SHG) or third harmonic generation (THG) of the incoming frequency or in sum (SFG) or difference frequency generation (DFG). For general reviews, including theoretical considerations, see [763, 764]. Surface second harmonic generation is more simple and versatile, whereas SFG is more powerful because it allows identification of adsorbed species at the interface via their vibrational resonances.

5.7.1 Second Harmonic Generation

Fundamentals. The optical response of certain media upon illumination with light of sufficiently high intensity can become nonlinear, i.e. in the reflected or transmitted light second, third or higher harmonics of the fundamental frequency of the illuminating light can be observed. This example of nonlinear optics is particularly fascinating and helpful in interfacial studies because the SHG is governed by selection rules that localize the origin to regions (or more generally media) lacking inversion symmetry. The interface region between two centrosymmetric media provides this particular asymmetry quite naturally, thus the method is by itself surface specific. The generated harmonic provides information about electronic and structural properties of the interface, but only to a limited extent of adsorbed species on the surface. Depth of penetration of the incoming probe and depth of information of the obtained signal are about 2 to 3 atomic layers of the solid. In single crystal surfaces and systems where the surface nonlinear susceptibility is dominated by molecular contributions from the adsorbate, dependencies on the angle of rotation of the surface may be observed. Theoretical treatments have been provided elsewhere [765, 766]. The theory of the generation of second harmonics via plasmon surface polaritons has been described [767]. A theoretical treatment of the combined effects of wavelength of illuminating light and rotational anisotropy useful in investigations of single crystal/solution interfaces has been given [768].

Instrumentation. An experimental setup is depicted in Fig. 5.131. Laser light of sufficiently high intensity is filtered to remove light at harmonic frequencies and reflected off the surface under investigation. From the reflected light that appears at an angle[45] determined by the Fresnel factors at the given wavelength, the fundamental and the third harmonic are filtered out and the second harmonic intensity is detected with a CCD camera.

Numerous studies employing both polycrystalline and single crystal metal electrodes and semiconductor surfaces have been reported; for an overview, see [772]. The possible advantages of a simultaneous (double beam) application of electroreflectance spectroscopy and second harmonic generation have been pointed out [769].

[45] This angle is generally close to the angle of specular reflection, e.g. 60° in and 58° out at 578-nm incident light wavelength and 289-nm SHG. The difference in the index of refraction of water between these wavelengths causes the change of the angle.

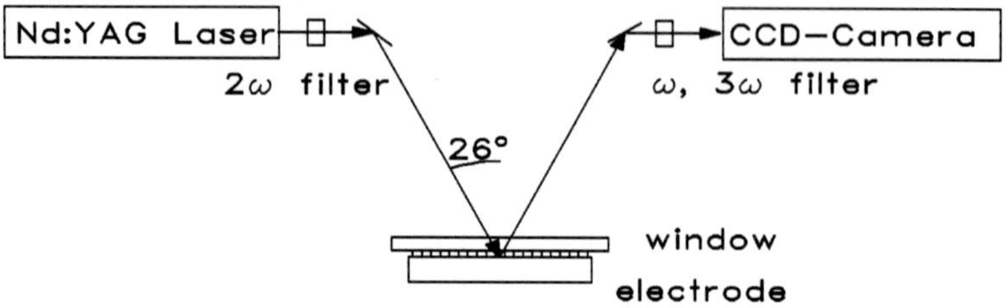

Fig. 5.131. Schematic setup for SHG experiments

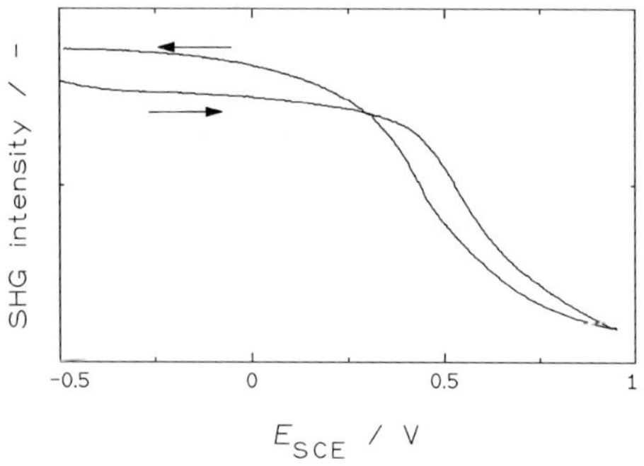

Fig. 5.132. SHG signal of a polycrystalline platinum electrode in contact with a solution of 0.5 **M** H_2SO_4, based on data in [770]

In a typical study of a polycrystalline platinum surface in contact with a solution of 0.5 **M** H_2SO_4, an electrode potential dependent intensity as shown in Fig. 5.132 was recorded [770].

The intensity increase in the positive-going potential scan starting at minimum intensity around the potential of zero charge E_{pzc} (around approx. $E_{SCE} = -0.1$ V) was assigned to bisulfate adsorption in the double layer region. The subsequent drop at more positive potentials was attributed to surface oxide formation. The increase of SHG intensity at potentials negative to the E_{pzc} was explained by invoking hydrogen adsorption [776]. The nonlinear surface susceptibility was found to correlate linearly with the surface coverage with hydrogen species. With single crystal electrodes, further structural information about the interface can be obtained from measurements of the SHG intensity on the angle of rotation of the surface; for a theoretical treatment and an overview, see [771, 772]. The combined effect of a wavelength ranging from the UV into the infrared region and the rotational anisotropy observed at the Ag(111)/solution interface has been explained by invoking the electronic band structure of the metal [774]. Extension of the study to single crystal silver electrode surfaces during charging and adsorption as well as modification by upd-layers of lead and thallium has been reported [773]; for a review, see [774].

During a slow electrode potential scan, a change in the SHG has been observed with a Au(111) electrode (Fig. 5.133) [775, 776].

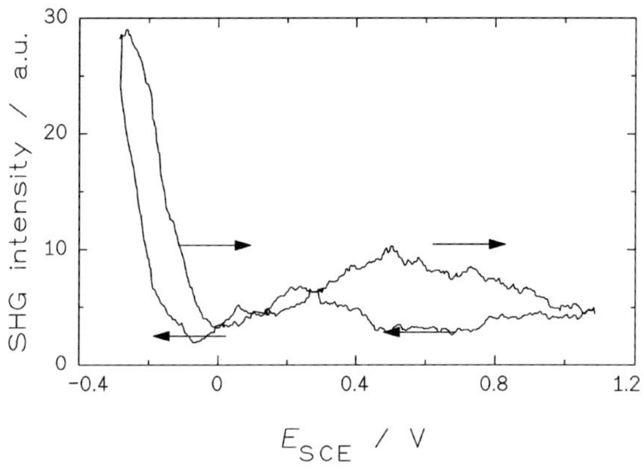

Fig. 5.133. SHG signal as a function of electrode potential for a Au(111) surface in contact with a solution of 0.01 **M** $HClO_4$, $dE/dt = 0.1$ mV s^{-1}, based on data in [775]

It has been explained by invoking a slow reconstruction of the gold surface from the (1 × 23) surface structure into the unreconstructed (1 × 1) surface. The observed hysteresis implies a slow surface reconstruction, which is complete on a minute time scale. Rotation of the samples revealed a threefold symmetry pattern for the unreconstructed Au(111)-(1 × 1) surface, whereas the Au(100)–(1 × 1) surface did not show rotational anisotropy of SHG intensity as expected for C_{4v} symmetry.

Second harmonic generation has also been used to study the semiconductor/ solution interface during the deposition of gold on Si(111) [777]. In a setup combining SHG and the electrochemical quartz crystal microbalance (EQCM), the underpotential deposition of copper on a polycrystalline gold surface has been studied; a decrease of the SHG signal by 60% upon formation of the upd-layer was found [778]. A study of the electrochemical liquid/liquid interface between two immiscible solutions where adsorption of surfactants occurred has been reported [779].

5.7.2 Sum and Difference Frequency Generation

Fundamentals. Sum frequency generation (SFG)[46] and the closely related difference frequency generation (DFG) are second-order non-linear optical processes. Light of a fixed frequency ω_1 interacts with light of a frequency ω_2 in space and time at an interface where the local susceptibility $\chi^{(2)}$ is non-zero, thereby generating a coherent beam with the frequency $\omega_s = \omega_1 + \omega_2$ in SFG and $\omega = \omega_1 - \omega_2$ in DFG, which is superimposed on a non-resonant background signal. This is basically the result of the non-linear susceptibility of the interface and the non-linear polarizability of adsorbed species. The rather complicated mathematics for the interaction occurring at a metal/solution interface have been treated elsewhere [780–783]. The

[46] Because mostly vibrational modes at interfaces are investigated, the method has also been called vibrational sum frequency generation (VSFG).

interface region between two centrosymmetric media provides this particular asymmetry quite naturally, thus the method is by itself surface-specific [784]. With one light frequency that is tunable, the resulting sum frequency can be tuned accordingly. According to theory, the SFG intensity is enhanced when either ω_s, ω_1 or ω_2 are in resonance with a vibrational transition in an adsorbate that is present at the interface or an electronic transition in the metal of the electrode. In addition the vibrational mode of the adsorbate has to be both infrared and Raman-active. Because of the large reduction in molecular symmetry upon adsorption on a surface, this condition is met in most cases. With a light source of tunable frequency emitting in the infrared (e.g. a tunable infrared laser diode, a tunable infrared laser (CO laser, dye laser) or a free-electron laser), the vibrational properties of the interface (i.e. of adsorbates and their bonding to the metal surface) can be probed. Besides their conceptual similarity, DFG differs from SFG in two details. The more stringent phase matching requirements for the two light beams pose narrower limits for the experimental design. A more effective filtering of the light emitted from the interface is necessary because of the competition between the desired DFG signal and the undesired fluorescence that is sometimes induced by the visible light beam [785]. For selected adsorbates, the observed SFG data have been modeled using multiconfigurational self-consistent field calculations [786]. Sum frequency generation at metal/solution and other interfaces has been reviewed extensively [787–791].

Instrumentation. For electrochemical studies two laser light beams—one operating in the visible and one that is tunable operating in the infrared—are required. The advent of tunable Ti:sapphire-based lasers might provide an additional source of infrared radiation [792]. The sum frequency intensity coming from the electrochemical interface at an angle of emission somewhere between the respective angles for the reflected beams of frequencies ω_1 and ω_2 is detected with a photomultiplier. The electrochemical cell is equipped with a prismatic window (e.g. CaF_2) to avoid refraction and unwanted reflection of the laser beams. Single crystal and polycrystalline electrodes can be investigated. In a broad study the structure of water at the silver/water interface has been investigated with SFG [793]. Peaks correlated with low-order water, water interacting with electrolyte ions, water specifically adsorbed on the silver surface and hydronium species were observed.

CO_{ad} formed by chemisorption of methanol on a platinum electrode has been studied with SFG (see Fig. 5.134) [786].

The spectra show two resonances located around 2065 cm^{-1} and 1965 cm^{-1}. The first corresponds to linearly bound CO, the latter was assigned to CO in a bridge-bound position [786]. The conversion of CO in a bridging location into species atop has been identified with SFG [794]. The combination of the high light intensity (e.g. free-electron lasers) and the possibility to obtain absolute spectra without any modulation technique has allowed the detection of generally weak bands like those of over- and underpotential deposited hydrogen on a platinum electrode [795, 796]. Studies of coadsorption of cyanide anions and cetylpyridinium cations on Au(111) and Au(210) revealed marked differences [797]. CN^- is bound

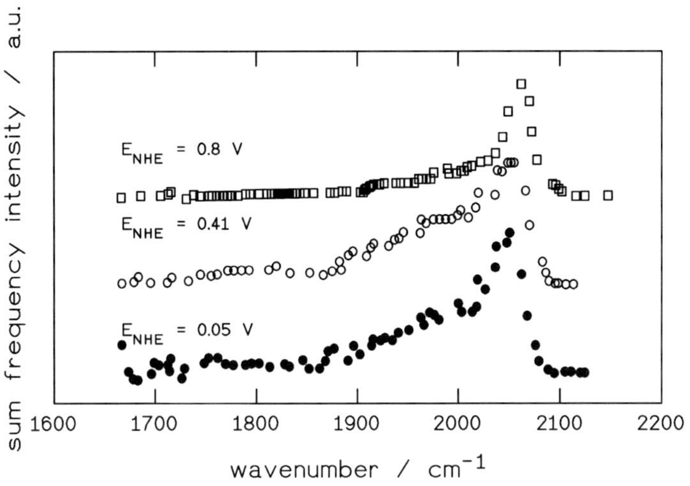

Fig. 5.134. SFG spectra of adsorbed CO on a polycrystalline platinum electrode in an aqueous solution of 0.1 **M** $HClO_4$ and 0.5 **M** methanol at various electrode potentials as indicated; data taken from [786]

more strongly on the more open Au(210) surface; the presence of cetylpyridinium[47] shows no significant effect. On the denser Au(111) surface, CN^- is less strongly adsorbed; the observed features, including surface ion pair formation, imply coadsorption of cetylpyridinium. Vibrational lifetimes of CN^- adsorbed on Pt(111) and on polycrystalline gold and silver electrode surfaces have been measured with SFG [798]. Lifetimes on platinum and gold ranging from 3 to 8 ps and 10 to 19 ps, respectively, are electrode potential dependent; at more positive electrode potentials, longer lifetimes were observed. Significantly longer lifetimes were observed with a silver electrode. In a solution of 0.1 **M** $NaClO_4$ in H_2O a value of 28 ps was found and, in D_2O, a lifetime of 60 ps was found. These values are close to those found for the free, solvated cyanide anions. The values are surprisingly large since lifetimes in the range of 1.5 to 3 ps are observed for the isoelectronic CO. Time-resolved SFG has been applied in studies of various species, including CO at the metal/solution interface [799].

In a study of thiocyanate adsorption on a polycrystalline gold surface and on electrodeposited cobalt and nickel films, changes in band frequencies were observed when moving from the naked gold to the electroplated surface. In the case of cobalt electrodeposition, an anion-induced underpotential deposition was proposed. Further changes of spectral feature were interpreted assuming differences in strength of interaction related to d-band filling of the substrate and in different adsorbate geometries [800]. Measurements of vibrational dynamics, i.e. vibrational lifetimes, are possible [801]. Vibrational sum frequency generation (VSFG) has been employed in a study of the atmospheric corrosion of zinc by various organic com-

47 Cetylpyridinium is a popular electrolyte solution additive in electroplating baths.

pounds [802]. In the case of acetic acid, various species formed by hydrogen bonding were identified; acetaldehyde underwent a hydration process resulting in the presence of a gem-diol at the interface. Rapid acquisition of spectra (5 s^{-1}) is possible using a femtosecond broad-band multiplex SFG apparatus; SFG spectra can be recorded during a slow electrode potential scan [803, 804]. Studies reported so far with this setup deal with CO-adsorption on platinum and Pt/Ru electrode surfaces. The strictly interfacial sensitivity of SFG as compared to the dependency of surface enhancement on active sites has been illuminated in a comparative study of cyanide adsorption on gold with SERS and SFG [805].

Approaches to interpretation of results based on quantum mechanical calculations have been described [806]. The interface between a platinum electrode surface and a room temperature ionic liquid (1-butyl-3-methylimidazolium tetrafluoroborate) in the presence of adsorbed CO has been probed with SFG [807]. The position of the electrode potential of zero charge E_{pzc} could be estimated and reorientation of adsorbed ions as a function of applied electrode potential could be deduced.

5.8 Mass Spectrometry[48]

Mass spectrometry has been applied in electrochemical investigations predominantly as an *ex situ* method because of the obvious incompatibility of the high vacuum needed for all types of mass spectrometry and the presence of a liquid electrolyte solution. Because of the amount of information provided in a mass spectrum, there have been various attempts to couple mass spectrometers with electrochemical cells as described below.

5.8.1 Differential Electrochemical Mass Spectrometry (DEMS)

Fundamentals. Mass spectrometry (MS) is an extremely powerful spectroscopy suitable not only for determination of molar masses, but also for identification of chemical species and molecular structures. It works only under ultrahigh vacuum conditions. Combining mass spectrometry with electrochemistry directly requires a special experimental setup matching the seemingly incompatible requirements. A first attempt was reported by Bruckenstein and Gadde [808, 809]. Using a hydrophobic porous membrane of polytetrafluoroethylene (PTFE), an electrochemical cell was coupled to a mass spectrometer. Gaseous reaction products dissolved in the electrolyte solution were sucked into the inlet system of the mass spectrometer. The mass intensity recorded with a time delay (rise time) of about 20 seconds was found to be proportional to the charge consumed during the electrochemical reaction. A similar setup using a silicone rubber membrane has been described by Brockman and Anderson [810]. In a further development of the experimental setup,

[48] The terms mass spectrometry and mass spectroscopy are used synonymously; because of the usage in another major field of application, the mass spectrometers used in the applications reviewed here are sometimes called residual gas analysers (RGA).

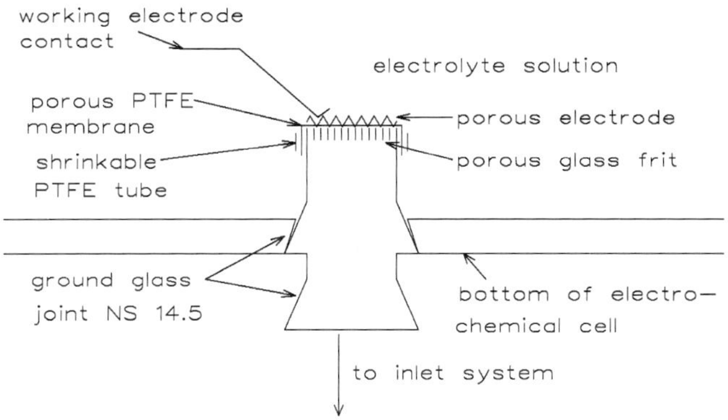

Fig. 5.135. Hydrophobic porous electrode and inlet system for DEMS [811]

Wolter and Heitbaum shortened the delay time significantly. In addition, they applied a different and more effective method of evacuation [811, 812]. The porous electrode and the inlet system connecting the electrochemical cell and the mass spectrometer are shown schematically in Fig. 5.135.

The hydrophobic porous membrane made frequently from PTFE serves as a mechanical separator between the liquid (aqueous and nonaqueous electrolyte solutions) inside the electrochemical cell and the vacuum in the ionization chamber of the mass spectrometer. Because of the nonwetting surface properties of the polymer and the small pore size, no liquid will penetrate the membrane. Water vapour (or the vapour of any other liquid used in the electrolyte solution) will pass the pores and establish a fairly high partial pressure in the inlet system. Powerful turbomolecular pumps keep the pressure at sufficiently low levels in this chamber. A narrow slit is inserted between the inlet system and analyser section. This pumping technique is generally called differential pumping. The working electrode is also porous in order to enable passage of any species to be analysed through the electrode towards the porous membrane. Any other volatile constituent of the electrolyte solution (e.g. dissolved alcohol) will show up in the ionization chamber. During interpretation of mass spectra, the high vapour pressure of these species has to be kept in mind as it may interfere with signals from other species in the search.

As a result, Wolter and Heitbaum obtained a mass signal proportional to the rate of the electrochemical reaction, i.e. to the current flowing through the porous electrode. Because of the applied pumping technique and because of the proportionality between electrochemical current (i.e. derivative of consumed charge) and mass signal, the method was called differential electrochemical mass spectrometry (DEMS).[49]

Instrumentation. The experimental setup includes a mass spectrometer (initially this was a standard type using magnetic mass separation; in subsequent versions and

[49] The name PERMS (permeable membrane mass spectrometry) suggested by Brockman and Anderson [810] has not entered the standard terminology.

in setups reviewed elsewhere [813], quadrupole mass spectrometers have frequently been used), at least two turbomolecular pumps and an electrochemical cell incorporating a porous working electrode, a counterelectrode and a reference electrode. The use of a porous electrode limits the scope of electrode materials to noble metals and other materials that can be prepared to be painted lacquer-like on the PTFE membrane or on other suitably porous materials [814, 815]. Modifications of the inlet system have enabled the use of compact electrode materials (including single crystals [816]) with the porous PTFE membrane mounted in close distance [817, 820] to the solid bulk electrode under investigation or with the material to be investigated mounted as a rotating electrode close to the inlet system [818–820]. An improved setup with significantly shorter response time, enabling the application of potential scan rates up to 1 $V \cdot s^{-1}$, has been described by Wasmus et al. [821]. The combination of DEMS with an electrochemical quartz microbalance (EQMB), allowing simultaneous detection of volatile products via DEMS and solid deposits via EQMB, has been described by Jusys et al. [822]. Systems investigated include fuel oxidation reactions (identification of volatile intermediates and reaction products) [824, 823], carbon dioxide reduction [826, 824, 825], electrolyte solution decomposition (i.e. of organic solvent as used in lithium batteries yielding CO_2, which is presumably involved in the formation of a solid electrolyte interface (SEI) [826, 827]), gas evolution reactions (oxygen evolution during water electrolysis, gassing of lead acid batteries) [820, 828] and systems of general technological interest, like hydroxylamine [829] or the electrooxidation of *N*,*N*-dimethyl-*p*-phenylenediamine [830]. Gas diffusion electrode-employing materials and processes as used for the fabrication of fuel cell electrodes have been studied in a comparative investigation of alcohol oxidation [831]. In the case of the C2 alcohols (ethanol and ethylene glycol), a strong dependency of carbon dioxide yield on the electrode potential was found. Formation of further oxidation products showing different capabilities of being further oxidized depending on the composition of the gas diffusion electrode was reported. The electrooxidation of ammonia at the three basal planes of monocrystalline platinum in contact with an alkaline solution has been studied with a thin layer flow cell [832]. Ammonia is oxidized almost exclusively at the Pt(100) surface towards N_2, NO and N_2O; prior to surface oxidation, a poison is formed that inhibits N_2 formation specifically. Electrochemical reduction of CO_2 and water at TiO_2 electrodes has been studied with DEMS [833].

A scanning type of DEMS (named SDEMS) has been described [834, 835]. A capillary with a porous membrane separating the inner vacuum line to the quadrupole mass spectrometer from the surrounding electrolyte solution is scanned over the electrode surface under investigation. Local electrocatalytical activities can be resolved on a mm scale.

5.8.2 Electrospray Mass Spectrometry

Fundamentals. Mass spectroscopy of species showing only low vapour pressures has been made possible using electrospray ionization coupled with a mass selector and detector [836, 837]. The liquid to be analysed is pumped through a capillary placed at a high electric potential of a few keV. The resulting large electric field results in rapid dispersion of the liquid, which is electrically charged. After evaporation of the solvent, the sample molecules are present as charged species, which are extracted into the UHV system of the mass spectrometer.

Instrumentation. Basically, a small-volume (preferably thin layer) electrochemical cell is coupled to a mass spectrometer (most frequently a quadrupole or a time-of-flight instrument) with an electrospray ionization source as the interfacing device (ESI–MS). A typical setup has been described elsewhere [838, 839] and a critical review of various combinations of different electrochemical cells with an ES-MS is available [840]. The influence of the design of the TLC on the mass spectra has been discussed [841]. In particular, interfering contributions from the processes at the counter electrode and incomplete conversion of the species under investigation can result in artifacts, which may change intensities and distributions of observed peaks and, in the worst case, these effects may result in a complete loss of the mass spectrum of the species formed at the working electrode. Suggestions for suitable cell design and adaptations of already available cells are described. Electrolyte solution is continuously pumped through the electrochemical cell into the ionization source. Reported studies deal with the electroreduction of 7,7′,8,8′-tetracyanoquinodimethane (TCNQ) and the electrooxidation of triphenylamine (TPA) [846]. In the former case, the initial radical anion, single- and double-charged charge transfer complexes and the double-charged anion were observed. Nearly complete transformation of the initial radical anion into an adduct $[(TCNQ_2)^{2-} + Li^+]$ was deduced from the experimental results. This adduct was also found as the result of the second electron transfer via a different route. In the case of the oxidation of TPA, monomer and dimer radical cations as well as neutral dimers were found. Formation of molecular dications (nickel and cobalt octaethylporphyrin) in the gas phase from neutral organics was assigned to the inherent electrolytic process occurring during electrospray ionization [842]. Cyanide ion oxidation from a nonaqueous acetonitrile-based electrolyte solution has been investigated and $[C_5N_5]^-$, $[C_{12}N_{12}]^{2-}$ and $[C_6N_6]^{*-}$ have been identified as intermediates [843]. Cesium ion uptake into and release from hexacyanoferrate films has been studied [844]. Electrooxidation of ruthenium cyclopentadienyl complexes at a platinum electrode was investigated with ESI-MS coupled to an electrochemical cell [845]. The use of ESI-FTMS (electrospray ionization Fourier transform mass spectrometry) in investigations of conducting polymers has been proposed [846]. In a study of oxidation of substituted polyamides at a glassy carbon electrode ESI-MS was used [847].

The combination of ESI–MS with electrochemically modulated liquid chromatography (EMLC) has been described [848]. In EMLC, an electrically conductive phase (e.g. porous carbon) is used as the stationary phase in an LC instrument. By modulating the electrode potential of this phase the retention of analytes is altered. Reported applications mainly involved separation of complex mixtures of organic compounds, including aromatic sulfonates, carboxylic acids, dansylated amino acids, corticosteroids, benzodiazepines and inorganic and metal ions.

5.8.3 Thermospray Mass Spectrometry

Fundamentals. In order to identify nonvolatile species in the effluent of a liquid chromatography (LC) setup with mass spectroscopy, the thermospray interface has been developed [849]. The effluent of the LC is pumped through a heated capillary into the ionization chamber of a mass spectrometer (typically a quadrupole mass spectrometer). With adequate pumping, a pressure low enough for chemical ionization could be maintained therein. The mechanism of thermospray ionization starts with almost complete evaporation of the solvent upon entrance into the ionization chamber, resulting in a supersonic jet of a superheated mist [850]. The nonvolatile components, including the ions of the electrolyte that is required for further ionization in the liquid phase, are kept in the droplets, which are charged positively or negatively corresponding to their composition of anions and cations. Assisted by the large local electric field around the droplets, molecular ions of nonvolatile components with a few solvent molecules evaporate from the droplets. After thermal equilibration with the solvent vapor in the ionization chamber, the ions diffuse towards the sampling aperture into the mass spectrometer.

Instrumentation. Hambitzer and Heitbaum described a coupling of the thermospray inlet system with an electrochemical cell [851, 852]. The electrolyte solution is pumped into the cell. The working electrode was initially mounted as a coil of platinum wire around the exit bore connecting the cell to the interface. Thus the liquid in which the reaction products are present was transferred directly to the thermospray interface. An advanced version that could be used with flat electrodes instead of a wire is depicted in Fig. 5.136.

This setup has been used in studies of electrooxidation of aniline [853–855]. Higher aniline oligomers were identified. In an improved version described by Volk et al. [856] the dead time between electrochemical generation of a species (i.e. current signal) and the appearance of the corresponding mass signal was reduced to 0.5 s. Results obtained during the electroreduction of uric acid at a graphite electrode are displayed in Fig. 5.137.

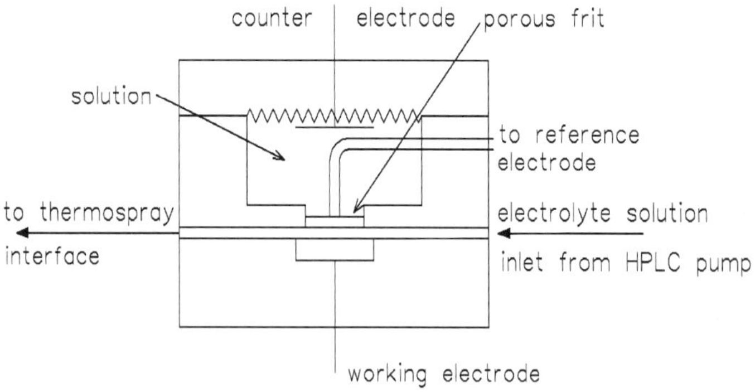

Fig. 5.136. Electrochemical cell for thermospray mass spectrometry [851]

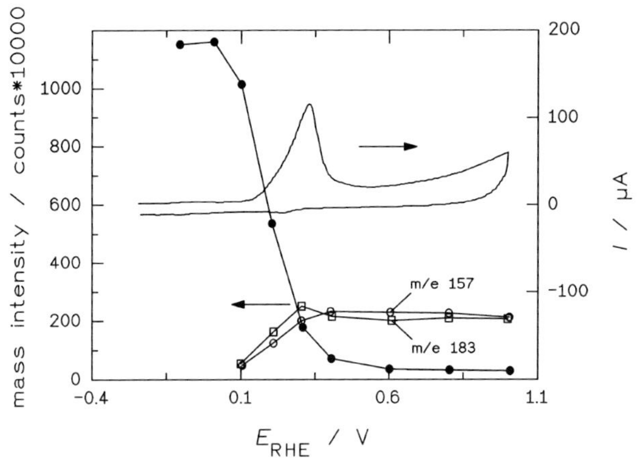

Fig. 5.137. Cyclic voltammogram and selected mass intensity vs. electrode potential curves measured during electrooxidation of uric acid at pyrolytic graphite; m/e 157 ≡ allantoine, m/e 167 ≡ (uric acid - H)$^-$, m/e 183 ≡ imine alcohol, based on data in [856]

5.8.4 Inductively Coupled Plasma Mass Spectrometry (ICPMS)

Fundamentals. Mass spectroscopy of species released from a suitable electrode by anodic stripping voltammetry (ASV) can be detected online by moving the electrolyte solution from the electrode into the ionization chamber of an inductively coupled plasma mass spectrometry. In the electrically generated plasma with a temperature of several thousand Kelvin, both metals and nonmetals are simply ionized. Using ASV, both preconcentration of the species to be measured (mostly heavy ions like Cd^{2+} or Cu^{2+}) and removal over other interfering cations can be afforded [857].

Instrumentation. Application in determination of As(III) and Se(IV) species with ICPMS coupled directly with a porous gold-plated carbon electrode (reticulated vitreous carbon) that is operated in the anodic stripping mode has been reported [858]; similar studies of Cr(IV) and V(V) [859] and Cu and Cd [860] determination are available.

5.8.5 Thermodesorption Mass Spectrometry (TDMS)

Fundamentals. Besides introduction of gaseous species into the ionization chamber, other techniques have been employed to transfer species to be analysed [861]. For analysis of materials adsorbed or deposited as solids on an inert support, thermally induced desorption is a possible approach. The flow of particles desorbing from the surface, which is heated with the temperature increasing linearly with time, is analysed with respect to intensity and particle mass. Results yield information about the strength of bonding and/or adsorptive interaction and about the chemical composition of the species. An alternative approach is the use of secondary ion mass spectrometry (SIMS) [862, 863]. Obviously both approaches are feasible only under

UHV conditions. As has been demonstrated in selected cases, the electrochemical double layer may adhere to an electrode surface upon emersion of the electrode from an aqueous electrolyte solution [864] making this method a valid approach.

Instrumentation. A mass spectrometer equipped for thermal desorption is interfaced with an electrochemical cell [865]. The working electrode, which is made of metal foil, is mounted on a manipulator. It can be immersed into the electrolyte solution for electrochemical measurements. After emersion, it is transferred into the ultrahigh vacuum part (ionization chamber) of the mass spectrometer. Care must be taken to exclude contact of the electrode with ambient air. Because any solvent adhering to the electrode and the manipulator will be removed by vigorous pumping, the loss of species only weakly bound to the electrode surface must be taken into account during interpretation of the measured mass spectra. Heating of the electrode can be done with a high intensity halogen lamp. The small number of reports deals mostly with investigations of fuel cell reactions, particularly with strongly adsorbed reaction intermediates and reaction products [820, 872, 866].

5.9 Miscellaneous Spectroscopies and Methods

5.9.1 Probe beam deflection (PBD)[50]

Fundamentals. When a narrow beam of light (e.g. a laser) passes a condensed phase of inhomogeneous density (gradients of density caused by local changes in concentration) it is deflected (mirage effect, see Fig. 5.138). The actual deflection Ψ of the beam depends for small angles and considering geometric optics only on the path length w of the light beam in the inhomogeneous medium and on the dependence of the refractive index n on the concentration c according to

$$\Psi = \frac{w}{n}\frac{\partial n}{\partial c}\frac{\partial c}{\partial z}.$$

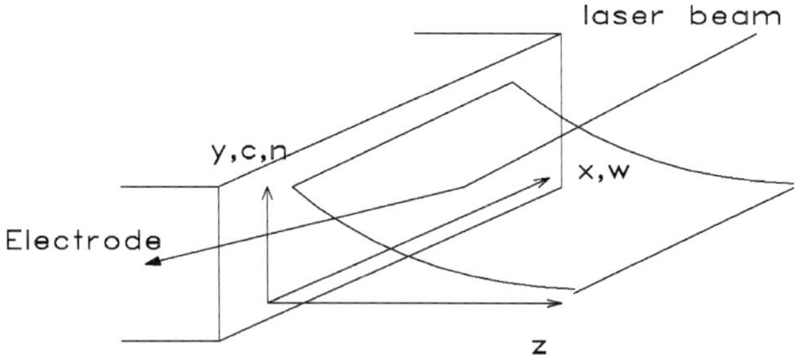

Fig. 5.138. Principle of probe beam deflection

[50] The term “photothermal deflection” is used as an equivalent to “probe beam deflection”; the term “optical beam deflection” has been used only infrequently.

For most binary electrolyte solutions $\partial n/\partial c$ is constant. Since both the length of the pathway where the laser beam is influenced by the concentration gradient and the refractive index n of the solution are constant, the deflection Ψ depends practically only on $\partial c/\partial z$ [867].

In general, the beam is deflected towards a region of higher concentration. A positive deflection (away from the electrode surface) thus indicates a lower concentration at the interface and a higher concentration inside the bulk of the solution. Accordingly, the concentration gradient is caused by an ion flux towards the interface. An ion flux in the opposite direction causes a negative beam deflection. Strictly speaking, this assignment is correct only for simple binary electrolytes and positive values of $\partial c/\partial z$. This beam deflection effect has been applied to study the mechanism and kinetics of electrode processes accompanied by significant changes in the concentration of species in the solution phase involved in this electrochemical reaction. Measurements of the probe beam deflection as a function of the wavelength of light illuminating the electrochemical interface are also possible; they are called photothermal deflection spectroscopy and are treated elsewhere (Sect. 5.2.7) [868]. (Actually this effect is used when aligning the setup: A choppered beam of light is directed perpendicularly towards the surface and the signal at the position-sensitive detector is subsequently maximized by adjustment of the cell and the measuring beam.) A combination of both methods has been treated in [869]. Theoretical modelling has been described [870] and a comprehensive overview has been provided [871]. Changes of mass, composition and other properties of the interphase that are closely connected to the flux of the ion are conveniently studied with ellipsometry (see p. 192) or the electrochemical quartz microbalance. Laser interferometry as a somewhat related technique has been applied infrequently in studies yielding information similar to that obtained with PBD [872, 873].

Instrumentation. The experimental setup is simple and straightforward. The electrochemical interphase to be investigated is mounted inside an electrochemical cell that allows a beam of light (usually laser light, which is provided most commonly by HeNe gas ion lasers adjusted to a typical diameter of 80 µm) to enter the cell and pass the solution phase parallel and as close to the electrode surface as possible. The exciting beam arrives at a position sensitive detector that indicates any deflection of the beam as a function of electrode potential. The setup is shown in Fig. 5.139.

A scanning refractometer that is basically very similar to the setup described in detail above has been reported by Tamor and Zanini [874]. The laser beam scans the electrolyte solution volume close to the electrode surface. Data on beam position and refracted beam incidence were used to calculate local refractive indices as a function of distance from the electrode, current density, etc.

Probe beam deflection can be observed during metal deposition and dissolution. The method is sensitive towards changes in the concentration gradient originating from dissolution or deposition of submonolayer amounts of material on the electrode surface [875]. A typical result of PBD measurements is displayed in Fig. 5.140. The negative PBD response observed during anodic stripping of the silver indicates concentration decay into the solution phase. The positive response

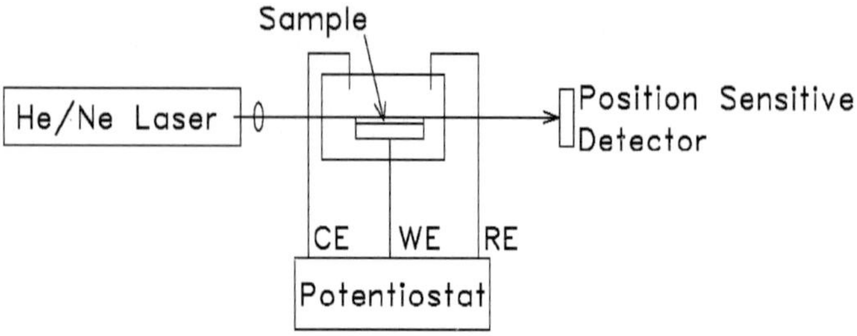

Fig. 5.139. Experimental setup for probe beam deflection studies

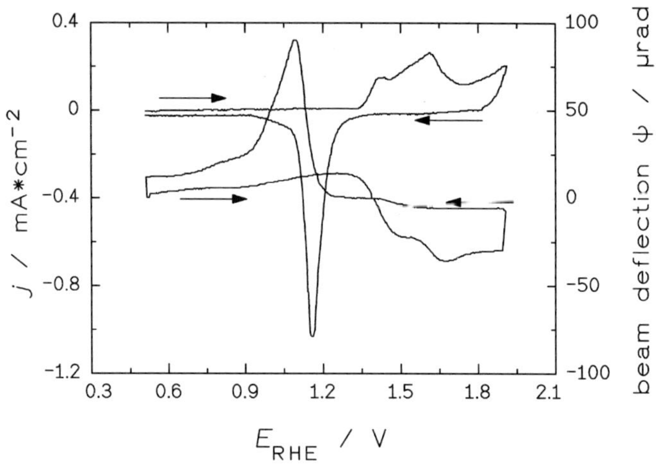

Fig. 5.140. Cyclic voltammogram (*top*) and PBD response (*bottom*) observed with a silver electrode exposed to an electrolyte solution of 2 **M** $HClO_4$ + 0.0015 **M** $AgClO_4$, dE/dt = 20 mV s^{-1} [875]

during cathodic silver deposition is caused by the concentration decrease close to the electrode surface.

In a study of silver dissolution into an aqueous perchlorate solution, $(Ag^+)_{solv}$ ions were identified as oxidation products [876]. Probe beam deflection studies of underpotential deposition have been described [877]. Probe beam deflection has frequently been employed in investigations of doping and dedoping phenomena of intrinsically conducting polymers [878–905]. During the former process, a considerable number of anions are incorporated into the polymer for charge compensation; alternatively, protons may leave the polymer for charge compensation. During dedoping, the reverse processes may take place. Considerable changes in concentration of the species involved in the solution are observed. In a study of the effect of the concentration of HCl on the exchange of protons and anions during oxidative doping/reductive dedoping of polyaniline, PBD responses that varied significantly from acid concentration were found (see Fig. 5.141).

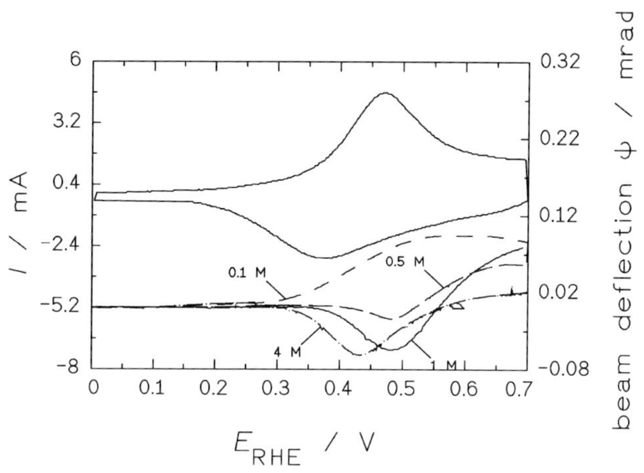

Fig. 5.141. Cyclic voltammogram (*top*) and PBD responses (*bottom*) at various concentrations of HCl observed with a polyaniline film on a gold electrode exposed to an aqueous electrolyte solution of x **M** HCl, $dE/dt = 50\ \mathrm{mV\,s^{-1}}$ (based on data in [906])

Both protons and chloride ions are exchanged. At low electrode potentials, proton exchange dominates the charge compensation equilibria. At higher potentials the type of anion (sulfate ions were used in addition to chloride) determines the relative contribution of protons and anions.

The influence of the molecular structure of the intrinsically conductive polymer and, consequently, the chemical identity of the redox reaction center has a dominant influence on the PBD response, indicating again different contributions of cations and anions in the charge compensation (compare Figs. 5.142 and 5.143).

Poly(*N*-methylaniline) shows a considerably closer spacing between the two redox processes associated with the transition from the fully reduced leucoemeraldine state into the semioxidized emeraldine and finally into the fully oxidized pernigraniline state. Both electrode potential shifts can be explained by invoking structural changes effected by the introduction of the methyl group. Further assignment of the movement of ionic species that are associated with the redox processes and necessary for charge compensation are possible based on the PBD response. A positive sign of the beam deflection indicates ion insertion. A negative sign implies ion expulsion. During the first oxidation peak both polymers show a similar positive response indicating anion insertion; the zero deflection range has been explained by assuming protonation of the reduced state of the polymer. During the second peak, the response differs considerably. Insertion is followed by ion (proton) expulsion with the polyaniline ion (anion), whereas in case of poly(*N*-methylaniline) the response indicates ion (anion) insertion at all potentials in the peak range. This qualitative difference in ion exchange mechanisms is related to the structural modification at the nitrogen atom, which is effected by methyl substitution. The flat response at low potentials might be caused by a compensating effect of simultaneous proton egress and anion ingress.

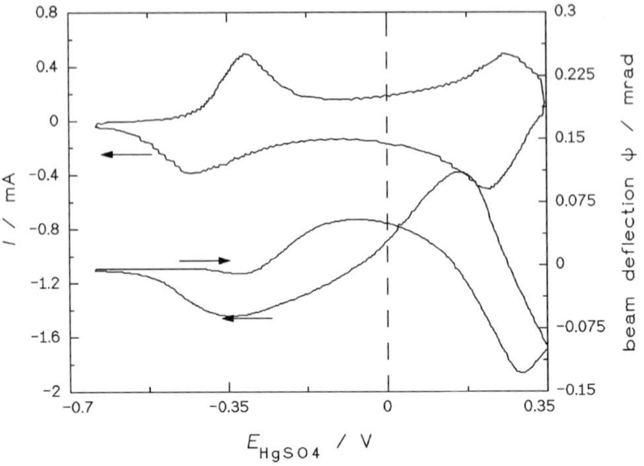

Fig. 5.142. Cyclic voltammogram and PBD response observed with a polyaniline film on a gold electrode exposed to an aqueous electrolyte solution of 1 **M** $HClO_4$, $dE/dt = 50\ mV\,s^{-1}$ (based on data in [907])

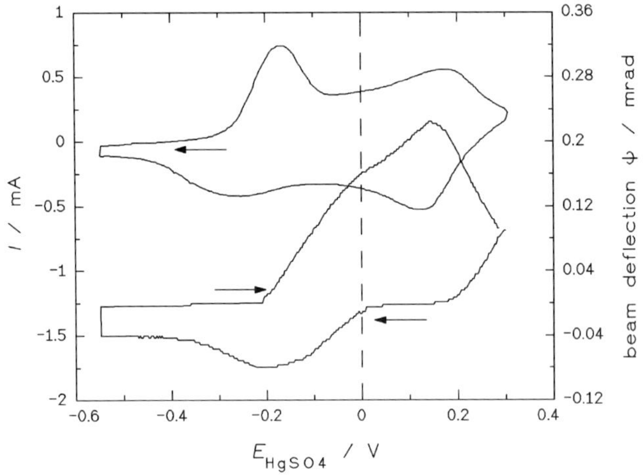

Fig. 5.143. Cyclic voltammogram (*top*, $dE/dt = 100\ mV\,s^{-1}$) and PBD response (*bottom*, $dE/dt = 50\ mV\,s^{-1}$) observed with a poly(*N*-methylaniline) film on a gold electrode exposed to an aqueous electrolyte solution of 1 **M** $HClO_4$, $dE/dt = 100\ mV\,s^{-1}$ (based on data in [907])

Reported studies include various deposition and dissolution processes [876, 908] and mechanistic studies of electrode processes at redox-active polymer-coated surfaces [909–912]. In a typical study of a polymer film containing ruthenium ions as redox-active centers, a cyclic voltammogram indicative of the one electron transfer at this center is observed as depicted in Fig. 5.144. The movement of ions needed for charge compensation in the film after a change of the redox state can be detected with PBD. The positive-going response during oxidation of the ruthenium center

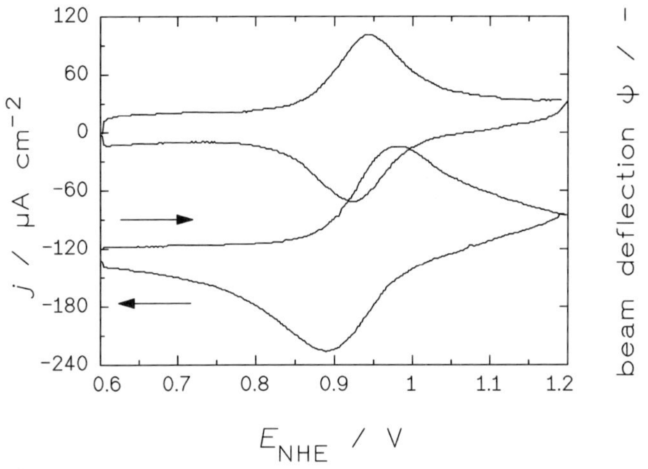

Fig. 5.144. Cyclic voltammogram (*top*) and PBD response (*bottom*) of a redox-active film of [Ru(pby)$_2$ClPVP]Cl exposed to an aqueous electrolyte solution of 1 **M** HCl, dE/dt = 50 mV·s^{-1} (based on data in [909])

and the corresponding negative response upon reduction imply anion movement. The slight delay between the signals is mainly caused by the fact that the laser beam probes changes in refractive index, i.e. in local changes of concentration gradient, which are about 50 to 100 μm away from the electrode. The concentration changes effected by anion ingress at the film/solution interface need some time to extend so far into the solution.

Although the mechanism of the electron transfer itself is on a time-scale that is not accessible with PBD associated processes, the movement of ionic and solvent species can be detected. Separation of thermal and ionic effects is possible [913] and the Peltier effects of redox couples could be measured [920].

5.9.2 Light Reflection Method

Fundamentals. Reflection of light (monochromatic or white) at an electrode surface will depend basically on two influences: Absorption of light by colored surface layers and scattering in case of rough or otherwise not ideally reflecting surfaces. Both influences may depend on electrode potential.

Instrumentation. A beam of light from a suitable source (e.g. a laser or white light source) is directed at the surface under investigation after passage through a chopper. The reflected light is analysed with respect to its intensity. An application in studies of the kinetics of redox transformations of intrinsically conducting polymers (utilizing the rather fast response time of the employed optical components) has been reported and rate constants in the range of 0.05 . . . 8 s^{-1} were deduced [914].

5.9.3 Phase-Shift Interferometry

Fundamentals. A beam of monochromatic light is polarized and split into two beams. One beam passes an optical medium, e.g. the electrolyte solution in an electrochemical cell close to the solution/electrode interface. After passage it interferes with the other beam; for a review of two-beam interferometry; see [915, 916]. This technique was limited both with respect to spatial resolution and sensitivity; for further details, see [917]. Temporal or spatial modulation of the reference beam enabled the use of very thin cells (i.e. the sensitivity was increased) and the two-dimensional reconstruction of concentration maps.

Instrumentation. A thin layer cell consisting of two high-quality glass windows and two wires acting as electrodes and spacers in a galvanostatic two-electrode arrangement is inserted into the sample beam of a two-beam interferometer. The laser light is polarized and adjusted in intensity by a half-wave retardation plate and split by a polarizing cube beamsplitter. The sample beam passes the electrochemical cell close to the solution/electrode interface to be investigated. The reference beam passes an additional half-wave retardation plate, the phase modulation mirror operated by a piezo actuator and a mechanical chopper enabling the measurement of temporal changes in concentration distribution. With a beamsplitter, both beams are mixed and the resulting interference patterns are recorded with a camera. From the obtained fringe pattern, the concentration map is calculated [924]. Reported results dealing with metal deposition processes, deposition of conducting polymers and magnetic field effects during electrodeposition have been reviewed elsewhere [924].

5.9.4 Photoacoustic Methods

Fundamentals. Illumination of a surface or interface both with monochromatic or with white light can result in considerable temperature fluctuations because of the photothermal effect. As a consequence of heating the illuminated surface and its support or the adjacent phase (solution), pressure changes can be induced. These pressure changes can be detected using a piezoelectric element or a microphone brought into mechanical contact with the substrate that supports the illuminated surface. Changes in the piezoelectric signal (the photoacoustic signal) can be correlated with processes taking place at the electrode surface or with structural or further features of the interface/interphase [918]. Measurements taken as a function of wavelength of the illuminating light are called photoacoustic spectroscopy (see Sect. 5.1.9, see also Sect. 5.1.10).

Instrumentation. Metal sheets are used as working electrodes. Piezoelectric elements were coupled mechanically with this electrode by using a thin layer of grease [919–921]. Figure 5.145 shows the experimental setup. For increased sensitivity, the incident light beam was modulated by means of a mechanical chopper and the photoacoustic signal was detected via a lock-in amplifier.

Detection of heat generated by absorbed radiation converted nonradiatively has also been done using pyroelectric materials like β-poly(vinylidene) fluoride onto

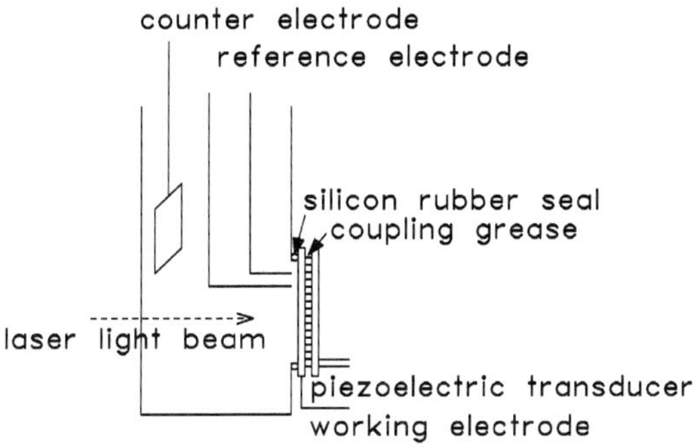

Fig. 5.145. Experimental setup for photoacoustic measurements [921]

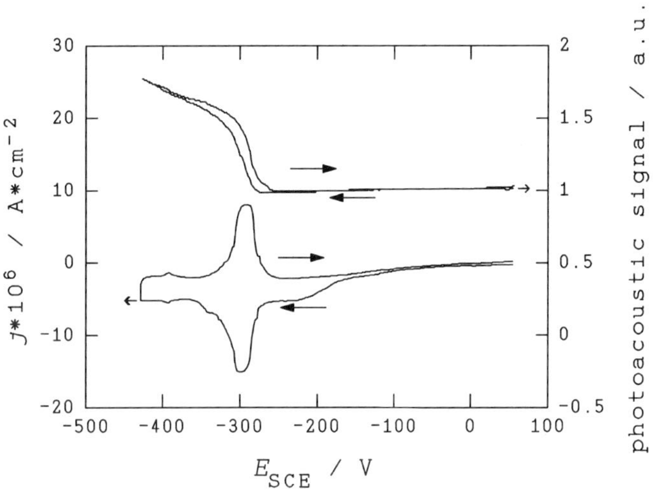

Fig. 5.146. Cyclic voltammogram (*bottom*) and potential dependent photoacoustic response (*top*) of a silver electrode in a solution of 0.1 **M** $NaClO_4$ + 0.01 **M** $Pb(ClO_4)_2$ + 0.005 **M** $HClO_4$, scan rate 1 $mV{\cdot}s^{-1}$, modulation frequency 125 Hz (based on data in [921])

which the material under investigation is deposited [922]. This principle has not been applied to electrochemical *in situ* studies.

A typical set of results as observed during the underpotential deposition and dissolution of lead onto a silver electrode is shown in Fig. 5.146. The increased absorptivity of the silver electrode covered with lead results in an increased photoacoustic signal. As the electrode potential is scanned only in the underpotential deposition region, the deposited amount of lead should correspond to a monolayer of atoms or a fraction thereof. Measurements of the photoacoustic signal as a function of time at various electrode potentials illustrate this point. At electrode potentials in the underpotential region the signal rises rapidly to a constant value.

At more negative potentials, the signal rises continuously as a function of time, indicating bulk deposition. Accordingly, this method is capable of indicating deposition of metal monolayers or similar deposits provided that their light absorption is different from the absorptivity of the substrate. A review of applications is available [925]. A variation of this method is known as photoacoustic spectroscopy (see p. 60). The incident light beam is composed not of white light as described above (p. 190), but is of a well-defined wavelength. Recording the photoacoustic signal as a function of the applied wavelength provides the desired absorption spectrum.

5.9.5 Ellipsometry

Fundamentals. Light emitted from a glowing wire (an incandescent or tungsten halogen lamp), or from most other natural and artificial sources, shows no preferred polarization, i.e. the electric and the magnetic vector (both are coupled, their planes of oscillation enclose an angle of 90°) of emitted electromagnetic waves are randomly oriented. When such a beam of light is reflected at a surface, the actual orientation of the electric vector with respect to the plane of reflection (defined by the incoming and the outgoing beams) has considerable influence on the properties of the outgoing wave. This influence can be studied more precisely with polarized light. The latter type of electromagnetic radiation is generated by certain types of lasers (gas ion) employing windows mounted at the Brewster angle[51] (so-called Brewster windows) at the ends of the plasma tube. Polarized light can be obtained from other sources by means of a polarizer[52] [923].

Upon reflection of polarized light, both the amplitude (i.e. the magnitude of the electric field vector) and the phase might undergo changes. This depends on the complex refractive index N of the material designated 1 according to

$$N_1 = n_1 - ik_1,$$

with the refractive index n_1 and the extinction coefficient k_1. The refractive index n_1 can be measured using an Abbé refractometer; the extinction coefficient k_1 is related to the absorption coefficient α_1 according to

$$\alpha_1 = 4\pi k_1/\lambda.$$

Further understanding of the interaction is most straightforward when only incident light with its electromagnetic vector parallel with the plane of reflection (E_p) and with its vector perpendicular (E_s[53]) are treated. After superposition of the electric field vectors of these two waves the resulting vector describes a circle, provided they were in phase and of the same amplitude. After reflection the mentioned changes

[51] The Brewster angle is the angle at which reflected light is completely linearly polarized; the reflected and the diffracted beams are perpendicular to each other.

[52] There are numerous polarizing optical components available. Their operation is based on birefringence, polarization at mirrors at the Brewster angle, optical dichroism, etc. Precision achieved with sheet polarizers is inadequate, thus devices like Glan–Thompson, Glan–Foucault or Rochon prisms have to be used.

[53] The symbol s derives from the German term for perpendicular, “senkrecht”.

result in a superposition wherein the resulting electric field vector describes an ellipse, hence the name of the method. The changes caused by the reflection can be expressed in various ways. A very convenient and popular description using the physical parameters gives the amplitude ratio as

$$\tan \boldsymbol{\Psi} = \frac{|E_\mathrm{p}|}{|E_\mathrm{s}|}.$$

The difference Δ of the time-independent phases ε of the two components is

$$\Delta = \varepsilon_\mathrm{p} - \varepsilon_\mathrm{s}.$$

Both parameters are combined in the basic equation of ellipsometry

$$\rho = (\tan \boldsymbol{\Psi})\mathrm{e}^{\mathrm{i}\Delta}.$$

These parameters are related to the optical properties of the investigated films. Textbooks on ellipsometry are available [924–926]. Thorough reviews of the fundamentals, including selected applications related to electrochemistry, have been published [927–930].

Instrumentation. An ellipsometer is used to determine the change of the polarization of light effected by reflection at a surface, as expressed with ρ. The principal optical elements are polarizer and compensators. The former device has been introduced on p. 192. The latter devices (also known as retarders) introduce a defined phase difference between two orthogonal components of a passing wave. The difference may be fixed (e.g. 90° or quarter-wave) or variable. Compensators are manufactured from crystalline birefringent materials like mica, calcite or quartz; their operation has been treated in detail elsewhere [931]. Measurements can be performed at a fixed angle of incidence or at variable angles; the wavelength used can be a fixed one or can be variable. In the latter case, the instrument is called a spectroscopic ellipsometer. Instruments can be grouped into two types: compensating and photometric (noncompensating). The basic components of a commercially designed spectroscopic ellipsometer of the former type are depicted in Fig. 5.147.

In the compensating instrument the phase change caused by the reflection is counteracted by a compensator yielding, in effect, linearly polarized light. The required information is extracted from the position of the fixed polarizer, the compensator in the incoming beam and the polarizer in the reflected beam. Further details of the instrumental setup and measurement procedure have been described in detail elsewhere [934]. This setup is also called a nulling ellipsometer because the correct adjustment of the three optical components resulting in an intensity minimum at the detector is searched. The sensitivity of the photomultipliers frequently used as detectors has to be taken into account. Obviously this approach is too slow for practical application. By adding Faraday modulators[54] in incoming and reflected beams,

[54] A Faraday modulator is a glass or fused silica cylinder mounted axially in an electrical coil powered by an electrical (AC or DC) current.

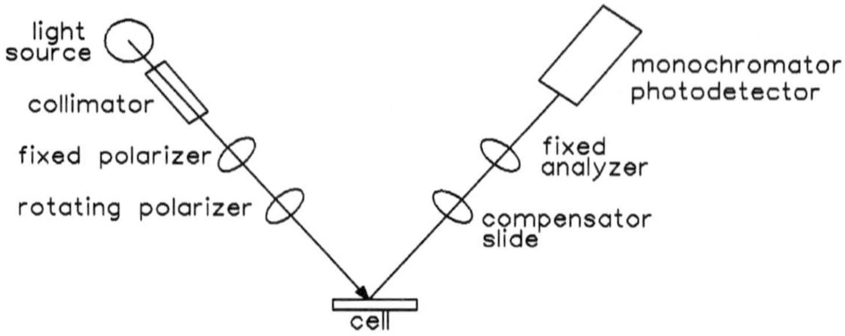

Fig. 5.147. Components of a spectrometric rotating polarizer ellipsometer

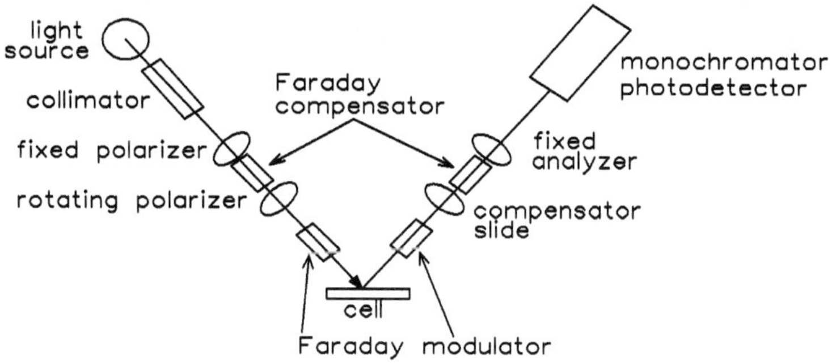

Fig. 5.148. Components of an automatic nulling ellipsometer

an oscillation of the plane of polarization is possible. Phase-sensitive detection of the signal at the detector with respect to the modulation allows faster determination of the minimum and the relevant position of the corresponding settings of the three optical devices. Further compensation of mechanical adjustment errors is possible by adding two further Faraday modulators that are operated with a constant current. The technical status of this type of instrument has been reviewed together with further types of instruments [932] and the setup of the former type is shown schematically in Fig. 5.148.

A further type of fast automatic ellipsometer for electrochemical investigations has been described [933] and an experimental approach to observe fast transients with ellipsometry was reported [934]. In the photometric mode, the intensity of the reflected light is measured as a function of the position of polarizer and sometimes compensator in the incoming beam; for further details and an overview, see [934]. A general overview of instrumental developments has been provided [935].

The process of an ellipsometric experiment can be separated into five steps [935]:

1. Providing an incident beam of a known state of polarization
2. Allowing this beam to interact with the interface/interphase under investigation
3. Measuring the state of the emerging light
4. Determination of parameters, like Ψ and Δ or their equivalents
5. Inferring properties and changes from these parameters

The final step almost always involves formulation of a model that describes the system under investigation. The parameters of this model are compared with those measured. By successive approximation, the parameters of the model are optimized until agreement is reached. In the most simple case, a clean, homogeneous and isotropic surface with N_2 is covered with a medium of a known value of N_1 and a two-layer model can be applied. From a single ellipsometric measurement the value of N_2 can already be determined [937]. When the surface is covered with a layer of a foreign material, a three-layer model must be used. Although the values of N_1 and N_3 can be determined in many cases, the number of parameters is considerably larger in a separate experiment where the intermediate layer is absent, which indicates a need for multiple experiments [937]. Numerous models combining information about optical parameters of ideal films with their real morphology and composition have been used to evaluate ellipsometric data [934]; nevertheless, in many studies, ellipsometry is used mostly to determine film thickness. A modification of an ellipsometer enabling microscopically resolved measurements (ellipsomicroscopy EMSI) has been described [936] and applied to studies of the spreading of corrosion on stainless steel.

Ellipsometry at noble metal electrode/solution interfaces has been used to test theoretically predicted microscopic parameters of the interface [937]. Investigated systems include numerous oxide layer systems [934–943], metal deposition processes [934], adsorption processes [934, 944] and polymer films on electrodes [945–947]. Submonolayer sensitivity has been claimed. Expansion and contraction of polyaniline films was monitored with ellipsometry by Kim et al. [948]. Film thickness as a function of the state of oxidation of redox active polyelectrolyte layers has been measured with ellipsometry [949]. The deposition and electroreduction of MnO_2 films has been studied [950]; below a thickness of 150 nm, the anodically formed film behaved like an isotropic single layer with optical constants independent of thickness. Beyond this limit, anisotropic film properties had to be assumed. Reduction was accompanied by an increase in thickness, which started at the oxide/solution interface.

5.9.6 Surface Plasmon Resonance Spectroscopy[55]

Fundamentals. Surface plasmon polaritons (SPP)[56] are electromagnetic charge density waves[57] (plasma oscillations) that collectively propagate along the metal/solution (or any other dielectric medium) interface [951–953]. They can be generated by electromagnetic radiation (e.g. light in the visible range of the electromag-

[55] The method is sometimes called surface plasmon spectroscopy (SPS); the term "resonance" refers to the peculiarities of the reflectivity vs. angle of incidence plots obtained with the method (minimum of reflectivity).

[56] SPPs are sometimes also called PSPs (plasmon surface polaritons).

[57] The transversal electromagnetic modes are either photon-like or similar to optical phonons, if their nature is a mix between these forms, the modes are sometimes called polaritons.

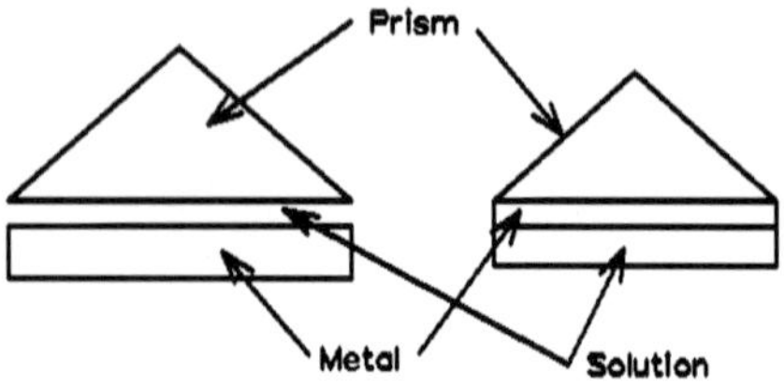

Fig. 5.149. Otto (*left*) and Kretschmann (*right*) configuration for SPP excitation

netic spectrum) coupling with oscillating modes of the electron density at this interface. Surface plasmon polaritons are generated only with p-polarized light waves with electric field vectors that are oriented parallel to the plane of incidence. Their propagation or wave vector lies in the plane of the metal surface. The field amplitudes associated with SPPR have a maximum at the interface and decay exponentially into both adjacent phases, making their investigation an extremely surface sensitive technique. Because of the electrooptical peculiarities SPPs cannot be caused by simple external reflection of a light beam at a smooth metal surface. Instead, prism coupling is needed, which uses either the Otto or the Kretschmann configuration depicted in Fig. 5.149. In the former configuration an electrolyte solution layer with the thickness of about the wavelength of the used light is inserted between the prism and the solid metal. In the latter configuration, the metal layer of about this thickness is deposited directly onto the prism. Details are discussed in the following section.

Surface plasmon polaritons have been used to enhance the surface sensitivity of various spectroscopic techniques like surface enhanced Raman spectroscopy and second harmonic generation. The generation of SPPs at an interface causes a lower intensity of the reflected light beam. This intensity reaches a minimum at a typical angle of incidence. This angle is called a surface plasmon resonance angle (θ_{sp}). Because the shape of the reflectivity vs. the angle of incidence curve depends on the electrooptical properties of the involved materials on both sides of the interface and can be calculated using Fresnel calculations, measurements of these curves provide a possibility to investigate the interface and any changes occurring therein by adsorption on the metal surface. Further details are provided elsewhere [954]. An overview of SPS in electrochemistry has been provided [955].

Instrumentation. For the measurement of the surface plasmon resonance, a suitable electrochemical cell incorporating the crucial elements needed for SPP excitation is required. A simple design based on the Kretschmann configuration is shown in Fig. 5.150. Because the metal under investigation has to be deposited by vapor deposition, only polycrystalline surfaces can be studied. The Otto configuration requires a well-defined, narrow gap between the base of the prism and the metal film. Depending on the employed wavelength, this distance ranges from 500 to 1000 nm. Invoking a typical distance of about 200 nm, dust particles have been noted as possible disturbances, making this arrangement practically meaningless [952]. Nevertheless, successful applications were previously reported, wherein the proper distance

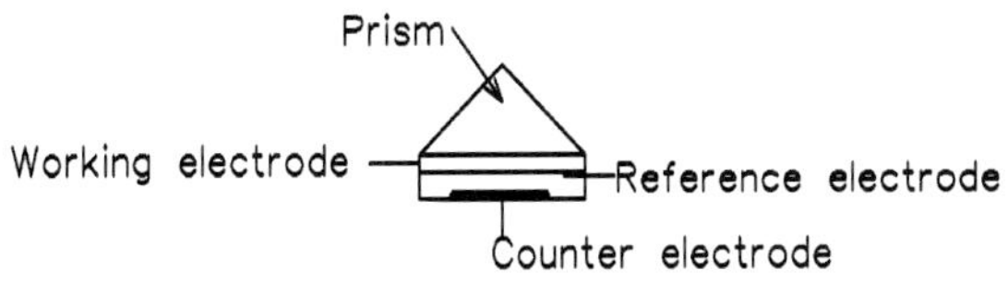

Fig. 5.150. Electrochemical cell for SPR measurements based on the Kretschmann configuration

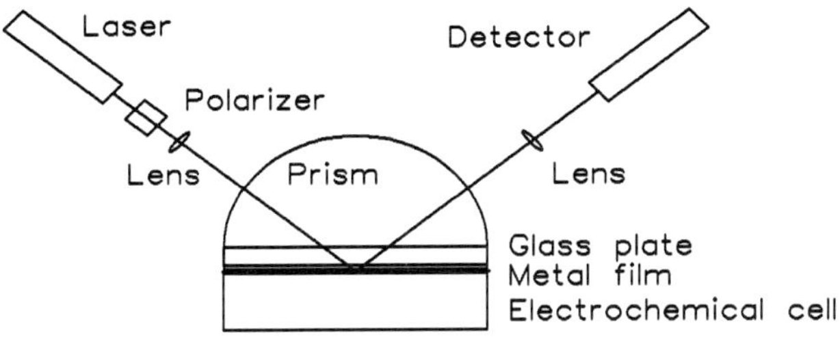

Fig. 5.151. Experimental setup for detection of SPR

was adjusted with a micrometer screw [956]. The Otto arrangement allows investigations of single crystal surfaces or of polymer coats on metal substrates.

The obvious advantage of the Kretschmann design is a simple setup without any specific problems related to the arrangement of the electrodes. A drawback is the need to have the working electrode under investigation deposited as a thin film on the prism. As already discussed in relation to various other internal reflection mode techniques, this might cause problems because of the poor electronic conductivity of thin metal films. This may result in uneven electrode potential distribution within the film and a poor response to fast changes of the electrode potential. In addition, the preparation of the metal layer itself might be a problem because of poor adhesion or uneven thickness. With the Otto configuration, this problem is readily avoided. Instead, the usual problems encountered with thin electrolyte films (TLC) have to be expected. The SPR is detected as a minimum in the reflected light intensity. The setup needed for this measurement includes a monochromatic light source (usually a laser), a polarizer for selection of the p-polarized light and a detector (e.g. photodiode matched to the wavelength of the incident light). Cell and detector are mounted on a θ-2θ-stage, as shown schematically in Fig. 5.151. Additional lenses are needed for focusing the incident beam and a chopper is used to simplify and improve light intensity measurements. Instead of a simple prism, a hemispherical prism is used, this results in a constant position of the incident beam on the electrode surface.

Surface plasmon resonance has been used to study condensed films at metal/solution interfaces [957, 958]. A typical result is displayed in Fig. 5.152.

The SPR is found around $\varphi_0 = 46°$. From data taken at various electrode potentials, the number of organic species in the condensed thymine film could be calculated [957]. With a combination of SPR and EQCMB, the adsorption of a perfluo-

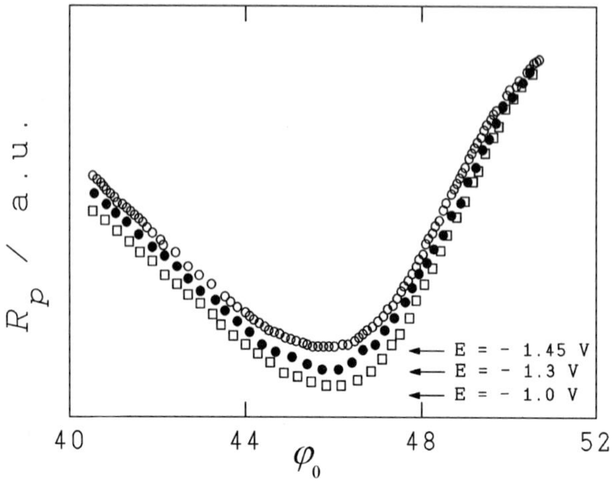

Fig. 5.152. Reflectivity of the interface mercury/aqueous solution of 1 **M** NaCl + 1 m**M** thymine as a function of the angle of incidence φ_0 and of the electrode potential as indicated, pH = 6; λ = 600 nm; data taken from [957]

ropolyether surfactant on a gold electrode has been investigated [959]. In studies of the potential dependency of the SPR as a function of electrode potential for a silver electrode in contact with aqueous solutions containing various anions, strong effects of specifically adsorbed halide ions were found [960]. Time-dependent measurements of SPR enabled separation of various stages of adsorption and surface oxide film formation on a gold electrode [961]. Oxidation and reduction of mercury and its ions at an ultrathin gold electrode was monitored with SPR [962]. Results imply a partially reversible formation of a gold amalgam with optical properties between those of gold and mercury; the mercury content was estimated to be about 12%. Surface plasmon resonance studies of a gold surface coated for corrosion protection with a thin layer of silica indicate that silica provides effective protection without completely inhibiting electrochemical processes [963]. These multilayer systems are under development with respect to analytical applications.

Surface plasmon resonance has been applied to studies of intrinsically conducting polymers. Electrochromic properties and optical conductivity were measured [964]. Ion ingress and egress processes could be related to doping/dedoping processes of polypyrrole; accumulation of neutral salt and solvent in the film were observed with a combination of SPR and the quartz crystal microbalance [965]. The mechanism of copolymerization of terthiophene with a carbazole and the properties of the obtained product were studied with a combination of various electrochemical methods, including the quartz crystal microbalance and SPR [966]. The advantageous effects of a micellar solution of a monomer (3,4-ethylene dioxythiophene EDOT, aqueous micellar solution) could be verified [967]. Self-assembled layer-by-layer films of polyaniline and sulfonated polyaniline (self-doped PANI) were investigated with a combination of SPR and surface plasmon field-enhanced light scattering (see below) [968]. Morphological changes and changes of the dielectric properties could be observed during electrochemical cycling of the ultrathin films. Kinetic studies of various adsorption processes have been reviewed [64].

5.9.7 Surface Plasmon Excitation and Related Methods

Fundamentals. Surface plasmons (SP) can be used to monitor optical properties of metal and semiconductor surfaces. For introductory overviews, see [969, 970]. Surface plasmon field-enhanced light scattering (SPFELS) is observed when an interface is illuminated by light under conditions stimulating surface plasmon excitation as reported [971].

Applications in electrochemistry similar to surface plasmon (resonance) spectroscopy are treated in the previous sections. Photoluminescence associated with surface plasmons is treated in Sect. 5.6.4.

Instrumentation. The experimental setup is similar to the one used in SPR. An additional photomultiplier mounted at fixed angle with respect to the base of the prism at the solution side of the electrode films is used to detect light scattered from the rough electrode surface and/or the polymer film deposited on the metal film (see Fig. 5.153).

With a combination of SPR and SPFELS, changes of optical (dielectric constant) and electrochemical properties during doping/dedoping of layer-by-layer self-assembled conducting films of polyaniline and sulfonated polyaniline were measured; results are displayed in Figs. 5.154 and 5.155 [968]. In a study of polyaniline deposition by electropolymerization, changes of both reflectivity and scattered light intensity were observed [971].

The observed change in reflectivity can be explained when taking into account ion movement (both cation egress and anion ingress) during oxidative doping. Changes in the dielectric constant of the film associated with doping/dedoping are evident in the SPFELS curves below.

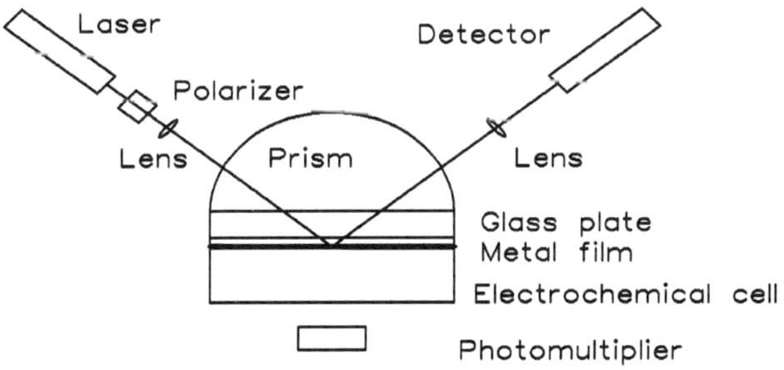

Fig. 5.153. Experimental setup for detection of SPFELS

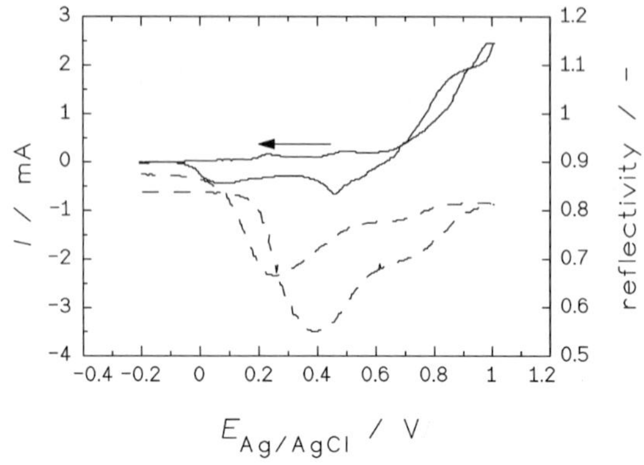

Fig. 5.154. Cyclic voltammogram of polyaniline and surface plasmon resonance reflectivity with a solution of 0.5 **M** H_2SO_4, second scan, $dE/dt = 20$ mV·s^{-1}, based on data in [971]

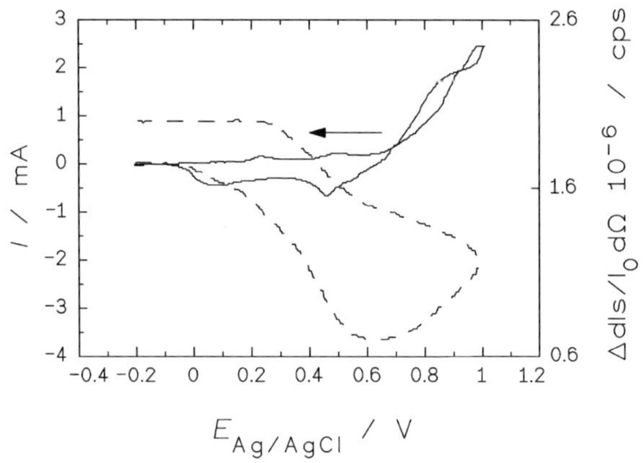

Fig. 5.155. Cyclic voltammogram of polyaniline and surface plasmon field-enhanced light scattering with a solution of 0.5 **M** H_2SO_4, second scan, $dE/dt = 20$ mV·s^{-1}, based on data in [971]

5.9.8 Inductively Coupled Plasma Atomic Emission Spectroelectrochemistry

Fundamentals. The composition of liquids with respect to both identity and concentration of dissolved species can be determined with inductively coupled plasma atomic emission spectrometry (ICP-AES) [972]. The employed spectrometer can be coupled directly with an electrochemical cell wherein processes like corrosion or anodic dissolution occur. Continuous aspiration of very small liquid volumes transferred into the spectrometer allows determination of rates of dissolution as a function of various experimental parameters like electrode potential [973].

Instrumentation. An electrochemical flow cell (flow rate approx. 2–3 mL·min^{-1}) is connected to an electrolyte solution reservoir. The solution is supplied via a peristaltic pump. The outlet is connected to the ICP-AES. Depending on the type and setup of the spectrometer, several metal ions can be monitored. In a study of the anodic dissolution of 304 stainless steel, the dissolution rates of numerous metallic components could be determined [980].

5.9.9 Positron Annihilation Spectroscopy (PAS)[58]

Fundamentals. Vacancies, vacancy clusters, nanometer-sized voids, dislocations and other open volume defects of atomic size can be detected and studied with positron annihilation spectroscopy [974, 975]. Positrons (p^+ or e^+) are nuclear species with the same mass and spin, but the opposite charge and magnetic momentum of the electron. The positron is the antiparticle of the electron as postulated by Dirac (1930) and finally detected by Anderson (1932). Positrons are formed during nuclear decay reactions of various isotopes (e.g. ^{58}Co, ^{22}Na, ^{64}Cu) or in linear accelerators. Monochromatic positrons of desired energy are obtained by using moderators (e.g. mono- or polycrystalline nickel or tungsten crystals). The positron is implanted in the solid under investigation. Its annihilation with an electron in the solid produces gamma radiation (about 511 keV) according to:

$$p^+(e^+) + e^- \rightarrow \text{positronium} \rightarrow 2h \cdot \nu(511\ \text{keV}).$$

The gamma radiation is Doppler shifted because of the momentum that is transferred to the photon during annihilation. This momentum is in turn controlled by the momentum of the annihilating electron, which is related to the nature of the annihilation site. In voids, valence electrons dominate with a narrow distribution of momentum, whereas in the bulk of a crystal a much broader distribution of electron momenta is observed. The Doppler broadening is wider in the latter case. The presence of defects can thus be detected by measuring the line shape of the annihilation photopeak. Positrons of variable energy and correspondingly different depth of penetration can be used for depth profiling. Thus this method can be used to study defects close to a surface, i.e. the metal/solution interface, as formed during open circuit dissolution of aluminum in caustic solutions [976].

Instrumentation. The experimental setup includes a positron source with a suitable converter (e.g. Cu(111)+S crystal) that is used for moderation of decay positrons from the radioactive source (e.g. 200–300 mCi ^{58}Co) into thermal positrons [977]. A NaI(Tl) or Ge(Li) crystal is used to detect the annihilation photons produced during interaction of positrons with the material under investigation. Light from these crystals is guided via a light pipe to a photomultiplier. Samples to be investigated have to be transferred from the electrochemical cell into the PASCA setup. Investigations reported so far have dealt mainly with metal dissolution and oxide layer formation processes [983], metal swelling upon hydrogen sorption [978] and intrinsically conducting polymers [979, 980].

---

[58] This method has also been called PASCA (Positron annihilation spectroscopy for chemical analysis).

5.9.10 Neutron Reflectivity

Fundamentals. A collimated beam of neutrons generated in a nuclear reactor or a spallation source with wavelength λ impinges onto an interface at an angle θ (close to glancing angle). The reflected intensity R is measured as a function of momentum transfer $\boldsymbol{Q}$

$$\boldsymbol{Q} = \frac{4\pi}{\lambda} \sin\theta .$$

The recorded reflectivity $R(\boldsymbol{Q})$ depends on the neutron refractive index perpendicular to the interface. This index depends on the scattering length density Nb, which is composed of the number density N of nuclei per cm^3 and the sum of the neutron scattering lengths b of the various nuclei. Because values of b vary for different nuclei, isotopic exchange causes variation of the neutron reflectivity without changing the electrochemical properties to a significant extent.[59] The reflectivity of an interface with a film attached is described by the Fresnel coefficients. Even in the approximate form [981], the recorded reflectivity profile cannot be deconvoluted into a scattering length density profile because of the lack of phase information. Instead, a calculated reflectivity profile based on a model of the Nb profile is fitted to the measured profile. Thus the desired structural information along the z-axis (perpendicular to the solution/electrode interface) is obtained. Using isotope exchange, the scattering length density gradients in the system can be varied systematically. Further details of the fundamentals and of the data interpretation have been reviewed elsewhere [982, 988]. A major advantage of the method is the high spatial (depth) resolution down to nanometers.

Instrumentation. Films to be investigated are deposited on a material (e.g. single crystal quartz) that is transparent for thermal neutrons ($0.1\ \text{nm} < \lambda < 0.6\ \text{nm}$). A typical cell is displayed in Fig. 5.156. The electrochemically inert metal used as

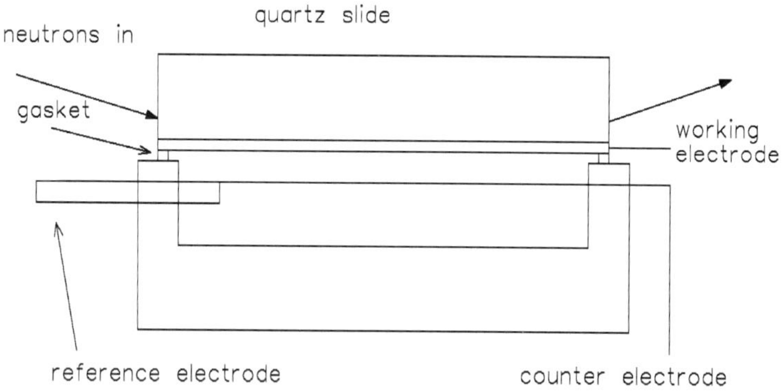

Fig. 5.156. Spectroelectrochemical cell for *in situ* neutron reflectivity measurements [989]

[59] This may be different in kinetic studies, where the kinetic isotope effect might come into play.

a conductive layer between the quartz plate and the film under investigation has to be selected with respect to matching the scattering density of quartz and metal (for an example, see [988]).

Investigated examples include the determination of the spatial distribution of a polymer, solvent and mobile species in poly(*o*-toluidine) [983, 984] and polybithiophene [989] films, film swelling and solvent content in electroactive films containing transition metal complexes [988, 985], postdeposition modified electroactive polymers [986] and organic adsorbate layers [987]. The method allows also the investigation of buried interfaces in bilayer systems of various polymers [988].

5.9.11 Neutron Scattering

Fundamentals. Neutrons can interact with matter in several ways. Depending on the neutron–nucleus interaction, they can be scattered coherently or incoherently and both processes can occur elastically or inelastically. For structural studies in electrochemical systems, diffraction, i.e. elastic coherent scattering, is of particular interest. Fundamentals of these modes of interaction, including spectroscopic aspects relevant for mobility studies, have been reviewed [989].

Instrumentation. The basic setup of a neutron scattering experiment is depicted in Fig. 5.157.

In experiments reported so far, which mostly pertain to electrode materials showing intercalation processes like those of transition metal chalcogenides or graphite (for a review, see [989]), an angular dispersive setup with position-sensitive detectors has been used. In this arrangement, the position of the detector is moved from angles corresponding to $2\theta_1$ to $2\theta_2$. Further studies of processes in molten electrolytes and during surface oxidation of various metals have been reported [989]. In inelastic scattering experiments, the energy change of the neutron is determined by moving the crystal serving as a monochromator. The resulting Doppler effect causes a modulation of the beam energy (in a typical case, by about ± 30 μeV).

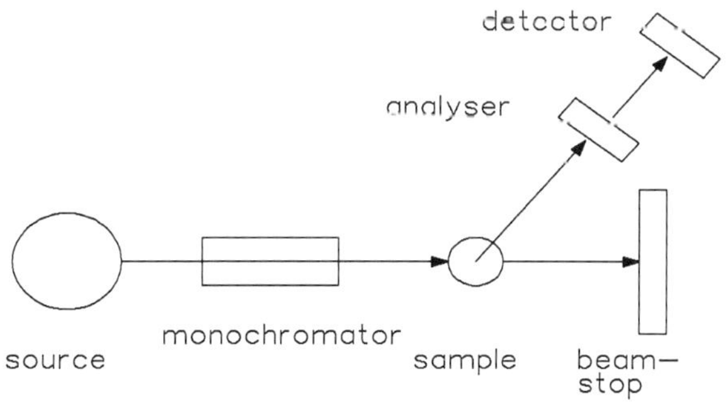

Fig. 5.157. Simplified setup for neutron scattering measurements

References

1. N. Sheppard, H.A. Willis, J.C. Rigg, Pure Appl. Chem. **57**, 105 (1985)
2. X.G. Zhang, *Electrochemistry of Silicon and its Oxide* (Kluwer/Plenum, New York, 2001)
3. J.B. Wagner Jr., in *Techniques for Characterization of Electrodes and Electrochemical Processes*, ed. by R. Varma, J.R. Selman (Wiley, New York, 1991), p. 3
4. H. Ohno (ed.), *Electrochemical Aspects of Ionic Liquids* (Wiley, Hoboken, 2005)
5. C. Aliaga, C.S. Santos, S. Baldelli, Phys. Chem. Chem. Phys. **9**, 3683 (2007)
6. K.E. Johnson, Interface **16**, 38 (2007)
7. T. Tsuda, C.L. Hussey, Interface **16**, 42 (2007)
8. V.E. Norvell, G. Mamantov, in *Molten Salt Techniques*, ed. by D.G. Lovering, R.J. Gale (Plenum, New York, 1983), p. 151
9. I. Galasiu, R. Galasiu, J. Thonstad, in *Nonaqueous Electrochemistry*, ed. by D. Aurbach (Dekker, New York, 1999), p. 461
10. F. Scholz, U. Schröder, R. Gulaboski, *Electrochemistry of Immobilized Particles and Droplets* (Springer, Heidelberg, 2005)
11. C.K. Mann, in *Electroanalytical Chemistry*, vol. 3, ed. by A.J. Bard (Dekker, New York, 1969), p. 57
12. W.R. Heineman, F.M. Hawkridge, H.N. Blount, in *Electroanalytical Chemistry*, vol. 13, ed. by A.J. Bard (Dekker, New York, 1984), p. 1
13. T. Kuwana, W.R. Heineman, Acc. Chem. Res. **9**, 241 (1976)
14. W.R. Heinemann, Extended Abstracts of the 201st Meeting of the Electrochemical Society, Philadelphia, USA, 12–17.05.2002, Ext. Abstr. #1301
15. A. Yildiz, P.T. Kissinger, C.N. Reilley, Anal. Chem. **40**, 1018 (1968)
16. J. Lippe, PhD Thesis. Universität Oldenburg, 1991
17. R. Holze, Collect. Czech. Chem. Comm. **65**, 899 (2000)
18. P.D. Boyle, D.C. Boyd, A.M. Mueting, L.H. Pignolet, Inorg. Chem. **27**, 4424 (1988)
19. R.A. Hatton, S.R. Day, M.A. Chesters, M.R. Willis, Thin Solid Films **394**, 292 (2001)
20. E. Matveeva, J. Electrochem. Soc. **152**, H138 (2005)
21. M. Hupert, A. Muck, R. Wang, J. Stotter, Z. Cvackova, S. Haymond, Y. Show, G.M. Swain, Diam. Relat. Mater. **12**, 1940 (2003)
22. H. Tian, A.J. Bergren, R.L. McCreery, Appl. Spectrosc. **61**, 1246 (2007)
23. E. Steudel, J. Posdorfer, R.N. Schindler, Electrochim. Acta **40**, 1587 (1995)
24. J.P. Bullock, K.R. Mann, Inorg. Chem. **28**, 4006 (1989)
25. R. Holze, Spectroelectrochemistry of Conducting Polymers, in *Handbook of Electronic and Photonic Materials and Devices*, vol. 8, ed. by H.S. Nalwa (Academic Press, San Diego, 2000), p. 209
26. R. Holze, in *Handbook of Advanced Electronic and Photonic Materials*, vol. 2, ed. by H.S. Nalwa (OPA N.V. and Gordon&Breach, Singapore, 2001), p. 171
27. A.M. Bond, F. Marken, E. Hill, R.G. Compton, H. Hügel, J. Chem. Soc. Perkin Trans. II **1997**, 1735 (1997)
28. M.H. Barley, C.E. Anson, J. Lewis, B.F.G. Johnson, J. Organometal. Chem. **339**, 151 (1988)
29. E.W. Becker, W. Ehrfeld, P. Hagmann, A. Maner, D. Münchmeyer, Microelec. Eng. **4**, 35 (1986)
30. W. Ehrfeld, D. Münchmeyer, Nucl. Instr. Meth. Phys. Res. **A303**, 523 (1991)
31. L. Dunsch, A. Neudeck, P. Rapta, Fresenius J. Anal. Chem. **367**, 314 (2000)
32. A. Neudeck, L. Dunsch, Ber. Bunsenges. Phys. Chem. **97**, 407 (1993)

33. A. Neudeck, L. Dunsch, J. Electroanal. Chem. **386**, 135 (1995)
34. M.B.G. Pilkington, B.A. Coles, R.G. Compton, Anal. Chem. **61**, 1787 (1989)
35. F. Hartl, H. Luyten, H.A. Nieuwenhuis, G.C. Schoemaker, Appl. Spectrosc. **48**, 1522 (1994)
36. M. Kummer, J.R. Kirchhoff, Anal. Chem. **65**, 3720 (1993)
37. M. Kummer, J.R. Kirchhoff, Electroanal. **8**, 524 (1996)
38. S.C. Paulson, C.M. Elliott, Anal. Chem. **68**, 1711 (1996)
39. T.B. Jarbawi, M.T. Stankovich, Anal. Chim. Acta **292**, 71 (1994)
40. W.R. Heinemann, B.J. Norris, J.F. Goelz, Anal. Chem. **47**, 79 (1975)
41. R.W. Murray, W.R. Heineman, G.W. O'Dom, Anal. Chem. **39**, 1666 (1967)
42. J. Salbeck, Anal. Chem. **65**, 2169 (1993)
43. J. Zak, M.D. Porter, T. Kuwana, Anal. Chem. **55**, 2219 (1983)
44. M.J. Simon, W.R. Heineman, G.P. Kreishman, Anal. Chem. **54**, 2382 (1982)
45. J.D. Brewster, J.L. Anderson, Anal. Chem. **54**, 2560 (1982)
46. Y.F. Lee, J.R. Kirchhoff, Anal. Chem. **65**, 3430 (1993)
47. G. Mamantov, C.L. Hussey, R. Marassi, in *Techniques for Characterization of Electrodes and Electrochemical Processes*, ed. by R. Varma, J.R. Selman (Wiley, New York, 1991), p. 471
48. C.Y. Chung, T.C. Wen, A. Gopalan, Electrochim. Acta **47**, 423 (2001)
49. A.A. Nekrasov, V.F. Ivanov, A.V. Vannikov, Electrochim. Acta **46**, 3301 (2001)
50. F. Alakhras, R. Holze, Synth. Met. **157**, 109 (2007)
51. P. Marque, J. Roncali, J. Phys. Chem. **94**, 8614 (1990)
52. W.T. Yap, G. Marbury, E.A. Blubaugh, R.A. Durst, J. Electroanal. Chem. **271** 325 (1989)
53. J. Salbeck, J. Electroanal. Chem. **340**, 169 (1992)
54. S. Mho, S.N. Hoier, B.-S. Kim, S.-M. Park, Bull. Kor. Chem. Soc. **15**, 739 (1994)
55. F. Gaillard, E. Levillain, J. Electroanal. Chem. **398**, 77 (1995)
56. D.-H. Han, B.-S. Kim, S.-J. Choi, Y. Jung, J. Kwak, S.-M. Park, J. Electrochem. Soc. **151**, E283 (2004)
57. I.S. Zavarine, C.P. Kubiak, J. Electroanal. Chem. **495**, 106 (2001)
58. R.S. Robinson, C.W. McCurdy, R.L. McCreery, Anal. Chem. **54**, 2356 (1982)
59. R.S. Robinson, R.L. McCreery, Anal. Chem. **53**, 997 (1981)
60. C.D. Batich, H.A. Laitinen, H.C. Zhou, J. Electrochem. Soc. **137**, 883 (1990)
61. J. López-Palacios, A. Colina, A. Heras, V. Ruiz, L. Fuente, Anal. Chem. **73**, 2883 (2001)
62. O. Orcajo, E. Ventosa, A. Martinez, A. Colina, A. Heras, V. Ruiz, J. Lopez-Palacios, J. Electroanal. Chem. **596**, 95 (2006)
63. W. Knoll, Ann. Rev. Phys. Chem. **49**, 569 (1998)
64. W.J. Doherty III, C.L. Donley, N.R. Armstrong, S.S. Saavedra, Appl. Spectrosc. **56**, 920 (2002)
65. J.N. Richardson, Z. Aguilar, N. Kaval, S.E. Andria, T. Shtoyko, C.J. Seliskar, W.R. Heineman, Electrochim. Acta **48**, 4291 (2003)
66. A.J. Bard, L.R. Faulkner, *Electrochemical Methods*, 2nd edn. (Wiley, New York, 2001), p. 736
67. H.G.O. Becker, H. Böttcher, F. Dietz, D. Rehorek, G. Roewer, K. Schiller, H.-J. Timpe, in *Einführung in die Photochemie*, ed. by H.G.O. Becker (Wissenschaften, Berlin, 1991)
68. A.B. Ellis, in *Chemistry and Structure at Interfaces*, ed. by R.B. Hall, A.B. Ellis (VCH, Deerfield Beach, 1986), p. 245

69. R.G. Compton, A.C. Fisher, R.G. Wellington, Electroanalysis **3**, 27 (1991)
70. R.G. Compton, A.C. Fisher, R.G. Wellington, J. Winkler, J. Phys. Chem. **96**, 8153 (1992)
71. R.G. Compton, A.C. Fisher, R.G. Wellington, J. Winkler, J. Chem. Soc. Perkin Trans. **1992**, 1359 (1992)
72. J.S. Littig, T.A. Nieman, Anal. Chem. **64**, 1140 (1992)
73. L. Sun, R.M. Crooks, J. Electrochem. Soc. **152**, E371 (2005)
74. B. Valeur, J.-C. Brochon (eds.), *New Trends in Fluorescence: Application to Chemical and Life Sciences* (Springer, Berlin–Heidelberg, 2001)
75. M. Dias, P. Hudhomme, E. Levillain, L. Perrin, Y. Sahin, F.-X. Sauvage, C. Wartelle, Electrochem. Commun. **6**, 325 (2004)
76. J. Salbeck, J. Electroanal. Chem. **340**, 169 (1992)
77. P. Jennings, A.C. Jones, A.R. Mount, Phys. Chem. Chem. Phys. **2**, 1241 (2000)
78. H. Nagatani, R.A. Iglesias, D.J. Fermin, P.-F. Brevet, H.H. Girault, J. Phys. Chem. B **104**, 6869 (2000)
79. H. Nagatani, T. Ozeki, T. Osakai, J. Electroanal. Chem. **588**, 99 (2006)
80. N. Nishi, K. Izawa, M. Yamamoto, T. Kakiuchi, J. Phys. Chem. B **105**, 8162 (2001)
81. J.D.E. McIntyre, in *Advances in Electrochemistry and Electrochemical Engineering*, vol. 9, ed. by P. Delahay, C.W. Tobias (Wiley, New York, 1973), p. 61
82. D.M. Kolb, in *Spectroelectrochemistry*, ed. by R.J. Gale (Plenum, New York, 1988), p. 87
83. W. Plieth, in *Spectroscopic and Diffraction Techniques in Interfacial Electrochemistry*, ed. by C. Gutiérrez, C. Melendres. NATO ASI Series C, vol. 320 (Kluwer Academic, Dordrecht, 1990), p. 223
84. D.E. Aspnes, Surf. Sci. **37**, 418 (1973)
85. L.M. Abrantes, R. Peat, L.M. Peter, A. Hamnett, Ber. Bunsenges. Phys. Chem. **91**, 369 (1987)
86. A. Tadjeddine, D.M. Kolb, R. Kötz, Surf. Sci. **101**, 277 (1980)
87. C. Gutierrez, in *Spectroscopic and Diffraction Techniques in Interfacial Electrochemistry*, ed. by C. Gutiérrez, C. Melendres. NATO ASI Series C, vol. 320 (Kluwer Academic, Dordrecht, 1990), p. 261
88. R.A. Castelli, P.D. Persans, W. Strohmayer, V. Parkinson, Corr. Sci. **49**, 4369 (2007)
89. M. Zhao, H. Huang, B.D. Cahan, I.T. Bae, S. Klippenstein, D.A. Scherson, Extended Abstracts of the 179th Meeting of the Electrochemical Society, Washington DC, USA, 05.05–10.05.1991, Ext. Abstr. #654
90. T.E. Furtak, D.W. Lynch, Phys. Rev. Lett. **35**, 960 (1975)
91. T.E. Furtak, D.W. Lynch, J. Electroanal. Chem. **79**, 1 (1977)
92. K.L. Shaklee, F.M. Pollak, M. Cardona, Phys. Rev. Lett. **15**, 883 (1965)
93. L.M. Abrantes, R. Peat, L.M. Peter, A. Hamnett, Ber. Bunsenges. Phys. Chem. **91**, 369 (1987)
94. I. Fromondi, D.A. Scherson, Electrochemical Society Spring Meeting 209th, Denver, Colorado, USA, May 7–11, 2006, Ext. Abstr. #1098
95. D.M. Kolb, J. Schneider, Surf. Sci. **162**, 764 (1985)
96. S.H. Liu, C. Hinnen, C. Nguyen Van Huong, N.R. de Tacconi, K.M. Ho, J. Electroanal. Chem. **176**, 325 (1984)
97. D.M. Kolb, J. Schneider, Electrochim. Acta **31**, 929 (1986)
98. B. Pettinger, U. Wenning, D.M. Kolb, Ber. Bunsenges. Phys. Chem. **82**, 1326 (1978)
99. B.G. McMillan, E.A. Berlouis, F.R. Cruickshank, P.F. Brevet, Electrochim. Acta **53**, 1157 (2007)

100. P.H. Schmidt, W.J. Plieth, J. Electroanal. Chem. **201**, 163 (1986)
101. F. Henglein, J. Lipkowski, D.M. Kolb, J. Electroanal. Chem. **303**, 245 (1991)
102. F. Henglein, D.M. Kolb, L. Stolberg, J. Lipkowski, Surf. Sci. **291**, 325 (1993)
103. T. Sagara, H. Kawamura, N. Nakashima, Langmuir **12**, 4253 (1996)
104. F. Bedioui, C. Bied-Charreton, C. Bongars, J. Devynck, C. Hinnen, J. Electroanal. Chem. **207**, 87 (1986)
105. B.G. McMillan, L.E.A. Berlouis, F.R. Cruickshank, P.F. Brevet, J. Electroanal. Chem. **599**, 177 (2007)
106. K. Nogami, H. Ueno, K. Yoshino, J. Appl. Phys. **64**, 6460 (1988)
107. W. Plieth, W. Kozlowski, T. Twomey, in *Adsorption of Molecules at Metal Electrodes*, ed. by J. Lipkowski, P.N. Ross (VCH, New York, 1992), p. 239
108. F. Scholz, B. Meyer, in *Electroanalytical Chemistry*, vol. 20, ed. by A.J. Bard (Dekker, New York, 1998), p. 1
109. U. Schröder, F. Scholz, J. Solid State Electrochem. **1**, 62 (1997)
110. Y. Borensztein, W.L. Mochan, J. Tarriba, R. Barrera, A. Tadjeddine, Phys. Rev. Lett. **71**, 2334 (1993)
111. A. Tadjeddine, D.M. Kolb, R. Kötz, Surf. Sci. **101**, 277 (1980)
112. J. Tarriba, W.L. Mochan, Phys. Rev. B **46**, 12902 (1992)
113. W.L. Mochan, R. Barrera, Y. Borensztein, A. Tadjeddine, Physica A **207**, 334 (1994)
114. O. Acher, B. Drevillon, Rev. Sci. Instr. **63**, 5332 (1992)
115. D.E. Aspnes, J.P. Harbison, A.A. Stunda, L.T. Florez, J. Vac. Sci. Technol. A **6**, 1327 (1988)
116. D.E. Aspnes, J.P. Harbison, A.A. Studna, L.T. Florez, Phys. Rev. Lett. **59**, 1687 (1987)
117. P. Harrison, T. Farrell, A. Maunder, C.I. Smith, P. Weightman, Meas. Sci. Technol. **12**, 2185 (2001)
118. V. Mazine, Y. Borensztein, L. Cagnon, P. Allongue, Phys. Status Solidi A **175**, 311 (1999)
119. B. Sheridan, D.S. Martin, J.R. Power, S.D. Barrett, C.I. Smith, P. Weightman, C.A. Lucas, R.J. Nichols, Phys. Rev. Lett. **85**, 4618 (2000)
120. A.J. Motheo, E.R. Gonzalez, G. Tremilliosi, A. Rakotondrainibe, J.M. Leger, B. Beden, C. Lamy, J. Brazil. Chem. Soc. **9**, 31 (1998)
121. C.I. Smith, A.J. Maunders, C.A. Lucas, R.J. Nichols, P. Weightman, J. Electrochem. Soc. **150**, E233 (2003)
122. C.I. Smith, N.J. Almond, P. Weightman, J. Electrochem. Soc. **154**, F90 (2007)
123. C.E. Vallet, S. Berns, J.J. Hendrickson, C.W. White, J. Electrochem. Soc. **135**, 387 (1988)
124. R.E. Malpas, A.J. Bard, Anal. Chem. **52**, 109 (1980)
125. C.E. Vallet, in *Spectroscopic and Diffraction Techniques in Interfacial Electrochemistry*, ed. by C. Gutiérrez, C. Melendres. NATO ASI Series C, vol. 320 (Kluwer Academic, Dordrecht, 1990), p. 133
126. G.H. Brilmyer, A.J. Bard, Anal. Chem. **52**, 685 (1980)
127. A.C. Boccara, D. Fournier, J. Badoz, Appl. Phys. Lett. **36**, 130 (1980)
128. J.D. Rudnick, F.R. McLarnon, E.J. Cairns, in *Techniques for Characterization of Electrodes and Electrochemical Processes*, ed. by R. Varma, J.R. Selman (Wiley, New York, 1991), p. 127
129. R.E. Russo, F.R. McLarnon, J.D. Spaer, E.J. Cairns, J. Electrochem. Soc. **134**, 2783 (1987)
130. B.S.H. Royce, D. Voss, A. Bocarsley, J. Phys. (Orsay, Fr.) **44**, C6-325 (1983)
131. C. Barbero, R. Kötz, O. Haas, Synth. Met. **101**, 170 (1999)

132. J.E. de Albuquerque, W.L.B. Melo, R.M. Faria, Rev. Sci. Instr. **74**, 306 (2003)
133. K. Nakanashi, N. Berova, R.W. Woody (eds.), *Circular Dichroism – Fundamentals and Applications* (VCH, New York, 1994)
134. for an introduction to the related linear dichroism effect (LD) see: J. Michl, E.W. Thulstrup, Acc. Chem. Res. **20** (1987) 192 and references therein
135. E. Charney, *The Molecular Basis of Optical Activity: Optical Rotatory Dispersion and Circular Dichroism* (Wiley, New York, 1979)
136. J. Bindl, P. Seitz, U. Seitz, E. Salbeck, J. Salbeck, J. Daub, Chem. Ber. **120**, 1747 (1987)
137. J. Salbeck, I. Aurbach, J. Daub, in *DECHEMA-Monographie*, vol. 112, ed. by J. Russow (VCH Verlagsgesellschaft, Weinheim, 1988), p. 177
138. J. Daub, J. Salbeck, I. Aurbach, Angew. Chem. **100**, 278 (1988)
139. X. Liu, K. Liu, G. Cheng, S. Dong, J. Electroanal. Chem. **510**, 103 (2001)
140. H. Goto, J. Electrochem. Soc. **154**, E63 (2007)
141. A. Abd-Elwahed, R. Holze, Curr. Top. Electrochem. **9**, 93 (2003)
142. R. Holze, J. Solid State Electrochem. **8**, 982 (2004)
143. C.M. Duff, G.A. Heath, Inorg. Chem. **30**, 2528 (1991)
144. J. Salbeck, J. Electroanal. Chem. **340**, 169 (1992)
145. L. Yu, J.-I. Lee, K.-W. Shin, C.-E. Park, R. Holze, J. Appl. Polym. Sci. **88**, 1550 (2003)
146. A. Abd-Elwahed, R. Holze, Elektrokhimiya **39**, 431 (2003)
147. W. Suĕtaka, in *Surface Infrared and Raman Spectroscopy*, ed. by C.J. Powell, A.W. Czanderna, D.M. Hercules, T.E. Madey, J.T. Yates Jr. Methods of Surface Characterization, vol. 3 (Plenum, New York, 1995)
148. A.M. Bradshaw, N.V. Richardson, Pure Appl. Chem. **68**, 457 (1996)
149. J.C. Lindon, G.E. Tranter, J.L. Holmes (eds.), *Encyclopedia of Spectroscopy and Spectrometry* (Academic Press, New York, 2000)
150. H. Günzler, H.M. Heise, *IR-Spektroskopie* (VCH, Weinheim, 1996)
151. H. Günzler, H.-U. Gremlich, *IR Spectroscopy* (VCH-Wiley, Weinheim, 2002)
152. R.A. Nyquist, C.L. Putzig, M.A. Leugers, *Infrared and Raman Spectral Atlas of Inorganic Compounds and Organic Salts: Text and Explanations* (Academic Press, San Diego, 1996)
153. I.R. Lewis, *Handbook of Raman Spectroscopy* (Dekker, New York, 2001)
154. W. Suëtaka, J.T. Yates, *Surface Infrared and Raman Spectroscopy – Methods and Applications* (Plenum, New York, 1995)
155. M. Pemble, in *Surface Analysis – The Principal Techniques*, ed. by J.C. Vickerman (Wiley, Chichester, 1997), p. 267
156. M.W. Urban, *Vibrational Spectroscopy of Molecules and Macromolecules on Surfaces* (Wiley, New York, 1993)
157. R.L. McCreery, in *Raman Spectroscopy for Chemical Analysis*, ed. by J.D. Winefordner. Chemical Analysis, vol. 157 (Wiley, New York, 2000)
158. H. Baumgärtel, H.-W. Jochims, B. Brutschy, Z. Phys. Chem. NF **154**, 1 (1987)
159. A. Bewick, K. Kunimatsu, B.S. Pons, J.W. Russell, J. Electroanal. Chem. **160**, 47 (1984)
160. P.R. Griffiths, *Chemical Infrared Fourier Transform Spectroscopy* (Wiley, New York, 1974)
161. B.C. Smith, *Fourier Transform Infrared Spectroscopy* (CRC, Boca Raton, 1996)
162. D.A. Long, *The Raman Effect* (Wiley, Chichester, 2002)
163. for an introductory overview including analytical applications see: R. Kellner, B. Mizaikoff, M. Jakusch, H.D. Wanzenböck, N. Weissenbacher, Appl. Spectrosc. **51**, 495 (1997)

164. see also: N. Weissenbacher, B. Lendl, J. Frank, H.D. Wanzenböck, B. Mizaikoff, R. Kellner, J. Mol. Struct. **410**, 539 (1997)
165. H. Ibach, *Electron Energy Loss Spectroscopy and Surface Vibrations* (Academic Press, New York, 1982)
166. M. Krejcik, M. Danek, F. Hartl, J. Electroanal. Chem. **317**, 179 (1991)
167. S.P. Best, Coord. Chem. Rev. **249**, 1536 (2005)
168. H.R. Sun, S.Y. Zhang, J.Q. Xu, G.Y. Yang, T.S. Shi, J. Electroanal. Chem. **455**, 57 (1998)
169. J.P. Bullock, D.C. Boyd, K.R. Mann, Inorg. Chem. **26**, 3084 (1987)
170. C.A. Blaine, J.E. Ellis, K.R. Mann, Inorg. Chem. **34**, 1552 (1995)
171. K.S. Yun, S. Joo, H.J. Kim, J. Kwak, E. Yoon, Electroanalysis **17**, 959 (2005)
172. C.M. Burba, R. Frech, 209th Electrochemical Society Spring Meeting, Denver, Colorado, USA, May 7–11, 2006, Ext. Abstr. #248
173. C.M. Burba, R. Frech, Electrochim. Acta **52**, 780 (2006)
174. G.J. Hansen, W.N. Hansen, Ber. Bunsenges. Phys. Chem. **91**, 317 (1987)
175. K.P. Pang, J.B. Benziger, M.P. Soriaga, A.T. Hubbard, J. Phys. Chem. **88**, 4583 (1984)
176. V.K.F. Chia, J.L. Stickney, M.P. Soriaga, S.D. Rosasco, G.N. Salaita, A.T. Hubbard, J.B. Benziger, K.W.P. Pang, J. Electroanal. Chem. **163**, 407 (1984)
177. O. Hofmann, K. Doblhofer, H. Gerischer, J. Electroanal. Chem. **161**, 337 (1984)
178. R.G. Greenler, J. Chem. Phys. **44**, 310 (1966)
179. W.G. Golden, D.D. Saperstein, M.W. Severson, J. Overend, J. Phys. Chem. **88**, 574 (1984)
180. J.B.L. Harkness, PhD. Dissertation. Massachusets Institute of Technology, Cambridge, 1970
181. J.I. Apse, PhD. Dissertation. Massachusets Institute of Technology, Cambridge, 1973
182. S.A. Francis, A.A. Ellison, J. Opt. Soc. Am. **49**, 131 (1959)
183. N.J. Harrick, *Internal Reflection Spectroscopy* (Interscience, London, 1967)
184. K. Molt, Fresenius Z. Anal. Chem. **319**, 743 (1984)
185. Y. Nishikawa, K. Fujiwara, K. Ataka, M. Osawa, Anal. Chem. **65**, 556 (1993)
186. J.-T. Li, Q.-S. Chen, S.-G. Sun, Electrochim. Acta **52**, 5725 (2007)
187. J.-T. Li, S.-R. Chen, X.-Y. Fan, L. Huang, S.-G. Sun, Langmuir **23**, 13174 (2007)
188. S. Schomaker, Dissertation. Universität Oldenburg, 1992
189. D.B. Parry, J.M. Harris, K. Ashley, Langmuir **6**, 209 (1990)
190. I.T. Bae, X. Xing, C.C. Liu, E. Yeager, J. Electroanal. Chem. **284**, 335 (1990)
191. M.G. Samant, K. Kunimatsu, H. Seki, Anal. Chem. **66**, 1781 (1994)
192. R.G. Greenler, J. Chem. Phys. **44**, 310 (1966)
193. W.R. Fawcett, A.A. Kloss, J.J. Calvente, N. Marinkovic, Electrochim. Acta **44**, 881 (1998)
194. J.D. Mozo, M. Dominguez, E. Roldan, J.M.R. Mellado, Electroanal. **12**, 767 (2000)
195. R.J. Nichols, in *Adsorption of Molecules at Metal Electrodes*, ed. by J. Lipkowski, P.N. Ross (VCH, New York, 1992), p. 347
196. B. Beden, C. Lamy, in *Spectroelectrochemistry*, ed. by R.J. Gale (Plenum, New York, 1988), p. 189
197. D.S. Bethune, A.C. Luntz, J.K. Sass, D.K. Roe, Surf. Sci. **197**, 44 (1988)
198. P. Liu, B. Jin, F. Cheng, J. Electroanal. Chem. **603**, 269 (2007)
199. S.J. Borg, S.P. Best, J. Electroanal. Chem. **535**, 57 (2002)
200. F. Joho, P. Novak, Electrochim. Acta **45**, 3589 (2000)
201. M. Liu, M. Ye, Q. Yang, Y. Zhang, Q. Xie, S. Yao, Electrochim. Acta **52**, 342 (2006)
202. R. Holze, Habilitationsschrift. Universität Oldenburg, 1988

203. G. Orozco, C. Gutierrez, J. Electroanal. Chem. **484**, 64 (2000)
204. B. Westerhoff, R. Holze, Ber. Bunsenges. Phys. Chem. **97**, 418 (1993)
205. B. Beden, in *Spectroscopic and Diffraction Techniques in Interfacial Electrochemistry*, ed. by C. Gutiérrez, C. Melendres. NATO ASI Series C, vol. 320 (Kluwer Academic, Dordrecht, 1990), p. 103
206. W.G. Golden, D.S. Dunn, J. Overend, J. Catal. **71**, 395 (1981)
207. F. Hoffman, Surf. Sci. Rep. **3**, 107 (1983)
208. J.B. Benziger, R.E. Preston, G.R. Schoofs, Appl. Opt. **26**, 343 (1987)
209. I. Chabay, G. Holzwarth, Appl. Opt. **14**, 454 (1975)
210. A. Bewick, S. Pons, in *Advances in Infrared and Raman Spectroscopy*, vol. 12, ed. by R.J.H. Clark, R.E. Hester (Wiley/Heyden, London, 1985)
211. H. Seki, K. Kunimatsu, W.G. Golden, Appl. Spectrosc. **39**, 437 (1985)
212. J.B. Benziger, R.E. Preston, G.R. Schoofs, Appl. Opt. **26**, 343 (1987)
213. W.G. Golden, D.D. Saperstein, M.W. Severson, J. Overend, J. Phys. Chem. **88**, 574 (1984)
214. W.G. Golden, D.D. Saperstein, J. Electron Spectrosc. Relat. Phenom. **30**, 43 (1983)
215. Opt. Commun. **8**, 26 (1973)
216. J. Gast, L. Genzel, U. Zwick, IEEE Trans. Microwave Theory Tech. **22**(12), 1026 (1974)
217. E.I. Sácz, R.M. Corn, Electrochim. Acta **38**, 1619 (1993)
218. T. Iwasita, F.C. Nart, in *Advances in Electrochemical Science and Engineering*, vol. 4, ed. by H. Gerischer, C.W. Tobias (VCH, Weinheim, 1995), p. 123
219. W. Tornquist, F. Guilleaume, G.L. Griffin, Langmuir **3**, 477 (1987)
220. R.M. Hammaker, S.A. Francis, R.P. Eischens, Spectrochim. Acta **21**, 1295 (1965)
221. A. Crossley, D.A. King, Surf. Sci. **68**, 528 (1977)
222. B.E. Hayden, A.M. Bradshaw, Surf. Sci. **125**, 787 (1983)
223. B. Alvarez, A. Rodes, J.M. Perez, J.M. Feliu, J. Phys. Chem. B **107**, 2018 (2003)
224. F. Hoffman, Surf. Sci. Rep. **3**, 107 (1983)
225. S.A. Wasileski, M.J. Weaver, Faraday Discuss. **121**, 285 (2002)
226. B. Westerhoff, PhD. thesis. Universität Oldenbrug, 1994
227. B. Westerhoff, R. Holze, Bull. Electrochem. **12**, 560 (1996)
228. M. Bron, R. Holze, J. Electroanal. Chem. **385**, 105 (1995)
229. M. Bron, R. Holze, Fresenius J. Anal. Chem. **361**, 694 (1998)
230. M. Bron, R. Holze, Electrochim. Acta **45**, 1121 (1999)
231. J. Poppe, PhD thesis. Technische Universität Chemnitz, 2002. http://archiv.tu-chemnitz.de/pub/2002/0002
232. R.J. Nichols, I. Burgess, K.L. Young, V. Zamlynny, J. Lipkwoski, J. Electroanal. Chem. **563**, 33 (2004)
233. M.A. Habib, J.O'M. Bockris, Langmuir **2**, 388 (1986)
234. P. Zelenay, M.A. Habib, J.O'M. Bockris, Langmuir **2**, 393 (1986)
235. G.J. Edens, X.P. Gao, M.J. Weaver, J. Electroanal. Chem. **375**, 357 (1994)
236. Y. Ikezawa, A. Yoshida, T. Ariga, Electrochim. Acta **51**, 3752 (2006)
237. G.L.J. Trettenhahn, G.E. Nauer, A. Neckel, Electrochim. Acta **41**, 1435 (1996)
238. T. Doneux, C. Buess-Herman, M.G. Hosseini, R.J. Nichols, J. Lipkowski, Electrochim. Acta **50**, 4275 (2005)
239. K. Marquez, R. Ortiz, J.W. Schultze, O.P. Marquez, J. Marquez, G. Staikov, Electrochim. Acta **48**, 711 (2003)
240. V. Rosca, M.T.M. Koper, Phys. Chem. Chem. Phys. **8**, 2513 (2006)
241. C.M. Johnson, C. Leygraf, J. Electrochem. Soc. **153**, B547 (2006)

242. Z.-Y. Zhou, N. Tian, Y.-J. Chen, S.-P. Chen, S.-G. Sun, J. Electroanal. Chem. **573**, 111 (2004)
243. P.W. Faguy, W.N. Richmond, R.S. Jackson, S.C. Weibel, G. Ball, J.H. Payer, Appl. Spectrosc. **52**, 557 (1998)
244. B.J. Barner, M.J. Green, E.I. Saez, R.M. Corn, Anal. Chem. **63**, 55 (1991)
245. M.J. Green, B.J. Barner, R.M. Corn, Rev. Sci. Instr. **62**, 1426 (1991)
246. W.N. Richmond, P.W. Faguy, S.C. Weibel, J. Electroanal. Chem. **448**, 237 (1998)
247. E.I. Saez, R.M. Corn, Electrochim. Acta **38**, 1619 (1993)
248. F. Guillaume, G.L. Griffin, Langmuir **5**, 783 (1989)
249. I. Zawisza, X. Bin, J. Lipkowski, Langmuir **23**, 5180 (2007)
250. R.E. Wittrig, C.P. Kubiak, J. Electroanal. Chem. **393**, 75 (1995)
251. I.S. Zavarine, C.P. Kubiak, J. Electroanal. Chem. **495**, 106 (2001)
252. S.P. Best, R.J.H. Clark, R.P. Cooney, R.C.S. McQueen, Rev. Sci. Instr. **58**, 2071 (1987)
253. I.S. Zavarine, T. Richmond, C.P. Kubiak, Inorg. Chem. Commun. **2**, 519 (1999)
254. T. Ito, T. Hamaguchi, H. Nagino, T. Yamaguchi, H. Kido, I.S. Zavarine, T. Richmond, J. Washington, C.P. Kubiak, J. Am. Chem. Soc. **121**, 4625 (1999)
255. I.S. Zavarine, C.P. Kubiak, T. Yamaguchi, K. Ota, T. Matsui, T. Ito, Inorg. Chem. **39**, 2696 (2000)
256. S.P. Best, R.J.H. Clark, R.C.S. McQueen, S. Joss, J. Am. Chem. Soc. **111**, 548 (1989)
257. J. Washington, C.P. Kubiak, Can. J. Chem. **74**, 2503 (1996)
258. R.J. Haines, R.E. Wittrig, C.P. Kubiak, Inorg. Chem. **33**, 4723 (1994)
259. J.P. Bullock, K.R. Mann, Inorg. Chem. **28**, 4006 (1989)
260. J.P. Bullock, M.C. Palazotto, K.R. Mann, Inorg. Chem. **30**, 1284 (1991)
261. J.P. Bullock, M.C. Palazzotto, K.R. Mann, Inorg. Chem. **29**, 4413 (1990)
262. S.C. Cheng, C.A. Blaine, M.G. Hill, K.R. Mann, Inorg. Chem. **35**, 7704 (1996)
263. N. Nanbu, Y. Sasaki, F. Kitamura, Electrochem. Commun. **5**, 383 (2003)
264. for an introduction and overview see e.g.: W. Knoll, Ann. Rev. Phys. Chem. **49**, 569 (1998)
265. Z. Ping, H. Neugebauer, A. Neckel, Electrochim. Acta **41**, 767 (1996)
266. M. Öhman, D. Persson, Electrochim. Acta **52**, 5159 (2007)
267. Y.X. Chen, M. Heinen, Z. Juysy, R.J. Behm, Poster at 3rd Gerischer-Symposium, Berlin, 06–08.07.2005
268. M. Heinen, Y.X. Chen, Z. Jusys, R.J. Behm, Electrochim. Acta **52**, 5634 (2007)
269. I.T. Bae, J. Electrochem. Soc. **153**, A2091 (2006)
270. I.T. Bae, M. Sandifer, Y.W. Lee, D.A. Tryk, C.N. Sukenik, D.A. Scherson, Anal. Chem. **67**, 4508 (1995)
271. M. Osawa, K. Yoshii, K. Ataka, T. Yotsuyanagi, Langmuir **10**, 640 (1994)
272. H. Visser, A.E. Curtright, J.K. McCusker, K. Sauer, Anal. Chem. **73**, 4374 (2001)
273. S.-M. Moon, C. Bock, B. MacDougall, J. Electroanal. Chem. **568**, 225 (2004)
274. P.J. Kulesza, M.A. Malik, A. Denca, J. Strojek, Anal. Chem. **68**, 2442 (1996)
275. A. Hartstein, J.R. Kirtley, J.C. Tsang, Phys. Rev. Lett. **45**, 201 (1980)
276. A. Hatta, Y. Chiba, W. Suetaka, Appl. Surf. Sci. **25**, 327 (1986)
277. M. Osawa, M. Ikeda, J. Phys. Chem. **95**, 9914 (1991)
278. M. Osawa, K. Ataka, K. Yoshii, T. Yotsuyanagi, J. Electron Spectrosc. Relat. Phenom. **64–65**, 371 (1993)
279. M. Osawa, A. Hatta, M. Kuramitsu, W. Suetaka, Surf. Sci. **175**, L787 (1986)
280. M. Osawa, K. Ataka, K. Yoshii, Y. Nishikawa, Appl. Spectrosc. **47**, 1497 (1993)
281. M. Osawa, W. Suetaka, Surf. Sci. **186**, 583 (1987)
282. M. Osawa, Bull. Chem. Soc. Jpn. **70**, 2861 (1997)

283. R.F. Aroca, D.J. Ross, Appl. Spectrosc. **58**, 324 (2004) A
284. M. Osawa, in *Handbook of Vibrational Spectroscopy*, ed. by J.M. Chalmers, P.R. Griffiths (Wiley, Chichester, 2002), p. 785
285. R. Aroca, S. Rodriguez-Llorente, J. Mol. Struct. **408/409**, 17 (1997)
286. R. Aroca, J. Mol. Struct. **292**, 17 (1993)
287. R. Aroca, *Surface-Enhanced Vibrational Spectroscopy* (Wiley, Chichester, 2006)
288. M. Osawa, Near-Field Opt. Surf. Plasmon Polaritons **81**, 163 (2001)
289. M. Osawa, K. Ataka, K. Yoshii, Y. Nishikawa, Appl. Spectrosc. **47**, 1497 (1993)
290. A. Hatta, Y. Chiba, W. Suetaka, Surf. Sci. **158**, 616 (1985)
291. S.G. Sun, W.B. Cai, L.J. Wan, M. Osawa, J. Phys. Chem. B **103**, 2460 (1999)
292. M. Osawa, K. Ataka, Surf. Sci. **262**, L118 (1992)
293. M. Loster, K.A. Friedrich, Surf. Sci. **523**, 287 (2003)
294. S.-G. Sun, in *Catalysis and Electrocatalysis at Nanoparticle Surfaces*, ed. by A. Wieckowski, E.R. Savinova, C.G. Vayenas (CRC, Boca Raton, 2003), p. 785
295. M. Osawa, K. Ataka, K. Yoshii, T. Yotsuyanagi, J. Electron. Spectrosc. Relat. Phenom. **64–65**, 371 (1993)
296. A. Pucci, Phys. Stat. Solid. B **242**, 2704 (2005)
297. H. Miyake, S. Ye, M. Osawa, Electrochem. Commun. **4**, 973 (2002)
298. J.M. Delgado, J.M. Orts, A. Rodes, Electrochim. Acta **52**, 4605 (2007)
299. A. Hatta, Y. Sasaki, W. Suetaka, J. Electroanal. Chem. **215**, 93 (1986)
300. K. Ataka, M. Osawa, Langmuir **14**, 951 (1998)
301. Th. Wandlowski, K. Ataka, S. Pronkin, D. Diesing, Electrochim. Acta **49**, 1233 (2004)
302. J.M. Delgado, J.M. Orts, A. Rodes, Langmuir **21**, 8809 (2005)
303. M. Osawa, K. Yoshii, K. Ataka, T. Yotsuyanagi, Langmuir **10**, 640 (1994)
304. M. Osawa, K. Yoshii, Appl. Spectrosc. **51**, 512 (1997)
305. M. Osawa, Denki Kagaku **64**, 957 (1996)
306. W. Uhmann, A. Becker, C. Taran, F. Siebert, Appl. Spectrosc. **45**, 390 (1991)
307. A. Yamakata, T. Uchida, J. Kubota, M. Osawa, J. Phys. Chem. B **110**, 6423 (2006)
308. A. Yamakata, J. Kubota, M. Osawa, 210th Electrochemical Society Meeting, Cancun, Mexico, 29.10–03.11.2006, Ext. Abstr. #1915
309. K. Ataka, M. Osawa, J. Electroanal. Chem. **460**, 188 (1999)
310. S.G. Sun, W.B. Cai, L.J. Wan, M. Osawa, J. Phys. Chem. B **103**, 2460 (1999)
311. W.-B. Cai, T. Amano, M. Osawa, J. Electroanal. Chem. **500**, 147 (2001)
312. L.J. Wan, M. Terashima, H. Noda, M. Osawa, J. Phys. Chem. B **104**, 3563 (2000)
313. A. Miki, S. Ye, T. Senzaki, M. Osawa, J. Electroanal. Chem. **563**, 23 (2004)
314. M.H. Shao, R.R. Adzic, Electrochim. Acta **50**, 2415 (2005)
315. Y.X. Chen, A. Miki, S. Ye, H. Sakai, M. Osawa, J. Am. Chem. Soc. **125**, 3680 (2003)
316. H. Shiroishi, Y. Ayato, K. Kunimatsu, T. Okada, J. Electroanal. Chem. **581**, 132 (2005)
317. M. Osawa, K. Yoshii, Y. Hibino, T. Nakano, I. Noda, J. Electroanal. Chem. **426**, 11 (1997)
318. A. Miki, S. Ye, M. Osawa, Chem. Commun. **2002**, 1500 (2002)
319. M. Osawa, K. Ito, W. Suetaka, T. Nakane, J. Electroanal. Chem. **270**, 459 (1989)
320. Q.-X. Li, X.-K. Xue, Q.-J. Xu, W.-B. Cai, Appl. Spectrosc. **61**, 1328 (2007)
321. Th. Wandlowski, K. Ataka, S. Pronkin, D. Diesing, Electrochim. Acta **49**, 1233 (2004)
322. K. Ataka, T. Yotsuyanagi, M. Osawa, J. Phys. Chem. **100**, 10664 (1996)
323. M. Futamata, J. Electroanal. Chem. **550-551**, 93 (2003)
324. K. Kunimatsu, H. Uchida, M. Osawa, M. Watanabe, J. Electroanal. Chem. **587**, 299 (2006)
325. Y. Ayato, K. Kunimatsu, M. Osawa, T. Okada, J. Electrochem. Soc. **153**, A203 (2006)

326. D.H. Murgida, P. Hildebrandt, Phys. Chem. Chem. Phys. **7**, 3773 (2005)
327. K. Ataka, J. Heberle, J. Am. Chem. Soc. **126**, 9445 (2004)
328. Q.B. Fan, C. Pu, K.L. Lay, E.S. Smotkin, J. Electrochem. Soc. **143**, L21 (1996)
329. Q.B. Fan, C. Pu, E.S. Smotkin, J. Electrochem. Soc. **143**, 3053 (1996)
330. Z.Y. Deng, J.D. Spear, J.D. Rudnicki, F.R. McLarnon, E.J. Cairns, J. Electrochem. Soc. **143**, 1514 (1996)
331. X. Lu, P.W. Faguy, M. Liu, J. Electrochem. Soc. **149**, A1293 (2002)
332. J.L.J. Daschbach, J.J. Smith, M.D. Morse, J. Electroanal. Chem. **209**, 387 (1986)
333. G.P. Williams, C.J. Hirschmugl, E.M. Kneedler, E.A. Sullivan, D.P. Siddons, Y.J. Chabal, F. Hoffman, K.D. Moeller, Rev. Sci. Instr. **60**, 2176 (1989)
334. C.J. Hirschmugl, Surf. Sci. **500**, 577 (2002)
335. Considerations of sensitivity, noise and other aspects pertaining to detectors in the far infrared have been reported in: J. Piotrowski, Photonics Spectra, May 2004, p. 80
336. C.A. Melendres, B. Beden, G.A. Bowmaker, J. Electroanal. Chem. **383**, 191 (1995)
337. C.A. Melendres, G.A. Bowmaker, J.M. Leger, B. Beden, Nucl. Instrum. Meth. Phys. Res. B **133**, 109 (1997)
338. C.A. Melendres, G.A. Bowmaker, J.M. Leger, B. Beden, J. Electroanal. Chem. **449**, 215 (1998)
339. C.A. Melendres, F. Hahn, G.A. Bowmaker, Electrochim. Acta **46**, 9 (2000)
340. G.A. Bowmaker, J.M. Leger, A. Lerille, C.A. Melendres, A. Tadjeddine, J. Chem. Soc. Faraday Trans. **94**, 1309 (1998)
341. C.A. Melendres, F. Hahn, J. Electroanal. Chem. **463**, 258 (1999)
342. C.A. Melendres, in *Spectroscopic and Diffraction Techniques in Interfacial Electrochemistry*, ed. by C. Gutiérrez, C. Melendres. NATO ASI Series C, vol. 320 (Kluwer Academic, Dordrecht, 1990), p. 181
343. P. Matousek, I.P. Clark, E.R.C. Draper, M.D. Morris, A.E. Goodship, N. Everall, M. Towrie, W.F. Finney, A.W. Parker, Appl. Spectrosc. **59**, 393 (2005)
344. N.A. Macleod, C. Eliasson, P. Matousek, Spectrosc. Eur. **19**(5), 7 (2007)
345. T. Nomoto, H. Onishi, Phys. Chem. Chem. Phys. **9**, 5515 (2007)
346. S. Simard, M.S. Odziemkowski, D.E. Irish, L. Brossard, H. Menard, J. Appl. Electrochem. **31**, 913 (2001)
347. C.T. Lee, M.S. Odziemkowski, D.W. Shoesmith, J. Electrochem. Soc. **153**, B33 (2006)
348. M. Odziemkowski, J. Flis, D.E. Irish, Electrochim. Acta **39**, 2225 (1994)
349. G.V.M. Jacintho, P. Corio, J.C. Rubim, J. Electroanal. Chem. **603**, 27 (2007)
350. L.J. Hardwick, M. Hahn, P. Ruch, M. Holzapfel, W. Scheifele, H. Buqa, F. Krumeich, P. Novak, R. Kötz, Electrochim. Acta **52**, 675 (2006)
351. P. Mayer, R. Holze, Surf. Sci. **522**, 55 (2003)
352. M. Fleischmann, P.J. Hendra, A.J. McQuillan, Chem. Phys. Lett. **26**, 163 (1974)
353. D.L. Jeanmaire, R.P. van Duyne, J. Electroanal. Chem. **84**, 1 (1977)
354. D.L. Jeanmaire, M.R. Suchanski, R.P. van Duyne, J. Am. Chem. Soc. **97**, 1699 (1975)
355. C.L. Haynes, A.D. MacFarland, R.P. van Duyne, Anal. Chem. **77**, 339A (2005)
356. H. Seki, J. Electron Spectrosc. Relat. Phenom. **30**, 287 (1983)
357. H. Seki, J. Electron Spectrosc. Relat. Phenom. **39**, 289 (1986)
358. M. Fleischmann, P. Graves, I. Hill, A. Oliver, J. Robinson, Electrochemical Society Spring Meeting, Cincinnati, USA, 6–11 May 1984, Ext. Abstr. #427
359. for overviews see: P. Mayer, R. Holze, Surf. Sci. **522**, 55 (2003)
360. for overviews see: S. Kania, R. Holze, Surf. Sci. **408**, 252 (1998)
361. M.J. Weaver, S. Farquharson, M.A. Tadayyoni, J. Chem. Phys. **82**, 4867 (1985)

362. R.K. Chang, T.E. Furtak (eds.), *Surface Enhanced Raman Spectroscopy* (Plenum, New York, 1982)
363. R. Aroca, G.J. Kovacs, in *Vibrational Spectra and Structure*, vol. 19, ed. by H.R. During (Elsevier, Amsterdam, 1991), p. 55
364. A. Otto, Coll. Surf. **38**, 27 (1989)
365. A. Otto, Appl. Surf. Sci. **6**, 309 (1980)
366. R.K. Chang, J. Phys. C **10**(12), 283 (1983)
367. D.A. Weitz, M. Moskovits, J.A. Creighton, in *Chemistry and Structure at Interfaces*, ed. by R.B. Hall, A.B. Ellis (VCH, Deerfield Beach, 1986), p. 197
368. A. Campion, P. Kambhampati, Chem. Soc. Rev. **27**, 241 (1998)
369. R.K. Chang, in *Spectroscopic and Diffraction Techniques in Interfacial Electrochemistry*, ed. by C. Gutiérrez, C. Melendres. NATO ASI Series C, vol. 320 (Kluwer Academic, Dordrecht, 1990), p. 155
370. D.A. Weitz, M. Moskovits, J.A. Creighton, in *Chemistry and Structure at Interfaces*, ed. by R.B. Hall, A.B. Ellis (VCH, Deerfield Beach, 1986), p. 197
371. R.L. Birke, T. Lu, J.R. Lombardi, in *Techniques for Characterization of Electrodes and Electrochemical Processes*, ed. by R. Varma, J.R. Selman (Wiley, New York, 1991), p. 211
372. R. Holze, Bull. Electrochem. **10**, 45 (1994)
373. R. Holze, *Current Topics in Electrochemistry*, vol. 2 (Council of Scientific Information, Trivandrum, 1993), p. 293
374. A.G. Brolo, D.E. Irish, B.D. Smith, J. Mol. Struct. **405**, 29 (1997)
375. E.S. Brandt, T.M. Cotton, in *Physical Methods of Chemistry Series*, vol. 9B, ed. by B.W. Rossiter, R.C. Baetzold (Wiley, New York, 1993), ch. 8
376. J.E. Pemberton, in *Electrochemical Interfaces: Modern Techniques for in-situ Interface Characterization*, ed. by H.D. Abruna (VCH, New York, 1991), p. 195
377. B. Pettinger, in *Adsorption of Molecules at Metal Electrodes*, ed. by J. Lipkowski, P.N. Ross (VCH, New York, 1992), p. 285
378. M. Moskovits, Rev. Mod. Phys. **57**, 783 (1985)
379. M.J. Weaver, S. Zou, in *Advances in Spectroscopy, Spectroscopy of Surfaces*, vol. 26, ed. by R.J.H. Clark, R.E. Hester (Wiley, Chichester, 1998), Chap. 5
380. S.P. Mulvaney, C.D. Keating, Anal. Chem. **72**, 145 (2000) R
381. V.A. Markel, V.M. Shalaev, P. Zhang, W. Huynh, L. Tay, T.L. Haslett, M. Moskovits, Phys. Rev. B **59**, 10903 (1999)
382. P. Zhang, S. Smith, G. Rumbles, M.E. Himmel, Langmuir **21**, 520 (2005)
383. P. Kambhampati, C.M. Child, M.C. Foster, A. Campion, J. Chem. Phys. **108**, 5013 (1998)
384. B. Pettinger, U. Wenning, H. Wetzel, Chem. Phys. Lett. **67**, 192 (1979)
385. T.E. Furtak, J. Electroanal. Chem. **150**, 375 (1983)
386. J. Gersten, A. Nitzan, J. Chem. Phys. **73**, 3023 (1980)
387. M.J. Weaver, S. Zou, H.Y.H. Chan, Anal. Chem. **72**, 39A (2000)
388. Q. Ye, J.X. Fang, L. Sun, J. Phys. Chem. B **101**, 8221 (1997)
389. G. Compagnini, C. Galati, S. Pignataro, Phys. Chem. Chem. Phys. **1**, 2351 (1999)
390. C.S. Allen, G.C. Schatz, R.P. van Duyne, Chem. Phys. Lett. **75**, 201 (1980)
391. C.J. Orendorff, L. Gearheart, N.R. Jana, C.J. Murphy, Phys. Chem. Chem. Phys. **8**, 165 (2006)
392. C.J. Addison, A.G. Brolo, Langmuir **22**, 8696 (2006)
393. B.G. McMillan, E.A. Berlouis, F.R. Cruickshank, P.F. Brevet, Electrochim. Acta **53**, 1157 (2007)

394. E. Vogel, W. Kiefer, V. Deckert, D. Zeisel, J. Raman Spectrosc. **29**, 693 (1998)
395. B. Bozzini, B. Busson, G.P. De Gaudenzi, L. D'Urzo, C. Mele, A. Tadjeddine, J. Electroanal. Chem. **602**, 61 (2007)
396. P. Kambhampati, C.M. Child, A. Campion, J. Chem. Soc. Faraday Trans. **92**, 4775 (1996)
397. A. Campion, P. Kambhampati, Chem. Soc. Rev. **27**, 241 (1998)
398. R.K. Chang, Ber. Bunsenges. Phys. Chem. **91**, 296 (1987)
399. J.F. Arenas, D.J. Fernandez, J. Soto, I. Lopez-Tocon, J.C. Otero, J. Phys. Chem. B **107**, 13143 (2003)
400. A.G. Brolo, D.E. Irish, B.D. Smith, J. Mol. Struct. **405**, 29 (1997)
401. P. Hildebrandt, S. Keller, A. Hoffmann, F. Vanheckke, B. Schrader, J. Raman Spectrosc. **24**, 791 (1993)
402. T. Sueoka, J. Inukai, M. Ito, Electron Spectrosc. Relat. Phenom. **64–65**, 363 (1993)
403. A.G. Brolo, D.E. Irish, J. Lipkowski, J. Phys. Chem. B **101**, 3906 (1997)
404. A.G. Brolo, D.E. Irish, G. Szymanski, J. Lipkowski, Langmuir **14**, 517 (1998)
405. M. Moskovits, D.P. DiLella, K.J. Maynard, Langmuir **4**, 67 (1988)
406. M. Moskovits, J. Chem. Phys. **77**, 4408 (1982)
407. M. Moskovits, J.S. Suh, J. Phys. Chem. **88**, 5526 (1984)
408. R.L. Birke, J.R. Lombardi, in *Spectroelectrochemistry*, ed. by R.J. Gale (Plenum, New York, 1988), p. 263
409. X.P. Gao, J.P. Davies, M.J. Weaver, J. Phys. Chem. **94**, 6858 (1990)
410. R. Holze, Bull. Electrochem. **10**, 45 (1994)
411. R.P. Cooney, T.P. Mernagh, J. Electroanal. Chem. **168**, 67 (1984)
412. B. Pettinger, G. Picardi, R. Schuster, G. Ertl, J. Electroanal. Chem. **554-555**, 293 (2003)
413. B. Pettinger, G. Picardi, R. Schuster, G. Ertl, Electrochemistry **68**, 942 (2000)
414. B. Pettinger, B. Ren, G. Picardi, R. Schuster, G. Ertl, Phys. Rev. Lett. **92**, 096101-1 (2004)
415. C. Viets, W. Hill, J. Mol. Struct. **563**, 163 (2001)
416. A. Vivoni, R.L. Birke, R. Foucault, J.R. Lombardi, J. Phys. Chem. B **107**, 5547 (2003)
417. D.K. Lambert, Sol. State Commun. **51**, 297 (1984)
418. D.K. Lambert, Phys. Rev. Lett. **50**, 2106 (1983)
419. D.K. Lambert, Electrochim. Acta **41**, 623 (1996)
420. S. Venkatesan, G. Erdheim, J.R. Lombardi, R.L. Birke, Surf. Sci. **101**, 387 (1980)
421. Y.X. Chen, S.Z. Zou, K.Q. Huang, Z.Q. Tian, J. Raman Spectrosc. **29**, 749 (1998)
422. B. Pettinger, U. Wenning, H. Wetzel, Chem. Phys. Lett. **67**, 192 (1979)
423. U. Wenning, B. Pettinger, H. Wetzel, Chem. Phys. Lett. **70**, 49 (1980)
424. H. Wetzel, B. Pettinger, U. Wenning, Chem. Phys. Lett. **75**, 173 (1980)
425. B. Pettinger, U. Wenning, H. Wetzel, Surf. Sci. **101**, 409 (1980)
426. E.S. Brandt, Anal. Chem. **57**, 1276 (1985)
427. G. Niaura, A.K. Gaigalas, V.L. Vilker, J. Raman Spectrosc. **28**, 1009 (1997)
428. J.W.M. Nissink, J.H. van der Maas, Appl. Spectrosc. **53**, 528 (1999)
429. P. Hendra, C. Jones, G. Warnes, *Fourier Transform Raman Spectroscopy* (Ellis Horwood, New York, 1991)
430. A. Crookell, M. Fleischmann, P.J. Hendra, M. Hanniet, Chem. Phys. Lett. **149**, 123 (1988)
431. R. Wen, Y. Fang, J. Electroanal. Chem. **576**, 237 (2005)
432. P. Gao, M.L. Patterson, M.A. Tadayyoni, M.J. Weaver, Langmuir **1**, 173 (1985)
433. R. Holze, Rev. Sci. Instr. **60**, 3348 (1989)
434. R. Holze, Surf. Sci. **202**, L612 (1988)

435. R.L. Garrell, K.D. Beer, W. Tanner, The 1987 Pittsburgh Conference, Atlantic City, NJ, USA, March 9–13, 1987, Ext. Abstr. #1025
436. K.D. Beer, W. Tanner, R.L. Garrell, J. Electroanal. Chem. **258**, 313 (1989)
437. see also for an example I.C.N. Diogenes, I.M.M. de Carvalho, E. Longhnotti, L.G.F. Lopes, M.L.A. Temperini, G.F.S. Andrade, I.S. Moreira, J. Electroanal. Chem. **605**, 1 (2007)
438. A.G. Brolo, D.E. Irish, J. Chem. Soc. Faraday Trans. **93**, 419 (1997)
439. Landolt-Börnstein, Numerical data and functional relationships in science and technology, in *New Series, Group IV: Physical Chemistry, vol. 9: Electrochemistry, Subvolume A: Electrochemical Thermodynamics and Kinetics*, ed. by W. Martienssen, M.D. Lechner (Springer, Berlin, 2007)
440. R.L. Sobocinski, J.E. Pemberton, Langmuir **6**, 43 (1990)
441. J.E. Pemberton, D.D. Tuschel, R.L. Sobocinski, A.L. Guy, N.A. Cross, Appl. Surf. Sci. **32**, 33 (1988)
442. D.D. Tuschel, J.E. Cook, J.E. Pemberton, Langmuir **2**, 380 (1986)
443. K.M. Balss, T.C. Kuo, P.W. Bohn, J. Phys. Chem. B **107**, 994 (2003)
444. S. Cintra, M.E. Abdelsalam, P.N. Bartlett, J.J. Baumberg, T.A. Kelf, Y. Sugawar, A.E. Russell, Faraday Discuss. **132**, 191 (2006)
445. M.E. Abdelsalam, P.N. Bartlett, J.J. Baumberg, S. Cintra, T.A. Kell, Electrochem. Commun. **7**, 740 (2005)
446. S. Mahajan, J.J. Baumberg, A.E. Russell, P.N. Bartlett, Phys. Chem. Chem. Phys. **9**, 6016 (2007)
447. M. Moskovits, Chem. Phys. Lett. **98**, 498 (1983)
448. S. Kania, R. Holze, in *GDCh-Monographie*, vol. 14, ed. by J. Russow, G. Sandstede, R. Staab (GDCh, Frankfurt, 1998), p. 115
449. P. Gao, M.J. Weaver, J. Phys. Chem. **90**, 4057 (1986)
450. M.L. Patterson, C.S. Allen, Appl. Surf. Sci. **18**, 377 (1984)
451. C. Vericat, M.E. Vela, R.C. Salvarezza, Phys. Chem. Chem. Phys. **7**, 3258 (2005) (for an overview focussed on surface analytical techniques applied to SAMs)
452. M. Bron, R. Holze, J. Solid State Electrochem. (submitted)
453. X.-W. Cao, J. Raman Spectrosc. **36**, 250 (2005)
454. Y. Maeda, H. Kitano, Kobunshi Ronbunshu **52**, 737 (1995)
455. P. Matejka, Chem. Listy **86**, 875 (1992)
456. P. He, H.T. Liu, Z.Y. Li, Y. Liu, X.D. Xu, J.H. Li, Langmuir **20**, 10260 (2004)
457. C. Siemes, A. Bruckbauer, A. Goussev, A. Otto, M. Sinther, A. Pucci, J. Raman Spectrosc. **32**, 231 (2001)
458. A.G. Brolo, D.E. Irish, J. Electroanal. Chem. **414**, 183 (1996)
459. S. Zou, M.J. Weaver, Anal. Chem. **70**, 2387 (1998)
460. S. Zou, C.T. Williams, E.K.Y. Chen, M.J. Weaver, J. Am. Chem. Soc. **120**, 3811 (1998)
461. M.J. Weaver, S. Zou, H.Y. Chan, Anal. Chem. **72**, 38 (2000) A
462. M. Fleischmann, Z.Q. Tian, J. Electroanal. Chem. **217**, 411 (1987)
463. M. Fleischmann, Z.Q. Tian, L.J. Li, J. Electroanal. Chem. **217**, 397 (1987)
464. B. Ren, X.F. Lin, Y.X. Jiang, P.G. Cao, Y. Xie, Q.J. Huang, Z.Q. Tian, Appl. Spectrosc. **57**, 419 (2003)
465. S. Zou, C.T. Williams, E.K.Y. Chen, M.J. Weaver, J. Phys. Chem. B **102**, 9039 (1998), 9743
466. P. Gao, M.J. Weaver, J. Phys. Chem. **89**, 5040 (1985)
467. R. Holze, Electroanal. **5**, 497 (1993)
468. R. Holze, Electrochim. Acta **36**, 1523 (1991)

469. S. Kania, R. Holze, Electrochim. Acta **48**, 945 (2003)
470. V. Schroeder, T.M. Devine, J. Electrochem. Soc. **146**, 4061 (1999)
471. Z.-Q. Tian, B. Ren, D.-Y. Wu, J. Phys. Chem. B **106**, 9463 (2002)
472. C. Zuo, P.W. Jagodzinski, J. Phys. Chem. B **109**(109) 1788
473. J.L. Yao, G.P. Pan, K.H. Xue, D.Y. Wu, B. Ren, D.M. Sun, J. Tang, X. Xu, Z.Q. Tian, Pure Appl. Chem. **72**, 221 (2000)
474. R. Holze, B. Bittins-Cattaneo, Electrochim. Acta **33**, 353 (1988)
475. J.-Z. Zheng, B. Ren, D.-Y. Wu, Z.-Q. Tian, J. Electroanal. Chem. **574**, 285 (2005)
476. P.G. Cao, R.A. Gu, B. Ren, Z.Q. Tian, Chem. Phys. Lett. **366**, 440 (2002)
477. H. Kwon, A.A. Gewirth, J. Electrochem. Soc. **154**, D577 (2007)
478. S. Migge, G. Sandmann, D. Rahner, H. Dietz, J. Solid State Electrochem. **9**, 132 (2005)
479. H. Luo, M.J. Weaver, J. Electroanal. Chem. **502**, 141 (2001)
480. S. Lecomte, H. Wackerbarth, T. Soulimane, G. Buse, P. Hildebrandt, J. Am. Chem. Soc. **120**, 7381 (1998)
481. H. Wackerbarth, U. Klar, W. Günther, P. Hildebrandt, Appl. Spectrosc. **53**, 283 (1999)
482. T. Itoh, K. Abe, K. Dokko, M. Mohamedi, I. Uchida, A. Kasuya, J. Electrochem. Soc. **151**, A2042 (2004)
483. E. Koglin, M.J. Schwuger, Faraday Discuss. **1992**, 213
484. A.G. Brolo, M. Odziemkowski, D.E. Irish, J. Raman Spectrosc. **29**, 713 (1998)
485. F.J. Vidal-Iglesias, J. Solla-Gullón, J.M. Pérez, A. Aldaz, Electrochem. Commun. **8**, 102 (2006)
486. P. Etchegoin, R.C. Maher, L.F. Cohen, H. Hartigan, R.J.C. Brown, M.J.T. Milton, J.C. Gallop, Chem. Phys. Lett. **375**, 84 (2003)
487. K. Kneipp, H. Kneipp, I. Itzkan, R.R. Dasari, M.S. Feld, Curr. Sci. **77**, 915 (1999)
488. K. Kneipp, H. Kneipp, I. Itzkan, R.R. Dasari, M.S. Feld, J. Phys.-Condens. Matter. **14**, R597 (2002)
489. I. Nabiev, I. Chourpa, M. Manfait, J. Raman Spectrosc. **25**, 13 (1994)
490. I. Nabiev, M. Manfait, Rev. Instr. Fr. Petrol. **48**, 261 (1993)
491. N. Matsuda, T. Sawaguchi, M. Osawa, I. Uchida, Chem. Lett. **1995**, 145
492. T. Fukuda, H. Takezoe, K. Ishikawa, A. Fukada, H.S. Woo, S.K. Jeong, E.J. Oh, J.S. Suh, Synth. Met. **69**, 175 (1995)
493. S.J. Cyvin, J.E. Rauch, J.C. Decius, J. Chem. Phys. **43**, 4083 (1965)
494. W.H. Li, X.Y. Li, N.T. Yu, Chem. Phys. Lett. **305**, 303 (1999)
495. S.M. Nie, L.A. Lipscomb, N.-T. Yu, Appl. Spectrosc. Rev. **26**, 203 (1991)
496. J.T. Golab, J.R. Sprague, K.T. Carron, G.C. Schatz, R.P. van Duyne, J. Chem. Phys. **88**, 7942 (1988)
497. S. Nie, PhD thesis. Northwestern University, Evanston, Illinois, 1989
498. W.H. Yang, J. Hulteen, G.C. Schatz, R.P. Van Duyne, J. Chem. Phys. **104**, 4313 (1996)
499. W.H. Li, X.Y. Li, N.T. Yu, Chem. Phys. Lett. **327**, 153 (2000)
500. W.H. Li, X.Y. Li, N.T. Yu, Chem. Phys. Lett. **312**, 28 (1999)
501. R. Holze, Z. Phys. Chem. **185**, 1 (1994)
502. R. Holze, Electrochim. Acta **36**, 999 (1991)
503. R. Holze, Electrochim. Acta **33**, 1619 (1988)
504. S. Popovici, W. Leyffer, R. Holze, J. Porphyrins Phthalocyanins **2**, 249 (1998)
505. S. Popovici, W. Leyffer, R. Holze, J. Porphyrins Phthalocyanins **3**, 265 (1999)
506. M. Stelter, PhD thesis. Universität Chemnitz, 2001. http://archiv.tu-chemnitz.de/pub/2001/0065/index.html
507. R. Holze, in *Handbook of Advanced Electronic and Photonic Materials and Devices*, vol. 8, ed. by H.S. Nalwa (Academic Press, San Diego, 2001), p. 209

508. R. Holze, in *Advanced Functional Molecules and Polymers*, vol. 2, ed. by H.S. Nalwa (Gordon & Breach, Amsterdam, 2001), p. 171
509. B. Tian, G. Zerbi, J. Chem. Phys. **92**, 3886 (1990)
510. B. Tian, G. Zerbi, J. Chem. Phys. **92**, 3892 (1990)
511. G. Zerbi, C. Castiglioni, M. Gussoni, Synth. Met. **41–43**, 3407 (1991)
512. G. Zerbi, C. Castiglioni, B. Tian, M. Gussoni, in *Springer Series in Solid State Sci. Electronic Properties of Conjugated Polymers III*, vol. 91, ed. by H. Kuzmany, M. Mehring, S. Roth (Springer, Berlin, 1989), p. 106
513. G. Zerbi, M. Veronelli, S. Martina, A.D. Schlüter, G. Wegner, J. Chem. Phys. **100**, 978 (1994)
514. G. Zerbi, M. Veronelli, S. Martina, A.-D. Schlüter, G. Wegner, Adv. Mater. **6**, 385 (1994)
515. G. Zerbi, B. Chierichetti, O. Inganäs, J. Chem. Phys. **94**, 4637 (1991)
516. B. Tian, G. Zerbi, in *Springer Series in Solid State Sci. Electronic Properties of Conjugated Polymers III*, vol. 91, ed. by H. Kuzmany, M. Mehring, S. Roth (Springer, Berlin, 1989), p. 113
517. T. Lindfors, A. Ivaska, J. Electroanal. Chem. **580**, 320 (2005)
518. M. Minsky, Scanning **10**, 128 (1988)
519. T. Wilson, C.J.R. Sheppard, *Scanning Optical Microscopy* (Academic Press, London, 1984)
520. M. Delhaye, J. Barbillat, J. Aubard, M. Bridoux, E. Da Silva, in *Raman Microscopy: Developments and Applications*, ed. by G. Rurrell, J. Corset (Academic Press, San Diego, 1996)
521. T.G. Corle, G.S. Kino, *Confocal Scanning Optical Microscopy and Related Techniques* (Academic Press, New York, 1996)
522. T. Vosgröne, A.J. Meixner, A. Anders, H. Dietz, G. Sandmann, W. Plieth, Surf. Sci. **597**, 102 (2005)
523. K. Kneipp, H. Kneipp, Appl. Spectrosc. **60**, 322A (2006)
524. Z. Wang, S. Pan, T.D. Krauss, H. Du, L.J. Rothberg, Proc. Natl. Acad. Sci. U.S.A. **100**, 8638 (2003)
525. J. Nan, Y. Yang, Z. Lin, Electrochim. Acta **51**, 4873 (2006)
526. W. Plieth, H. Dietz, A. Anders, G. Sandmann, A. Meixner, M. Weber, H. Kneppe, Surf. Sci. **597**, 119 (2005)
527. D.A. Smith, S. Webster, M. Ayad, S.D. Evans, D. Fogherty, D. Batchelder, Ultramicroscopy **61**, 1 (1995)
528. R. Zenobi, V. Deckert, Angew. Chem. Int. Ed. **39**, 1747 (2000)
529. V. Deckert, C. Budich, E. Bailo-Mur, GIT Labor-Fachzeitschrift **50**, 1014 (2006)
530. D. Zeisel, V. Deckert, R. Zenobi, T. Vo Dinh, Chem. Phys. Lett. **283**, 381 (1998)
531. V. Deckert, D. Zeisel, R. Zenobi, T. Vo-Dinh, Anal. Chem. **70**, 2646 (1998)
532. H. Baumgärtel, H.-W. Jochims, B. Brutschy, Z. Phys. Chem. NF **154**, 1 (1987)
533. E.M. Rowe, Phys. Today **5/1981**, 28 (1981)
534. C.J. Sparks Jr., Phys. Today **5/1981**, 40 (1981)
535. P. Gütlich, R. Link, A. Trautwein, *Mössbauer Spectroscopy and Transition Metal Chemistry (Inorganic Chemistry Concepts 3)* (Springer, Berlin, 1978)
536. D.A. Scherson, in *Spectroelectrochemistry*, ed. by R.J. Gale (Plenum, New York, 1988), p. 399
537. A.G. Maddock, *Mössbauer Spectroscopy: Principles and Applications* (Ellis Horwood, Chichester, 1997)
538. J. Blomquist, U. Helgeson, L.C. Moberg, Electrochim. Acta **27**, 1453 (1982)

539. M.E. Kordesch, J. Eldridge, D. Scherson, J. Electroanal. Chem. **164**, 393 (1984)
540. J.G. Gordon, J. Phys. **44**, C10-171 (1983)
541. A. Vertes, H. Leidheiser Jr., M.L. Varsanyi, G.W. Simmons, L. Kiss, J. Electrochem. Soc. **125**, 1946 (1978)
542. M.L. Varsanyi, J. Jaen, A. Vertes, L. Kiss, Electrochim. Acta **30**, 529 (1985)
543. K. Itaya, I. Uchida, V. Neff, Acc. Chem. Res. **19**, 162 (1986)
544. N.R. de Tacconi, K. Rajeshwar, R.O. Lezna, Chem. Mater. **15**, 3046 (2003)
545. K. Itaya, T. Ataka, S. Toshima, T. Shinohara, J. Phys. Chem. **86**, 2415 (1982)
546. H. Leidheiser Jr., J. Electrochem. Soc. **135**, 5C (1988)
547. C. Fierro, R.E. Carbonio, D. Scherson, E.B. Yeager, J. Phys. Chem. **91**, 6579 (1987)
548. W.E. O'Grady, J.O'M. Bockris, Chem. Phys. Lett. **5**, 116 (1970)
549. W.E. Grady, J.O'M. Bockris, Surf. Sci. **38**, 249 (1973)
550. W.E. O'Grady, J. Electrochem. Soc. **127**, 555 (1980)
551. D.A. Corrigan, R.S. Conell, C.A. Fierro, A. Scherson, J. Phys. Chem. **91**, 5009 (1987)
552. W. Wen, S. Ghosh, R.C. Urian, C. Heath, W. Reiff, S. Mukerjee, V. Naschitz, S. Licht, Extended Abstracts of the 203rd Meeting of the Electrochemical Society, Paris, France, 27.04–02.05.2003, Ext. Abstr. #67
553. D.A. Scherson, S.L. Gupta, C. Fierro, E.B. Yeager, M.E. Kordesch, J. Eldridge, R.W. Hoffman, J. Blue, Electrochim. Acta **28**, 1205 (1983)
554. D.A. Scherson, C. Fierro, D. Tryk, S.L. Gupta, E.B. Yeager, J. Eldridge, R.W. Hoffman, J. Electroanal. Chem. **184**, 419 (1985)
555. E. Yeager, D.A. Scherson, C.A. Fierro, ACS Symp. Ser. **288** (Catal. Charact. Ser.), 535 (1985)
556. R. Alcántara, G.F. Ortiz, P. Lavela, J.L. Tirado, Electrochem. Commun. **8**, 731 (2006)
557. C.M. Ionica-Bousquet, P.E. Lippens, L. Aldon, J. Olivier-Fourcade, J.C. Jumas, Chem. Mater. **18**, 6442 (2006)
558. B.K. Agarwal, *X-Ray Spectroscopy* (Springer, Berlin, 1991)
559. K. Tsuji, J. Injuk, R. Van Grieken (eds.), *X-Ray Spectrometry: Recent Technological Advances* (Wiley, Chichester, 2004)
560. Electrochim. Acta **47**(Special issue) (2002) 3035 ff
561. R. O'Malley, T. Rayment, A. Vollmer, H. Headspith, I. Harvey, S. Fiddy, Extended Abstracts of the 203rd Meeting of the Electrochemical Society, Paris, France, 27.04–02.05.2003, Ext. Abstr. #2835
562. W.-S. Yoon, C.P. Grey, M. Balasubramanian, X.-Q. Yang, J. McBreen, Chem. Mater. **15**, 3161 (2003)
563. W.-S. Yoon, M. Balasubramanian, X.-Q. Yang, Z. Fu, D.A. Fischer, J. McBreen, J. Electrochem. Soc. **151**, A246 (2004)
564. P.A. Lee, P.H. Citrin, P. Eisenberger, B.M. Kincaid, Rev. Mod. Phys. **53**, 769 (1981)
565. M. Bauer, H. Bertagnolli, Bunsenmagazin **9**, 216 (2007)
566. J.A. Horsley, F.W. Lytle, in *Chemistry and Structure at Interfaces*, ed. by R.B. Hall, A.B. Ellis (VCH, Deerfield Beach, 1986), p. 1
567. J. Haase, Chem. unserer Zeit **26**, 219 (1983)
568. G.M. Bommarito, D. Acevedo, J.F. Rodriguez, H.D. Abruna, J. Electroanal. Chem. **379**, 135 (1994)
569. H.-H. Strehblow, Extended Abstracts of the 203rd Meeting of the Electrochemical Society, Paris, France, 27.04–02.05.2003, Ext. Abstr. #2833
570. L.R. Sharpe, W.R. Heinemann, R.C. Elder, Chem. Rev. **90**, 705 (1990)
571. H.D. Abruna, J.H. White, M.J. Albarelli, G.M. Bommarito, M.J. Bedzyk, M. McMillan, J. Phys. Chem. **92**, 7045 (1988)

572. J. McBreen, W.E. O'Grady, G. Tourillon, E. Dartyge, A. Fontaine, J. Electroanal. Chem. **307**, 229 (1991)
573. C. Roth, N. Martz, T. Buhrmester, J. Scherer, H. Fuess, Phys. Chem. Chem. Phys. **4**, 3555 (2002)
574. D. Lützenkirchen-Hecht, R. Frahm, J. Phys. Chem. **105**, 9988 (2001)
575. H. Imai, K. Izumi, Y. Kubo, K. Kato, Y. Imai, Extended Abstracts of the 209th Meeting of the Electrochemical Society Spring Denver, Colorado, USA, 07.05–11.05.2006, Ext. Abstr. #302
576. A. Tadjeddine, G. Tourillon, D. Guay, Electrochim. Acta **36**, 1859 (1991)
577. T.E. Furtak, L. Wang, E.A. Creek, P. Samanta, T.M. Hayes, G. Kendall, W. Li, G. Liang, C.-M. Lo, Electrochim. Acta **36**, 1869 (1991)
578. M.F. Toney, J. McBreen, Interface **2**, 22 (1993)
579. L.B. Lurio, J. Pant, T.M. Hayes, L. Wang, T.E. Furtak, Physica B **209**, 413 (1995)
580. W.E. O'Grady, D.E. Ramaker, Electrochim. Acta **44**, 1283 (1998)
581. S. Mukerjee, J. McBreen, J. Electrochem. Soc. **146**, 600 (1999)
582. F. Lu, G.N. Salaita, H. Baltruschat, A.T. Hubbard, J. Electroanal. Chem. **222**, 305 (1987)
583. J. McBreen, S. Mukerjee, J. Electrochem. Soc. **142**, 3399 (1995)
584. H. Prinz, H. Strehblow, Electrochim. Acta **47**, 3093 (2002)
585. M. Teliska, V.S. Murthi, S. Mukerjee, D.E. Ramaker, J. Electrochem. Soc. **152**, A2159 (2005)
586. C. Roth, N. Benker, T. Buhrmester, M. Mazurek, M. Loster, H. Fuess, D.C. Koningsberger, D.E. Ramaker, J. Am. Chem. Soc. **127**, 14607 (2005)
587. I.T. Bae, W.-S. Yoon, 210. Electrochemical Society Meeting, Cancun, Mexico, 29.10–03.11.2006, Ext. Abstr. #697
588. A.E. Russell, A. Rose, Chem. Rev. **104**, 4613 (2004)
589. G.G. Long, J. Kruger, in *Techniques for Characterization of Electrodes and Electrochemical Processes*, ed. by R. Varma, J.R. Selman (Wiley, New York, 1991), p. 167
590. X.Q. Yang, K.H. Xue, H.S. Lee, Y.H. Guo, J. McBreen, T.A. Skotheim, Y. Okamoto, F. Lu, J. Electrochem. Soc. **140**, 943 (1993)
591. T.A. Skotheim, X.Q. Yang, K.H. Xue, H.S. Lee, J. McBreen, F. Lu, Electrochim. Acta **37**, 1635 (1992)
592. W.-S. Yoon, C.P. Grey, M. Balasubramanian, X.-Q. Yang, J. McBreen, Chem. Mater. **15**, 3161 (2003)
593. O. Haas, A. Deb, E.J. Cairns, A. Wokaun, J. Electrochem. Soc. **152**, A191 (2005)
594. R.G. Compton, A.M. Waller, in *Spectroelectrochemistry*, ed. by R.J. Gale (Plenum, New York, 1988), p. 349
595. T.M. McKinney, in *Electroanalytical Chemistry*, vol. 19, ed. by A.J. Bard (Dekker, New York, 1996), p. 109
596. R. Holze, Recent Res. Devel. Electrochem. **6**, 111 (2003)
597. A. Wieckowski, Extended Abstracts of the 203rd Meeting of the Electrochemical Society, Paris, France, 27.04–02.05.2003, Ext. Abstr. #1249
598. A. Heinzel, PhD Dissertation. Universität Oldenburg, 1985
599. A.H. Maki, D.H. Geske, Anal. Chem. **31**, 1450 (1959)
600. A.H. Maki, D.H. Geske, J. Chem. Phys. **30**, 1356 (1959)
601. A.H. Maki, D.H. Geske, J. Chem. Phys. **33**, 825 (1960)
602. D.H. Geske, A.H. Maki, J. Am. Chem. Soc. **82**, 2671 (1960)
603. D.H. Geske, A.H. Maki, J. Am. Chem. Soc. **83**, 1852 (1961)
604. W. Kaim, S. Ernst, V. Kasack, J. Am. Chem. Soc. **112**, 173 (1990)

605. M.A.M. Ameer, A.A. Ghoneam, J. Electrochem. Soc. **142**, 4082 (1995)
606. L.H. Piette, R. Ludwig, R.N. Adams, J. Am. Chem. Soc. **83**, 3909 (1961)
607. L.H. Piette, R. Ludwig, R.N. Adams, J. Am. Chem. Soc. **84**, 4212 (1962)
608. A.M. Bond, D.A. Fiedler, J. Electrochem. Soc. **144**, 1566 (1997)
609. K. Möbius, Z. Naturf. **20A**, 1093 (1965)
610. K. Möbius, Angew. Physik **17**, 534 (1964)
611. L.H. Piette, R. Ludwig, R.N. Adams, Anal. Chem. **34**, 916 (1962)
612. L.H. Piette, R. Ludwig, R.N. Adams, Anal. Chem. **34**, 1587 (1962)
613. F. Marken, R.D. Webster, S.D. Bull, S.G. Davies, J. Electroanal. Chem. **437**, 209 (1997)
614. C. Frontana, I. Gonzalez, J. Electroanal. Chem. **603**, 155 (2007)
615. R. Koopman, H. Gerischer, Ber. Bunsenges. Phys. Chem. **70**, 118 (1966)
616. R.N. Bagchi, A.M. Bond, F. Scholz, J. Electroanal. Chem. **252**, 259 (1988)
617. X.H. Mu, K.M. Kadish, Electroanalysis **2**, 15 (1990)
618. D.A. Fiedler, M. Koppenol, A.M. Bond, J. Electrochem. Soc. **142**, 862 (1995)
619. J.K. Dohrmann, K.J. Vetter, J. Electroanal. Chem. **20**, 23 (1969)
620. J.K. Dohrmann, Ber. Bunsenges. Phys. Chem. **74**, 575 (1970)
621. J.K. Dohrmann, F. Galluser, Ber. Bunsenges. Phys. Chem. **75**, 432 (1971)
622. J.K. Dohrmann, F. Galluser, H. Wittchen, Faraday Discuss. Chem. Soc. **56**, 350 (1973)
623. I.B. Goldberg, A.J. Bard, J. Phys. Chem. **75**, 3281 (1971)
624. R.D. Allendoerfer, G.A. Martincheck, S. Bruckenstein, Anal. Chem. **47**, 890 (1975)
625. R.D. Allendoerfer, J. Am. Chem. Soc. **97**, 218 (1975)
626. A. Heinzel, R. Holze, C.H. Hamann, J.K. Blum, Z. Phys. Chem. NF **160**, 11 (1989)
627. R. Holze, Habilitation thesis. Universität Oldenburg, 1989
628. A. Heinzel, R. Holze, C.H. Hamann, J.K. Blum, Electrochim. Acta **34**, 657 (1989)
629. C.H. Hamann, R. Holze, F. Köleli, DECHEMA-Monographie **121**, 297 (1990)
630. R. Holze, F. Köleli, C.H. Hamann, DECHEMA-Monographie **117**, 315 (1989)
631. R. Holze, C.H. Hamann, Tetrahedron **47**, 737 (1991)
632. J. Lippe, PhD Dissertation. Universität Oldenburg, 1991
633. J. Lippe, R. Holze, Synth. Met. **41–43**, 2927 (1991)
634. R. Holze, J. Lippe, Bull. Electrochem. **8**, 516 (1992)
635. F.J. Dyson, Phys. Rev. **98**, 349 (1955)
636. R.N. Bagchi, A.M. Bond, F. Scholz, Electroanalysis **1**, 1 (1989)
637. W.J. Albery, R.G. Compton, C.C. Jones, J. Am. Chem. Soc. **106**, 469 (1984)
638. A.J. Wain, M. Thompson, O.V. Klymenko, R.G. Compton, Phys. Chem. Chem. Phys. **6**, 4018 (2004)
639. I. Streeter, A.J. Wain, M. Thompson, R.G. Compton, J. Phys. Chem. **109**, 12636 (2005)
640. A.J. Wain, R.G. Compton, J. Electroanal. Chem. **587**, 203 (2006)
641. A.J. Wain, L. Drouin, R.G. Compton, J. Electroanal. Chem. **589**, 128 (2006)
642. L.J. Nunez-Vergara, P. Santander, P.A. Navarrete-Encina, J. Valenzuela, J.C. Sturm, J.A. Squella, J. Electrochem. Soc. **153**, E144 (2006)
643. A. Petr, L. Dunsch, A. Neudeck, J. Electroanal. Chem. **412**, 153 (1996)
644. J. Friedrich, M. Baumgarten, Appl. Magn. Reson. **13**, 393 (1997)
645. J. Wu, J.B. Day, K. Franaszczuk, B. Montez, E. Oldfield, A. Wieckowski, P.-A. Vuissoz, J.-P. Ansermet, J. Chem. Soc., Faraday Trans. **93**, 1017 (1997)
646. J.B. Day, J. Wu, E. Oldfield, A. Wieckowski, in *Electrochemical Nanotechnology*, ed. by W.J. Lorenz, W. Plieth (Wiley-VCH, Weinheim, 1998), p. 291
647. Y.Y. Tong, H.S. Kim, P.K. Babu, A. Wieckowski, J. Am. Chem. Soc. **124**, 468 (2002)
648. Y.Y. Tong, J.J. van der Klink, in *Catalysis and Electrocatalysis at Nanoparticle Surfaces*, ed. by A. Wieckowski, E.R. Savinova, C.G. Vayenas (CRC, Boca Raton, 2003), p. 455

649. P.K. Babu, Y.Y. Tong, H.S. Kim, A. Wieckowski, J. Electroanal. Chem. **524–525**, 157 (2001)
650. P. McGrath, A.M. Fojas, B. Rush, J.A. Reimer, E.J. Cairns, Electrochemical Society Spring Meeting 209th, Denver, Colorad, USA, May 7–11, 2006, Ext. Abstr. #1097
651. J.B. Wagner Jr., in *Techniques for Characterization of Electrodes and Electrochemical Processes*, ed. by R. Varma, J.R. Selman (Wiley, New York, 1991), p. 3
652. S.B. Piepho, P.N. Schatz, *Group Theory in Spectroscopy with Applications to MCD* (Wiley, New York, 1983)
653. for a representative study see: P.M.A. Gadsby, A.J. Thomson, J. Am. Chem. Soc. **112**, 5003 (1990)
654. N. Kobayashi, Y. Nishiyama, J. Phys. Chem. **89**, 1167 (1985)
655. M. Zhao, S. Kim, I.T. Bae, C. Rosenblatt, D.A. Scherson, J. Electroanal. Chem. **297**, 507 (1991)
656. M. Zhao, S. Kim, I.T. Bae, C. Rosenblatt, D.A. Scherson, Anal. Chem. **63**, 2990 (1991)
657. S.D. Bader, J. Magn. Magn. Mater. **100**, 440 (1991)
658. E.R. Moog, C. Liu, S.D. Bader, J. Zak, Phys. Rev. B **39**, 6949 (1989)
659. H.A. Wierenga, W. de Jong, M.W.J. Prins, Th. Rasing, R. Vollmer, A. Kirilyuk, H. Schwabe, J. Kirschner, Phys. Rev. Lett. **74**, 1462 (1995)
660. J.R. Hampton, J.-L. Martinez-Albertos, H.D. Abruna, Rev. Sci. Instrum. **73**, 3018 (2002)
661. W. Schindler, J. Kirschner, Rev. Sci. Instr. **67**, 3578 (1996)
662. K.M. Poduska, S. Morin, Rev. Sci. Instrum. **74**, 4723 (2003)
663. W. Schindler, in *Solid–Liquid Interfaces: Macroscopic Phenomena – Microscopic Understanding*, ed. by K. Wandelt, S. Thurgate. Topics in Applied Physics, vol. 85 (Springer, Heidelberg, 2003), p. 243
664. W. Schindler, J. Kirschner, Phys. Rev. B **55**, r1989 (1997)
665. W. Schindler, T. Koop, J. Kirschner, Z. Physik. Chem. **208**, 93 (1999)
666. J.R. Hampton, J.-L. Martinez-Albertos, H.D. Abruna, Langmuir **19**, 4309 (2003)
667. W. Schindler, T. Koop, J. Kirschner, Z. Physik. Chem. **208**, 93 (1999)
668. A. Gündel, L. Cagnon, A. Morrone, J. Schmidt, P. Allongue, Phys. Chem. Chem. Phys. **3**, 3330 (2001)
669. A. Gündel, A. Morrone, J. Schmidt, L. Cagnon, P. Allongue, J. Magn. Magn. Mater. **226**, 1616 (2001)
670. Y.P. Ma, J.P. Wiksow, M. Samuleviciene, K. Leinartras, E. Juzeliunas, J. Phys. Chem. B **106**, 12549 (2002)
671. J.G. Bellingham, M.L.A. MacVicar, M. Nisenoff, P.C. Searson, J. Electrochem. Soc. **133**, 1753 (1986)
672. E. Juzeliunas, J.H. Hinken, Electrochim. Acta **45**, 3452 (2000)
673. E. Juzeliunas, M. Samuleviciene, A. Sudavicius, J.H. Hinken, Electrochem. Solid State Lett. **3**, 24 (2000)
674. E. Juzeliunas, J.H. Hinken, J. Electroanal. Chem. **477**, 171 (1999)
675. C. Andrieu, F. Dalard, J.J. Rameau, F. Alcouffe, M. Reboul, J. Mater. Sci. **33**, 3177 (1998)
676. A. Abedi, J.J. Fellenstein, A.J. Lucas, J.P. Wikswo Jr., Rev. Sci. Instrum. **70**, 4640 (1999)
677. H. Richter, A. Knecht, Materialprüfung **39**, 390 (1997)
678. G. Skennerton, A. Abedi, R.G. Kelly, J.P. Wikswo Jr., J. Corros. Sci. Eng. **3** (2000) paper 3 (online journal)
679. D. Li, Y.P. Ma, W.F. Flanagan, B.D. Lichter, J.P. Wikswo Jr., Corrosion **53**, 93 (1997)

680. D. Li, Y.P. Ma, W.F. Flanagan, B.D. Lichter, J.P. Wikswo Jr., Corrosion **52**, 219 (1996)
681. D. Li, Y.P. Ma, W.F. Flanagan, B.D. Lichter, J.P. Wikswo Jr., J. Miner. Metals Mater. **47**, 36 (1995)
682. J.G. Bellingham, M.L.A. MacVicar, M. Nisenoff, IEEE Trans. Magn. **23**, 477 (1987)
683. A.D. Hibbs, R.E. Saeger, D.W. Cox, T.H. Aukerman, T.A. Sage, R.S. Landis, Rev. Sci. Instrum. **63**, 3652 (1992)
684. A.D. Hibbs, IEEE Trans. Magn. **139**, 2447 (1992)
685. D.E. Williams, A.R.J. Kucernak, R. Peat, Electrochim. Acta **38**, 57 (1993)
686. D.E. Williams, A.R.J. Kucernak, R. Peat, Electrochim. Acta **38**, 71 (1993)
687. R. Hutton, D.E. Williams, Anal. Chem. **67**, 280 (1995)
688. R.S. Hutton, D.E. Williams, Chem. Soc. Faraday Trans. **90**, 345 (1994)
689. R.B. Severeyn, R.J. Gale, in *Spectroelectrochemistry*, ed. by R.J. Gale (Plenum, New York, 1988), p. 41
690. A. Brodsky, Yu. Gurevich, Electrochim. Acta **13**, 1245 (1968)
691. A. Brodsky, Yu. Pleskov, in *Progress in Surface Science*, vol. 2, ed. by S. Davidson (Pergamon, Oxford, 1972), p. 2
692. A.M. Brodsky, Yu.V. Pleskov, Yu.Ya. Gurevich, Z.A. Rotenberg, *Modern Photoelectrochemistry. Photoemission Phenomena (Sovremennaja Fotoelektrochimija. Fotoemissionnye Yavleniya)* (Nauka, Moscow, 1974)
693. Y.Y. Gurevich, Y.V. Pleskov, Z.A. Rotenberg, *Photoelectrochemistry* (Consultants Bureau, New York, 1980)
694. B. Feuerbacher, B. Fitton, R.F. Willis (eds.), *Photoemission* (Wiley, Chichester, 1978)
695. Yu.V. Pleskov, Z.A. Rotenberg, J. Electroanal. Chem. **20**, 1 (1969)
696. L.M. Peter, A.J. Rudge, Extended Abstracts of the 179th Meeting of the Electrochemical Society, Washington DC, USA, 05.05–10.05.1991, Ext. Abstr. #675
697. G.C. Barker, A.W. Gardner, D.C. Sammon, J. Electrochem. Soc. **113**, 1182 (1966)
698. A.N. Frumkin, O.A. Petrii, B.B. Damaskin, in *Comprehensive Treatise of Electrochemistry*, vol. 1, ed. by J'O.M. Bockris, B.E. Conway, E. Yeager (Plenum, New York, 1980), p. 221
699. A. Frumkin, *Zero Charge Potentials* (Nauka, Moscow, 1979)
700. Z.A. Rotenberg, Yu.B. Pleskov, Elektrokhimiya **9**, 1419 (1973)
701. V.V. Eletskii, V. Pleskov, Elektrokhimiya **10**, 167 (1974)
702. A.M. Brodsky, Yu.Ya. Gurevich, S.V. Sheberstov, J. Electroanal. Chem. **20**, 353 (1971)
703. V.A. Benderski, A.S. Kotkin, A.G. Krivenko, Russ. Electrochem. Engl. Tr. **29**, 449 (1993)
704. V.S. Bagotzky, A.M. Skundin, Electrochim. Acta **30**, 899 (1985)
705. P.A. Kilmartin, G.A. Wright, Electrochim. Acta **41**, 1677 (1996)
706. for an introduction and overview see: P.A. Kilmartin, G.A. Wright, Electrochim. Acta **41**, 1677 (1996)
707. P.A. Kilmartin, G.A. Wright, Synth. Met. **104**, 145 (1999)
708. for an introductory overview see: A.J. Bard, L.R. Faulkner, *Electrochemical Methods* (Wiley, New York, 2001)
709. H. Gerischer, *Advances in Electrochemistry and Electrochemical Engineering*, vol. 1 (1961), p. 139
710. V.A. Myamlin, Yu.V. Pleskov, *Electrochemistry of Semiconductors* (Plenum, New York, 1967)
711. P.J. Holmes, *The Electrochemistry of Semiconductors* (Academic Press, New York, 1961)

712. R. Memming, in *Electroanalytical Chemistry*, vol. 11, ed. by A.J. Bard (Dekker, New York, 1979), p. 1
713. A.J. Nozik, Annu. Rev. Phys. Chem. **29**, 189 (1978)
714. S.A. Campbell, L.M. Peter, J.S. Buchanan, J. Power Sources **40**, 137 (1992)
715. G. Waldner, R. Gómez, M. Neumann-Spallart, Electrochim. Acta **52**, 2634 (2007)
716. M.W. Peterson, B.A. Parkinson, J. Electrochem. Soc. **135**, 1424 (1988)
717. Yu.V. Pleskov, in *Advances in Electrochemical Science and Engineering*, vol. 8, ed. by R.C. Alkire, D.M. Kolb (VCH, Weinheim, 2003), p. 209
718. Y.V. Pleskov, Russ. J. Electrochem. **38**, 1275 (2002)
719. Yu.V. Pleskov, A.Ya. Sakharova, M.D. Krotova, L.L. Bouilov, B.V. Spitsyn, J. Electroanal. Chem. **228**, 19 (1987)
720. R. Tenne, C. Levy-Clement, Isr. J. Chem. **38**, 57 (1998)
721. A.Y. Sakharova, Y.V. Pleskov, F. Diquarto, S. Piazza, C. Sunseri, I.G. Teremetskaya, V.P. Varnin, J. Electrochem. Soc. **142**, 2704 (1995)
722. Yu.V. Pleskov, Yu.Ya. Gurevich, *Semiconductor Photoelectrochemistry* (Consultants Bureau, New York, 1986)
723. L. Boonma, T. Yano, D.A. Tryk, K. Hashimoto, A. Fujishima, J. Electrochem. Soc. **144**, L142 (1997)
724. E.A. Ponomarev, L.M. Peter, J. Electroanal. Chem. **396**, 219 (1995)
725. L.M. Peter, E.A. Ponomarev, D.J. Fermin, J. Electroanal. Chem. **427**, 79 (1997)
726. E. Barsoukov, J.R. Macdonald (eds.), *Impedance Spectroscopy* (Wiley, Hoboken, 2005)
727. R. Peat, L.M. Peter, J. Electroanal. Chem. **228**, 351 (1987)
728. E.A. Ponomarev, L.M. Peter, J. Electroanal. Chem. **397**, 45 (1995)
729. L. Bay, K. West, Sol. Energ. Mater. Sol. Cells **87**, 613 (2005)
730. D.J. Blackwood, A. Borazio, R. Greef, L.M. Peter, J. Stumper, Electrochim. Acta **37**, 889 (1992)
731. J. Li, L.M. Peter, J. Electroanal. Chem. **199**, 1 (1986)
732. L.M. Peter, Chem. Rev. **90**, 753 (1990)
733. D.J. Fermin, E.A. Ponomarev, L.M. Peter, J. Electroanal. Chem. **473**, 192 (1999)
734. P.M. DiCarmine, O.A. Semenikhin, Electrochim. Acta **53**, 3744 (2008)
735. L.M. Abrantes, L.M. Peter, J. Electroanal. Chem. **150**, 593 (1983)
736. J.S. Buchanan, N.P. Freestone, L.M. Peter, J. Electroanal. Chem. **182**, 383 (1985)
737. L.M. Peter, J. Electroanal. Chem. **144**, 315 (1983)
738. J.S. Buchanan, L.M. Peter, Electrochim. Acta **33**, 127 (1988)
739. S.M. Wilhelm, N. Hackerman, J. Electrochem. Soc. **128**, 1668 (1981)
740. J.F. McAleer, L.M. Peter, Faraday Disc. Chem. Soc. **70**, 67 (1980)
741. J. Nan, Y. Yang, Z. Lin, Electrochim. Acta **51**, 4873 (2006)
742. M.W. Peterson, B.A. Parkinson, J. Electrochem. Soc. **135**, 1424 (1988)
743. L.M. Peter, J. Electroanal. Chem. **98**, 49 (1979)
744. L.M. Peter, Electrochim. Acta **23**, 1073 (1978)
745. D.J. Fermin, E.A. Ponomarev, L.M. Peter, J. Electroanal. Chem. **473**, 192 (1999)
746. T.B. Cao, L.H. Wei, S.M. Yang, M.F. Zhang, C.H. Huang, W.X. Cao, Langmuir **18**, 750 (2002)
747. K. Nonomura, T. Loewenstein, E. Michaelis, D. Wöhrle, T. Yoshida, H. Minoura, D. Schlettwein, Phys. Chem. Chem. Phys. **8**, 3867 (2006)
748. H. Flaisher, R. Tenne, M. Halmann, J. Electroanal. Chem. **402**, 97 (1996)
749. Y. Yang, Z.G. Lin, Synth. Met. **64**, 43 (1994)
750. Y. Byun, B.W. Wessels, 173rd Meeting of the Electrochemical Society, Atlanta, GA, USA, May 15–20, 1988, Ext. Abstr. #323

751. Y. Byun, B.W. Wessels, J. Electrochem. Soc. **135**, 144c (1988)
752. Yu.V. Pleskov, V.P. Varnin, I.G. Teremetskaya, A.V. Churikov, J. Electrochem. Soc. **144**, 175 (1997)
753. J. Schefold, J. Electroanal. Chem. **362**, 97 (1993)
754. J. Schefold, J. Electrochem. Soc. **139**, 2862 (1992)
755. Y. Byun, B.W. Wessel, Appl. Phys. Lett. **52**, 1352 (1988)
756. Z.H. Yang, P. Zhang, D.J. Wang, D. Zhang, X.D. Chai, T.J. Li, Synth. Met. **85**, 1293 (1997)
757. P. Zhang, Z.H. Yang, D.J. Wang, S.H. Kan, X.D. Chai, J.Z. Liu, T.J. Li, Synth. Met. **84**, 165 (1997)
758. R. Kern, R. Sastrawan, J. Ferber, R. Stangl, J. Luther, Electrochim. Acta **47**, 4213 (2002)
759. E.M.M. Sutter, I. Gérard, A. Etcheberry, J. Electrochem. Soc. **146**, 1448 (1999)
760. A. Baba, W. Knoll, J. Phys. Chem. B **107**, 7733 (2003)
761. G.I. Pennarun, C. Boxall, D. O'Hare, Analyst **121**, 1779 (1996)
762. F.P.L. Andrieux, C. Boxall, D. O'Hare, J. Electroanal. Chem. **589**, 177 (2006)
763. A. Tadjeddine, A. Peremans, in *Spectroscopy for Surface Science*, ed. by R.J.H. Clark, R.E. Hester. Advances in Spectroscopy, vol. 26 (Wiley, Chichester, 1998), p. 159
764. C.T. Williams, D.A. Beattie, Surf. Sci. **500**, 545 (2002)
765. G.L. Richmond, J.M. Robinson, V.L. Shannon, Progr. Surf. Sci. **28**, 1 (1988)
766. G.L. Richmond, in *Electroanalytical Chemistry*, vol. 17, ed. by A.J. Bard (Dekker, New York, 1991), p. 87
767. R.M. Corn, M. Romagnoli, M.D. Levenson, M.R. Philpott, J. Chem. Phys. **81**, 4127 (1984)
768. R. Georgiadis, G.L. Richmond, J. Phys. Chem. **95**, 2895 (1991)
769. I. Fromondi, D.A. Scherson, Electrochemical Society Spring Meeting 209th, Denver, Colorado, USA, May 7–11, 2006, Ext. Abstr. #1098
770. D.J. Campbell, R.M. Corn, J. Phys. Chem. **92**, 5796 (1988)
771. G.L. Richmond, in *Electroanalytical Chemistry*, vol. 17, ed. by A.J. Bard (Dekker, New York, 1991), p. 87
772. G. Richmond, in *Advances in Electrochemical Science and Engineering*, vol. 2, ed. by H. Gerischer, C.W. Tobias (VCH, Weinheim), p. 142
773. D.A. Koos, V.L. Shannon, G.L. Richmond, J. Phys. Chem. **94**, 2091 (1990)
774. R.M. Corn, in *Adsorption of Molecules at Metal Electrodes*, ed. by J. Lipkowski, P.N. Ross (VCH, New York, 1992), p. 391
775. A. Friedrich, B. Pettinger, D.M. Lolb, G. Lupke, R. Steinhoff, G. Marowsky, Chem. Phys. Lett. **163**, 123 (1989)
776. see also: B. Pettinger, C. Bilgcr, J. Lipkowski, in *Topics Appl. Phys.*, vol. 85, ed. by K. Wandelt, S. Thurgate (Springer, Berlin, 2003), p. 223
777. R. Srinivasan, I.I. Suni, J. Electrochem. Soc. **146**, 570 (1999)
778. S. Lakkaraju, M.J. Bennahmias, G.L. Borges, J.G. Gordon, M. Lazaga, B.M. Stone, K. Ashley, Appl. Optics **29**, 4943 (1990)
779. D.A. Higgins, R.M. Corn, J. Phys. Chem. **97**, 489 (1993)
780. A. Tadjeddine, A. Peremans, P. Guyot-Sionnest, Surf. Sci. **335**, 210 (1995)
781. A. Tadjeddine, A. Le Rille, Electrochim. Acta **45**, 601 (1999)
782. A. Tadjeddine, Surf. Rev. Lett. **7**, 423 (2000)
783. G. Rupprechter, Phys. Chem. Chem. Phys. **3**, 4621 (2001)
784. Y.R. Shen, Nature **337**, 519 (1989)

785. A. Le Rille, A. Tadjeddine, W.Q. Zheng, A. Peremans, Chem. Phys. Lett. **271**, 95 (1997)
786. M. Tadjeddine, J.P. Flament, A. Tadjeddine, J. Electroanal. Chem. **408**, 237 (1996)
787. G.L. Richmond, Chem. Rev. **102**, 2693 (2002)
788. G.L. Richmond, Annu. Rev. Phys. Chem. **52**, 357 (2001)
789. K.B. Eisenthal, Chem. Rev. **96**, 1343 (1996)
790. Z.D. Schultz, S.K. Shaw, A.A. Gewirth, J. Am. Chem. Soc. **127**, 15916 (2005)
791. P. Guyot-Sionnest, A. Tadjeddine, Chem. Phys. Lett. **172**, 341 (1990)
792. D.K. Hore, J.L. King, F.G. Moore, D.S. Alavi, M.Y. Hamamoto, G.L. Richmond, Appl. Spectrosc. **58**, 1377 (2004)
793. Z.D. Schultz, S.K. Shaw, A.A. Gewirth, J. Am. Chem. Soc. **127**, 15916 (2005)
794. F. Vidal, B. Busson, A. Tadjeddine, Electrochim. Acta **49**, 3637 (2004)
795. A. Peremans, A. Tadjeddine, J. Chem. Phys. **103**, 7197 (1995)
796. A. Tadjeddine, A. Peremans, J. Electroanal. Chem. **409**, 115 (1996)
797. B. Bozzini, C. Mele, A. Fanigliulo, B. Busson, F. Vidal, A. Tadjeddine, J. Electroanal. Chem. **574**, 85 (2004)
798. C. Matranga, P. Guyot-Sionnest, J. Chem. Phys. **112**, 7615 (2000)
799. K. Uosaki, H. Noguchi, M. Ito, T. Okada, 210th Electrochemical Society Meeting, Cancun, Mexico, 29.10–03.11.2006, Ext. Abstr. #1910
800. C. Rambaud, L. Cagnon, J.-P. Levy, G. Tourillon, J. Electrochem. Soc. **151**, E352 (2004)
801. A. Peremans, A. Tadjeddine, P. Guyot-Sionnest, Chem. Phys. Lett. **247**, 243 (1995)
802. C. Magnus Johnsson, E. Tyrode, C. Leygraf, J. Electrochem. Soc. **153**, B113 (2006)
803. G.Q. Lu, A. Lagutchev, D.D. Dlott, A. Wieckowski, Surf. Sci. **585**, 3 (2005)
804. A. Wieckowski, G.-Q. Lu, A. Lagoutchev, D. Dlott, 209th Electrochemical Society Spring Meeting, Denver, Colorado, USA, 07–11.05.2006, Ext. Abstr. #1070
805. B. Bozzini, B. Busson, G.P. De Gaudenzi, L. D'Urzo, C. Mele, A. Tadjeddine, J. Electroanal. Chem. **602**, 61 (2007)
806. M. Tadjeddine, J.P. Flament, Chem. Phys. **240**, 39 (1999)
807. S. Baldelli, Electrochemical Society Spring Meeting 209th, Denver, Colorado, USA, May 7–11, 2006; Ext. Abstr. #1041
808. S. Bruckenstein, R.R. Gadde, J. Am. Chem. Soc. **93**, 793 (1971)
809. S. Bruckenstein, R.R. Gadde, J. Electroanal. Chem. **50**, 163 (1974)
810. T.J. Brockman, L.B. Anderson, Anal. Chem. **56**, 207 (1984)
811. O. Wolter, J. Heitbaum, Ber. Bunsenges. Phys. Chem. **88**, 2 (1984)
812. O. Wolter, J. Heitbaum, Ber. Bunsenges. Phys. Chem. **88**, 6 (1984)
813. B. Bittins-Cattaneo, E. Cattaneo, P. Königshoven, W. Vielstich, in *Electroanalytical Chemistry*, vol. 17, ed. by A.J. Bard (Dekker, New York, 1991), p. 181
814. M. Casciola, R. Clemetini, E. Skou, J. Munk, J. Mater. Chem. **5**, 991 (1995)
815. E. Skou, J. Munk, J. Electroanal. Chem. **367**, 93 (1994)
816. R. Stadler, Z. Jusys, H. Baltruschat, Electrochim. Acta **47**, 4485 (2002)
817. H. Baltruschat, in *Interfacial Electrochemistry*, ed. by A. Wieckowski (Dekker, New York, 1999), p. 577
818. D. Tegtmeier, A. Heindrichs, J. Heitbaum, Ber. Bunsenges. Phys. Chem. **93**, 201 (1989)
819. S. Wasmus, E. Cattaneo, W. Vielstich, Electrochim. Acta **35**, 771 (1990)
820. M. Fujihira, T. Noguchi, J. Electroanal. Chem. **347**, 457 (1993)
821. S. Wasmus, S.R. Samms, R.F. Savinell, J. Electrochem. Soc. **142**, 1183 (1995)
822. Z. Jusys, H. Massong, H. Baltruschat, J. Electrochem. Soc. **146**, 1093 (1999)

823. V. Rao, C. Cremers, U. Stimming, L. Cao, S. Sun, S. Yan, G. Sun, Q. Xin, J. Electrochem. Soc. **154**, B1138 (2007)
824. P. Friebe, P. Bogdanoff, N. Alonso-Vante, H. Tributsch, J. Catal. **168**, 374 (1997)
825. P. Dubé, G.M. Brisard, J. Electroanal. Chem. **582**, 230 (2005)
826. M. Winter, R. Imhof, F. Joho, P. Novak, J. Power Surces **82**, 818 (1999)
827. P. Novak, F. Joho, R. Imhof, J.C. Panitz, O. Haas, J. Power Sources **82**, 212 (1999)
828. Z. Jusys, A. Vaskelis, Ber. Bunsenges. Phys. Chem. **101**, 1865 (1997)
829. V. Rosca, G.L. Beltramo, M.T.M. Koper, J. Electroanal. Chem. **566**, 53 (2004)
830. A.D. Modestov, J. Gun, I. Savotine, O. Lev, J. Electroanal. Chem. **565**, 7 (2004)
831. V. Rao, C. Cremers, U. Stimming, 209th Electrochemical Society Spring Meeting, Denver, Colorado, USA, May 7–11, 2006, Ext. Abstr. #1102
832. F.J. Vidal-Iglesias, J. Solla-Gullón, J.M. Feliu, H. Baltruschat, A. Aldaz, J. Electroanal. Chem. **588**, 331 (2006)
833. B. Neumann, P. Bogdanoff, H. Tributsch, S. Sakthivel, H. Kisch, J. Phys. Chem. B **109**, 16579 (2005)
834. K. Jambunathan, A.C. Hillier, J. Electrochem. Soc. **150**, E312 (2003)
835. K. Jambunathan, S. Jayaraman, A.C. Hillier, Langmuir **20**, 1856 (2004)
836. H. Budzikiewicz, *Massenspektrometrie* (VCH, Weinheim, 1992)
837. E. Schröder, *Massenspektrometrie – Begriffe und Definitionen* (Springer, Berlin, 1991)
838. J.R. Pretty, G.J. van Berkel, Rapid Commun. Mass. Spectrom. **12**, 1644 (1998)
839. V. Kertesz, G.J. van Berkel, J. Solid State Electrochem. **9**, 390 (2005)
840. F.M. Zhou, G.J. van Berkel, Anal. Chem. **67**, 3643 (1995)
841. C. Zettersten, R. Lomoth, L. Hammarström, P.J.R. Sjöberg, L. Nyholm, J. Electroanal. Chem. **590**, 90 (2006)
842. G.J. van Berkel, F.M. Zhou, J. Am. Soc. Mass Spectrom. **7**, 157 (1996)
843. T. Guo, A. Illies, V. Cammarata, M. Arndt, W. Sonzogniy, J. Electroanal. Chem. **610**, 102 (2007)
844. V. Kertesz, N.M. Dunn, G.J. Van Berkel, Electrochim. Acta **47**, 1035 (2002)
845. V. Gutkin, J. Gun, P.V. Prikhodchenko, O. Lev, L. Gonsalvi, M. Peruzzini, A. Romerosa, T. Campos Malpartida, C. Lidrissic, J. Electrochem. Soc. **154**, F7 (2007)
846. A.R. Dolan, E.P. Maziarz, T.D. Wood, Eur. J. Mass Spectrom. **6**, 241 (2000)
847. Z.H. Ji, J.J. Li, G. Yuan, S.M. Cai, J. Liu, C. Liu, J. Electroanal. Chem. **570**, 265 (2004)
848. H.T. Deng, O.J. van Berkel, H. Takano, D. Gazda, M.D. Porter, Anal. Chem. **72**, 2641 (2000)
849. C.R. Blakley, M.L. Vestal, Anal. Chem. **55**, 750 (1983)
850. M.L. Vestal, Int. J. Mass Spectrom. Ion Phys. **46**, 193 (1983)
851. G. Hambitzer, J. Heitbaum, Anal. Chem. **58**, 1067 (1986)
852. G. Hambitzer, P.P. Heinz, I. Stassen, J. Heitbaum, Synth. Met. **55**, 1317 (1993)
853. G. Hambitzer, I. Stassen, Synth. Met. **55**, 1045 (1993)
854. I. Stassen, G. Hambitzer, J. Electroanal. Chem. **440**, 219 (1997)
855. G. Hambitzer, J. Heitbaum, I. Stassen, J. Electroanal. Chem. **447**, 117 (1998)
856. K.J. Volk, R.A. Yost, A. Brajter-Toth, Anal. Chem. **61**, 1709 (1989)
857. J.R. Pretty, E.H. Evans, E.A. Blubaugh, W.-L. Shen, J.A. Caruso, T.M. Davidson, J. Anal. Atom. Spetctrom. **5**, 437 (1990)
858. J.R. Pretty, E.A. Blubaugh, J.A. Caruso, Anal. Chem. **65**, 3396 (1993)
859. J.R. Pretty, E.A. Blubaugh, J.A. Caruso, T.M. Davidson, Anal. Chem. **66**, 1540 (1994)
860. J.R. Pretty, E.A. Blubaugh, E.H. Evans, J.A. Caruso, T.M. Davidson, J. Anal. Atom Spectrom. **7**, 1131 (1992)
861. J.J. Gross, *Mass Spectrometry* (Springer, Heidelberg, 2004)

862. H.W. Buschmann, S. Wilhelm, W. Vielstich, Electrochim. Acta **31**, 939 (1986)
863. R.A. De Souza, M. Martin, Bunsen-Magazin **8**, 109 (2006) (see for an overview)
864. W.N. Hansen, D.M. Kolb, J. Electroanal. Chem. **100**, 493 (1979)
865. S. Wilhelm, Dissertation. Universität Bonn, 1988
866. S. Wilhelm, W. Vielstich, H.W. Buschmann, T. Iwasita, J. Electroanal. Chem. **229**, 377 (1987)
867. C. Barbero, M.C. Miras, R. Kötz, Electrochim. Acta **37**, 429 (1992)
868. J.D. Rudnicki, F.R. McLarnon, E.J. Cairns, in *Techniques for Characterization of Electrodes and Electrochemical Processes*, ed. by R. Varma, J.R. Selman (Wiley, New York, 1991), p. 127
869. R.E. Russo, F.R. McLarnon, J.D. Spaer, E.J. Cairns, J. Electrochem. Soc. **134**, 2783 (1987)
870. M.F. Mathias, J. Electroanal. Chem. **407**, 115 (1996)
871. C.A. Barbero, Phys. Chem. Chem. Phys. **7**, 1885 (2005)
872. K.S.V. Santhanam, R.N. O'Brien, J. Electroanal. Chem. **160**, 377 (1984)
873. K.S.V. Santhanam, R.N. O'Brien, in *Techniques for Characterization of Electrodes and Electrochemical Processes*, ed. by R. Varma, J.R. Selman (Wiley, New York, 1991), p. 401
874. M.A. Tamor, M. Zanini, J. Electrochem. Soc. **133**, 1399 (1986)
875. R. Kötz, C. Barbero, O. Haas, Ber. Bunsenges. Phys. Chem. **97**, 427 (1993)
876. A.G. Brolo, S.D. Sharma, Electrochim. Acta **48**, 1375 (2003)
877. M.J. Henderson, E. Bitziou, A.R. Hillman, E. Vieil, J. Electrochem. Soc. **148**, E105 (2001)
878. E. Vieil, C. Lopez, J. Electroanal. Chem. **466**, 218 (1999)
879. C. Barbero, M.C. Miras, R. Kötz, O. Haas, Sol. State. Ionics **60**, 167 (1993)
880. C. Lopez, M.F. Mendes Viegas, G. Bidon, E. Vieil, Synth. Met. **63**, 73 (1994)
881. M. Kalaji, L. Nyholm, L.M. Peter, A.J. Rudge, J. Electroanal. Chem. **310**, 113 (1991)
882. E. Vieil, K. Meerholz, T. Matencio, J. Heinze, J. Electroanal. Chem. **368**, 183 (1994)
883. M. El Rhazi, C. Lopez, C. Deslouis, M.M. Musiani, B. Tribollet, E. Vieil, J. Chim. Phys. Phys. Chim. Biol. **92**, 859 (1995)
884. C.W. Liu, G.J. Cheng, J.H. Li, J.G. Jin, S.J. Dong, J. Electroanal. Chem. **407**, 243 (1996)
885. C. Barbero, M.C. Miras, R. Kötz, O. Haas, J. Electroanal. Chem. **437**, 191 (1997)
886. C. Barbero, M.C. Miras, O. Haas, R. Kötz, J. Electrochem. Soc. **144**, 4170 (1997)
887. H. Huang, C.W. Liu, B.F. Liu, G.J. Cheng, S.J. Dong, Electrochim. Acta **43**, 999 (1998)
888. M.J. Henderson, A.R. Hillman, E. Vieil, J. Electroanal. Chem. **454**, 1 (1998)
889. D. Dini, F. Decker, G. Zotti, Synth. Met. **101**, 22 (1999)
890. L.M. Abrantes, J.P. Correia, Electrochim. Acta **44**, 1901 (1999)
891. B. Piro, E.A. Bazzaoui, M.C. Pham, P. Novak, O. Haas, Electrochim. Acta **44**, 1953 (1999)
892. M.J. Henderson, H. French, A.R. Hillman, E. Vieil, Electrochem. Solid State Lett. **2**, 631 (1999)
893. M.J. Henderson, A.R. Hillman, E. Vieil, J. Phys. Chem. B **103**, 8899 (1999)
894. V.M. Schmidt, C. Barbero, R. Kötz, J. Electroanal. Chem. **352**, 301 (1993)
895. M.J. Henderson, A.R. Hillman, E. Vieil, Electrochim. Acta **45**, 3885 (2000)
896. C. Barbero, E.J. Calvo, R. Etchenique, G.M. Morales, M. Otero, Electrochim. Acta **45**, 3895 (2000)
897. C. Barbero, M.C. Miras, R. Kötz, O. Haas, Synth. Met. **55**, 1539 (1993)
898. C. Barbero, M.C. Miras, E.J. Calvo, R. Kötz, O. Haas, Langmuir **18**, 2756 (2002)

899. C. Barbero, M.C. Miras, R. Kötz, O. Haas, Synth. Met. **55**, 1539 (1993)
900. P. Novak, R. Kötz, O. Haas, J. Electrochem. Soc. **140**, 37 (1993)
901. C. Barbero, M.C. Miras, R. Kötz, O. Haas, Sol. State Ionics **60**, 167 (1993)
902. R. Kötz, C. Barbero, O. Haas, Ber. Bunsenges. Phys. Chem. **97**, 427 (1993)
903. V.W. Jones, M. Kalaji, G. Walker, C. Barbero, R. Kötz, J. Chem. Soc. Faraday Trans. **90**, 2061 (1994)
904. C. Barbero, O. Haas, M. Mostefai, M.C. Pham, J. Electrochem. Soc. **142**, 1829 (1995)
905. M.C. Miras, C. Barbero, R. Kötz, O. Haas, V.M. Schmidt, J. Electroanal. Chem. **338**, 279 (1992)
906. C. Barbero, M.C. Miras, O. Haas, R. Kötz, J. Electrochem. Soc. **138**, 669 (1991)
907. C. Barbero, M.C. Miras, O. Haas, R. Kötz, J. Electroanal. Chem. **310**, 437 (1991)
908. G.M. Brisard, J.D. Rudnicki, F. McLarnon, E.J. Cairns, Electrochim. Acta **40**, 859 (1995)
909. O. Haas, J. Rudnicki, F.R. McLarnon, E.J. Cairns, J. Chem. Soc. Faraday Trans. **87**, 939 (1991)
910. M. Vilas-Boas, I.C. Santos, M.J. Henderson, C. Freire, A.R. Hillman, E. Vieil, Langmuir **19**, 7460 (2003)
911. M. Vilas-Boas, M.J. Henderson, C. Freire, A.R. Hillman, E. Vieil, Chem. Eur. J. **6**, 1160 (2000)
912. H.J. Salavagione, J. Arias-Pardilla, J.M. Pérez, J.L. Vázquez, E. Morallón, M.C. Miras, C. Barbero, J. Electroanal. Chem. **576**, 139 (2005)
913. J.M. Rosolen, M. Fracastoro-Decker, F. Decker, J. Electroanal. Chem. **346**, 119 (1993)
914. Y. Harima, K. Kishimoto, H. Mizota, Electrochim. Acta **53**, 657 (2007)
915. R. Muller, in *Advances in Electrohemistry and Electrochemical Engineering*, vol. 9, ed. by P. Delahay, C. Tobias (Wiley, New York, 1991), p. 281
916. K. Santhanam, R. O'Brien, in *Techniques for Characterization of Electrodes and Electrochemical Processes*, ed. by R. Varma, J.R. Selman (Wiley, New York, 1991), p. 401
917. C. Léger, J. Elezgaray, F. Argoul, J. Electroanal. Chem. **486**, 204 (2000)
918. C.E. Vallet, in *Spectroscopic and Diffraction Techniques in Interfacial Electrochemistry*, ed. by C. Gutiérrez, C. Melendres. NATO ASI Series C, vol. 320 (Kluwer Academic, Dordrecht, 1990), p. 133
919. S. Yoshihara, M. Ueno, Y. Nagae, A. Fujishima, J. Electroanal. Chem. **243**, 475 (1988)
920. S. Yoshihara, M. Ueno, Y. Nagae, A. Fujishima, Electrochim. Acta **33**, 1685 (1988)
921. S. Yoshihara, R. Takahashi, M. Okamoto, E. Sato, A. Fujishima, Electrochim. Acta **36**, 1959 (1991)
922. J.E. De Albuquerque, W.L.B. Melo, R.M. Faria, J. Polym. Sci. B-Polym. Phys. **38**, 1294 (2000)
923. E. Kubacki, **40**(12), 74 (2006) (see for an introduction)
924. R.J. Archer, *Manual on Ellipsometry* (Gaertner, Chicago, 1968)
925. R.M.A. Azzam, N.M. Bashara, *Ellipsometry and Polarized Light* (North-Holland, Amsterdam, 1977)
926. W.E.J. Neal, *Ellipsometry. Theory and Applications* (Plenum, New York, 1982)
927. R. Muller, in *Techniques for Characterization of Electrodes and Electrochemical Processes*, ed. by R. Varma, J.R. Selman (Wiley, New York, 1991), p. 31
928. W. Plieth, W. Kozlowski, T. Twomey, in *Adsorption of Molecules at Metal Electrodes*, ed. by J. Lipkowski, P.N. Ross (VCH, New York, 1992), p. 239
929. S. Gottesfeld, in *Electroanalytical Chemistry*, vol. 15, ed. by A.J. Bard (Dekker, New York, 1989), p. 143

930. R. Greef, in *Comprehensive Treatise of Electrochemistry*, vol. 8, ed. by R.E. White, J.O'M. Bockris, B.E. Conway, E. Yeager (Plenum, New York, 1984), p. 339
931. D.A. Holmes, J. Opt. Soc. Am. **54**, 1115 (1964)
932. R. Muller, Surf. Sci. **56**, 19 (1976)
933. B.D. Cahan, R.F. Spanier, Surf. Sci. **16**, 166 (1969)
934. J.M.M. Droog, G.A. Bootsma, J. Electroanal. Chem. **105**, 261 (1979)
935. P.S. Hauge, Surf. Sci. **96**, 108 (1980)
936. M. Dornhege, C. Punckt, J.L. Hudson, H.H. Rotermund, J. Electrochem. Soc. **154**, C24 (2007)
937. F. Chao, M. Costa, J. Lecoeur, Electrochim. Acta **36**, 1839 (1991)
938. S. Gottesfeld, G. Maio, J.B. Floriano, G. Tremiliosi-Filho, E.A. Ticianelli, E.R. Gonzalez, J. Electrochem. Soc. **138**, 3219 (1991)
939. T. Ohtsuka, K.E. Heusler, J. Electroanal. Chem. **100**, 319 (1979)
940. R. Parsons, W.H.M. Visscher, J. Electroanal. Chem. **36**, 329 (1972)
941. J.M.M. Droog, P.T. Alderliesten, G.A. Bootsma, J. Electroanal. Chem. **99**, 173 (1979)
942. S. Gottesfeld, S. Srinivasan, J. Electroanal. Chem. **86**, 89 (1978)
943. J.L. Ord, J. Electrochem. Soc. **129**, 335 (1982)
944. K. Kunimatsu, R.H. Parsons, J. Electroanal. Chem. **100**, 335 (1979)
945. H. Arwin, D.E. Aspnes, R. Bjorklund, I. Ljundström, Synth. Met. **6**, 309 (1983)
946. L.A.A. Pettersson, T. Johansson, F. Carlsson, H. Arwin, O. Inganäs, Synth. Met. **101**, 198 (1999)
947. P. Christensen, A. Hamnett, Electrochim. Acta **45**, 2443 (2000)
948. D.R. Kim, W. Cha, W.K. Paik, Synth. Met. **84**, 759 (1997)
949. M. Tagliazucchi, D. Grumelliz, E.J. Calvo, Phys. Chem. Chem. Phys. **8**, 5086 (2006)
950. M. Hernández Ubeda, M.A. Pérez, H.T. Mishima, H.M. Villullas, J.O. Zerbino, B.A. López de Mishima, M. López Teijelo, J. Electrochem. Soc. **152**, A37 (2005)
951. N.W. Ashcroft, N.D. Mermin, *Festkörperphysik* (Oldenbourg, München, 2001)
952. W. Knoll, Annu. Rev. Phys. Chem. **49**, 569 (1998)
953. M. Mansuripur, A.R. Zakharian, J.V. Moloney, OPN Opt. Photonics News **18**(4), 44 (2007)
954. D.G. Hanken, C.E. Jordan, B.L. Frey, R.M. Corn, in *Electroanalytical Chemistry*, vol. 20, ed. by A.J. Bard (Dekker, New York, 1998), p. 141
955. N. Zhang, R. Schweiss, Y. Zong, W. Knoll, Electrochim. Acta **52**, 2869 (2007)
956. A. Tadjeddine, D.M. Kolb, R. Kötz, Surf. Sci. **101**, 277 (1980)
957. A. Tadjeddine, A. Rahmani, Electrochim. Acta **36**, 1855 (1991)
958. B.L. Frey, C.E. Jordan, S. Kornguth, R.M. Corn, Anal. Chem. **67**, 4452 (1995)
959. L.E. Bailey, D. Kambhampati, K.K. Kanazawa, W. Knoll, C.W. Frank, Langmuir **18**, 479 (2002)
960. J.G. Gordon II, S. Ernst, Surf. Sci. **101**, 499 (1980)
961. Y. Iwasaki, T. Horiuchi, M. Morita, O. Niwa, Electroanalysis **9**, 1239 (1997)
962. M. Vasjari, Y.M. Shirshov, A.V. Samoylov, V.M. Mirsky, J. Electroanal. Chem. **605**, 73 (2007)
963. S. Szunerits, R. Boukherroub, Electrochem. Commun. **8**, 439 (2006)
964. A. Baba, J. Lübben, K. Tamada, W. Knoll, Langmuir **19**, 9058 (2003)
965. A. Bund, A. Baba, S. Berg, D. Johannsmann, J. Lübben, Z. Wang, W. Knoll, J. Phys. Chem. B **107**, 6743 (2003)
966. P. Taranekar, T. Fulghum, A. Baba, D. Patton, R. Advincula, Langmuir **23**, 908 (2007)
967. R. Schweiss, J.F. Lübben, D. Johannsmann, W. Knoll, Electrochim. Acta **50**, 2849 (2005)

968. A. Baba, M.K. Park, R.C. Advincula, W. Knoll, Langmuir **18**, 4648 (2002)
969. A. Otto, in *Optical Properties of Solids – New Developments*, ed. by B.O. Seraphin (North-Holland, Amsterdam, 1976), p. 677
970. H. Raether, in *Physics of Thin Films*, vol. 9, ed. by G. Hass (Academic Press, New York, 1977), p. 145
971. A. Baba, R.C. Advincula, W. Knoll, J. Phys. Chem. B **106**, 1581 (2002)
972. P.W.J.M. Boumans, *Inductively Coupled Plasma Emission Spectrometry Part 1* (Wiley, New York, 1987)
973. K. Ogle, S. Weber, J. Electrochem. Soc. **147**, 1770 (2000)
974. P.J. Schultz, K.G. Lynn, Rev. Mod. Phys. **60**, 701 (1988)
975. X. Wu, P. Asoka-Kumar, K.G. Lynn, K.R. Hebert, J. Electrochem. Soc. **141**, 3361 (1994)
976. K. Hebert, M. Fomino, H. Wu, P. Asoka-Kumar, K. Lynn, ECS Fall Meeting, Boston 1–6.11.1998, Ext. Abstr. #281
977. K.G. Lynn, H. Lutz, Rev. Sci. Instrum. **51**, 977 (1980)
978. I.I. Bardyshev, Yu.M. Polukarov, Russ. J. Electrochem. **40**, 9 (2004)
979. V.S. Subrahmanyam, S.K. Das, B.N. Ganguly, A. Bhattacharya, A. De, Mater. Res. Bull. **32**, 1063 (1997)
980. B. Das, S. Kar, D. Sanyal, D. Banerjee, R. Bhattacharya, J. Appl. Polym. Sci. **89**, 930 (2003)
981. R.W. Wilson, R. Cubitt, A. Glidle, A.R. Hillman, P.M. Saville, J.G. Vos, J. Electrochem. Soc. **145**, 1454 (1998)
982. R.M. Richardson, M.J. Swann, A.R. Hillmann, S.J. Roser, Faraday Discuss. **94**, 295 (1992)
983. A.R. Hillman, L. Bailey, A. Glidle, J.M. Cooper, N. Gadegaard, J.R.P. Webster, J. Electroanal. Chem. **532**, 269 (2002)
984. L. Bailey, M.J. Henderson, A.R. Hillman, N. Gadegaard, A. Glidle, Physica B **276**, 373 (2000)
985. R.W. Wilson, R. Cubitt, A. Glidle, A.R. Hillman, P.M. Saville, J.G. Vos, Electrochim. Acta **44**, 3533 (1999)
986. A. Glidle, L. Bailey, C.S. Hadyoon, A.R. Hillman, A. Jackson, K.S. Ryder, P.M. Saville, M.J. Swann, J.R.P. Webster, Anal. Chem. **73**, 5596 (2001)
987. I. Burgess, V. Zamlynny, G. Szymanski, A.L. Schwan, R.J. Faragher, J. Lipkowski, J. Majewski, S. Satija, J. Electroanal. Chem. **550–551**, 187 (2003)
988. L. Bailey, A. Glidle, R.M. Richardson, S.J. Roser, P.M. Saville, M.J. Swann, J.R.P. Webster, J. Am. Chem. Soc. **120**, 12882 (1998)
989. C. Riekel, in *Techniques for Characterization of Electrodes and Electrochemical Processes*, ed. by R. Varma, J.R. Selman (Wiley, New York, 1991), p. 279

6 Diffraction and Other X-Ray Methods

A beam of X-ray (i.e. electromagnetic radiation with a wavelength between $\lambda = 0.01$ and 10 nm) interacts with matter in three different ways:

1. It may be transmitted basically unchanged
2. It may be scattered or reflected either elastically or inelastically
3. It may be absorbed.

Although the first case looks rather unimportant at first glance, the transparency of a given medium for X-rays is of considerable importance for the application of any method based on X-rays because the construction of a spectroelectrochemical cell and the influence of the medium around the interface or the interphase to be investigated by any X-ray method depends on this transparency. At a wavelength $\lambda = 0.154$ nm (this is copper K_α radiation, which is most commonly used in X-ray experiments), a layer of 0.1 mm water will absorb 20% of the incoming X-ray photons and a layer of 1 mm will absorb about 90%. Thus any cell design and experimental setup must take into account even this seemingly unattractive feature. Increasing the energy of the X-rays, i.e. decreasing the wavelength, results in lower absorption. With molybdenum K_α radiation ($\lambda = 0.071$ nm), even several centimeters of water pose no serious absorption problem. Unfortunately, X-rays with such high energies pose other challenges and problems related to the availability of suitable detectors.

X rays are basically scattered or absorbed by all atoms in their path, thus they are not inherently surface sensitive. Depending on the type of interaction employed in a given method, various approaches are possible to confer this surface sensitivity when desired, as outlined below.

6.1 X-Ray Diffraction Methods

Besides the use of X-ray absorption for the elucidation of adsorbates and adsorbate–surface interactions (see Chap. 5), X-ray diffraction is a tool that is useful particularly for the determination of structural data at a very high level of precision provided the presence of a minimum level of structuring. The advent of synchrotron radiation that provides an intense source of electromagnetic radiation in the range of X-rays has greatly stimulated the application of X-ray diffraction methods. A broad overview of experimental approaches and recent results has been provided elsewhere [1–4].

6.1.1 X-Ray Diffraction

Fundamentals. A wavefront of electromagnetic radiation arriving at an array of atoms that act as scattering centers causes these centers to emit (scatter) spheric waves of electromagnetic radiation of the same energy as the incoming radiation. Depending on the wavelength of the employed radiation and the angle of incidence and the spatial arrangement of the scatterers, the emitted radiation of the different scatterers will interfere constructively or destructively. With a crystalline sample containing a considerable number of scatterers, the conditions for constructive interference will be fulfilled only when the angle of incidence relative to the crystallographic plane in which the scatterers are located is matched very precisely. The relationship is given by Bragg's equation, $n\lambda = 2d_{\mathrm{hkl}} \sin \vartheta$, with n being the order of the reflection, λ the wavelength of the used radiation, d_{hkl} the crystallographic spacing and ϑ the angle of diffraction (i.e. the angle of incidence). From the number and intensities found at different values of values of ϑ, the crystallographic data of the sample under investigation can be derived. In order to obtain a diffractogram, a crystal is illuminated with X-rays of a fixed wavelength λ. The angle of incidence is varied by turning the crystal and the diffracted intensity is measured. For further details, see [5, 6].

X-ray diffraction, as described above, is done in an angular dispersive way (ADXD). The diffracted X-ray intensity is displayed as a function of the scattering angle 2ϑ. This diffractogram is correct only for the single wavelength used during its acquisition. Any change of the incident wavelength causes a nonlinear stretching of the diffractogram. A more general display is obtained by displaying the scattered intensity as a function of the scattering parameter q. Further details and advantages of this approach have been discussed elsewhere [7, 8]. Consequently, two different ways to obtain a diffractogram are possible:

1. Use of light with a fixed wavelength and measurement of the diffracted intensity as a function of the scattering angle
2. Use of polychromatic light (Bremsstrahlung) and detection of the diffracted radiation with a solid-state detector connected to a multichannel analyser

Instrumentation. A setup includes a source for X-rays (usually a fixed or a rotating anode source; more recently, synchrotron radiation has become an attractive choice as a source of radiation), a monochromator or at least a filter to select the desired wavelength, the spectroelectrochemical cell and the detector (for an overview on detectors see [3, 9]).

The cell has to fulfill three general requirements:

1. The path for the incident and the reflected beam should not be obstructed.

2. The number of scattering or absorbing particles beyond the sample under investigation in the beam path should be kept at a minimum.
3. The electrochemical characteristics of the cell (like even current distribution, reliable potential control, low *iR*-drop) should be as perfect as possible.

For X-ray diffraction experiments, two basic designs are possible: The Bragg (or reflection) and the Laue (or transmission) mode. In electrochemical investigations, the former is better suited for studies of adsorbates or of other features parallel to the electrode surface, whereas the second mode is suitable for thick films or layers. In both cases, a cell window as transparent as possible for X-rays with sufficient stability towards this radiation is needed. Thin polymer foils (Mylar® or Melinex®) are most commonly used. A typical design of a cell of the Laue type as depicted in Fig. 6.1 shows the X-ray passing through two polymer film windows and the electrolyte solution.

The working electrode is coated onto one of the windows. In order to keep the scattering from the electrolyte solution low, one window is mounted on the end of a hollow syringe barrel. For electrochemical measurements, it is retracted to provide acceptable current distribution; for X-ray measurements, the barrel is moved as close as possible towards the fixed window. The poor electrochemical properties of the cell in the latter position are an obvious drawback of the cell design.

A spectroelectrochemical cell of the Bragg (reflection) type, as used for the investigation of materials that are chemically or electrochemically deposited onto a gold film that was sputtered first onto a porous membrane foil, has been described [10]. The cross section as displayed in Fig. 6.2 shows the Prussian blue-coated membrane assembly with its coating towards a polyethylene film that is transparent and amorphous for X-rays.

The electrolyte penetrates the porous membrane via the porous glass body and the glass fiber paper connection is provided with the silver chloride coated silver plate that acts both as a reference and a counter electrode. The electrochemical behavior of the cell as demonstrated with cyclic voltammetry is fairly close to standard electrochemical cells. Results show changes of lattice constants of Prussian blue as

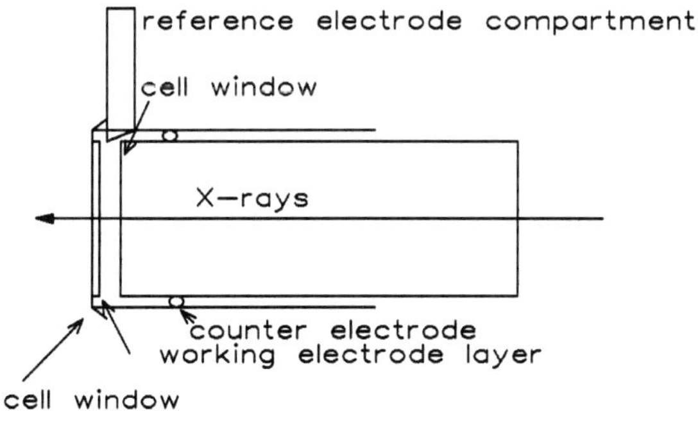

Fig. 6.1. Spectroelectrochemical cell for X-ray diffraction studies in transmission mode

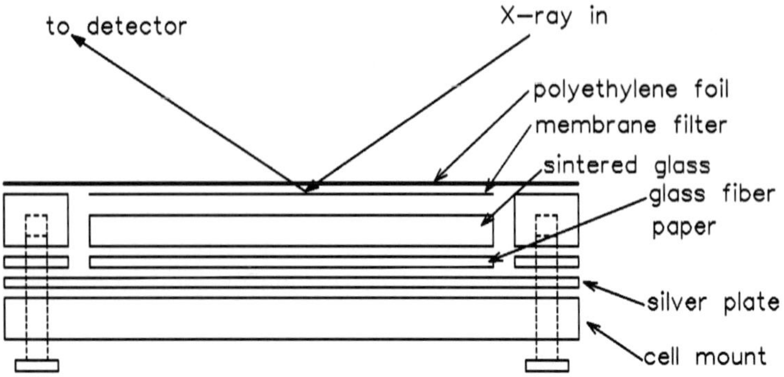

Fig. 6.2. Spectroelectrochemical cell (for better identification of components in exploded arrangement) for *in situ* X-ray diffraction studies according to [10]

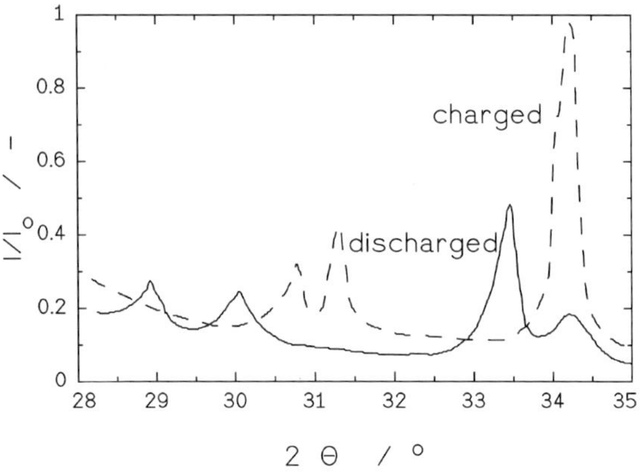

Fig. 6.3. *In situ* X-ray diffraction pattern of TiS_2 recorded during charging/discharging experiments (based on data in [12]; see also [13])

a function of the electrode potential and the transferred charge. Because of the penetration depth of X-rays and the need for a minimum of crystallinity, this approach is not exactly surface sensitive; instead, it probes the interphase between the current collector (i.e. the gold layer) and the electrolyte solution. Results obtained during charging/discharging of TiS_2 (as proposed for use in secondary lithium batteries battery [11]) that involve formation of $LiTiS_2$ are displayed in Fig. 6.3; the employed cell is depicted schematically in Fig. 6.4.

Upon discharge lithium is intercalated. This results in shifts of the (101), (002) and (100) Bragg peaks. Detailed studies of the various shifts as functions of the state of charge/discharge reveal further information on the mechanism of the different phase transitions [12]. Reactions of lithium with S_8 in a secondary Li/S cell have been tracked with *in situ* X-ray diffraction [15]. The electrochemical reaction

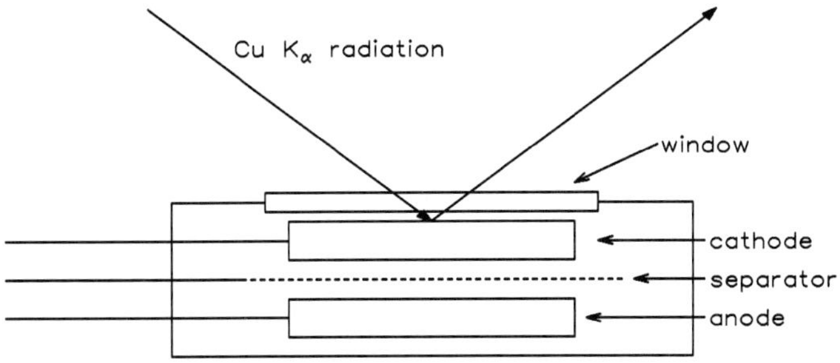

Fig. 6.4. *In situ* X-ray diffraction electrochemical cell (based on [14])

of lithium with crystalline silicon was monitored [16]. Various Li_xSi_y phases were identified. The transition from the crystalline into the amorphous state in the first lithium intercalation was observed [17]. For alternate cell constructions, see [18]. For a cell used in studies of ion insertion into electrode materials, see [19]. A complete cell that has not been modified in any way has been used with high energy X-rays from a synchrotron source for simultaneous measurement of lithium ion battery cell component diffraction patterns; recorded diffraction peaks were assigned to cathode and anode material [20]. A cell that is particularly suitable for measurements under ultrapure conditions has been described [21].

Basically the same design has been employed in a study of the solid state electrochemistry of PbO and Pb(OH)Cl (laurionite) [22]. The particles of red PbO (litharge) were attached to a paraffin-impregnated graphite (PIG) rod that was used as a working electrode. X-ray diffraction patterns obtained at various electrode potentials (see Fig. 6.5) show peaks indicative of PbO, a mix of PbO and Pb and finally of Pb as a function of reduction potential and time. Reflexes of graphite were also observed and could be distinguished easily from those of the electrochemically active material. The reduction proceeded entirely as a solid state reaction and no evidence of solution phase intermediates was observed.

Improved surface specificity can be obtained by making measurements at the grazing angle of incidence. The structure of small two-dimensional clusters and linear nanostructures of copper and cadmium deposits on Pt(533) at characteristic electrode potentials in the upd-range has been studied with GIXD combined with GIXAFS [23]. Using X-rays provided by a synchrotron (the method is now named SR-GIXRD), the corrosion of mild steel in the presence of carbon dioxide containing brine electrolyte solutions was studied. The mechanism of corrosion, particularly corrosion products that are present as thin films like $Fe_2O_2CO_3$, $Fe_2O_2CO_3$ and $Fe_2(OH)_2CO_3$, was identified with SR-GIXRD [24]. Investigations of iron chalcogenide glasses suitable for ion selective electrodes (ISE) with a combination of electrochemical impedance measurements and GIXRD have been reported together with a description of a suitable electrochemical cell [25, 26]. Selective dissolution of various crystallographic surfaces of iron chalcogenides associated with

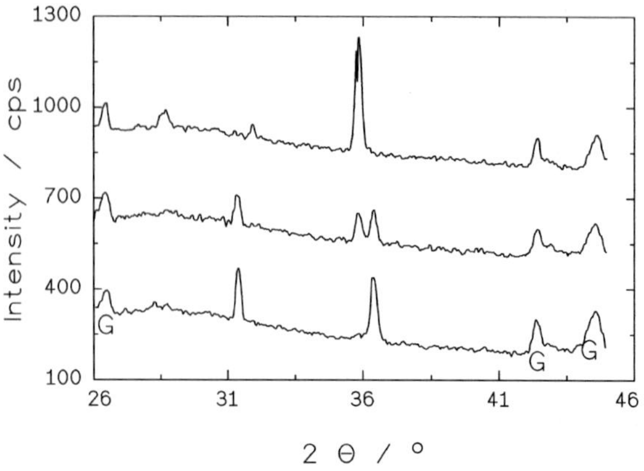

Fig. 6.5. *In situ* X-ray diffraction pattern of PbO and its reduction products: before reduction, at $E_{Ag/AgCl} = -0.5$ V (*top*); during reduction at $E_{Ag/AgCl} = -1$ V, after reduction, at $E_{Ag/AgCl} = -1.3$ V (based on data by [22])

the electrode potential shifts of these materials that are employed in ion selective electrodes have been monitored with SR-GIXRD [27]; results were found to be in agreement with those obtained with AFM.

The cell designs and examples discussed above pertain to X-ray diffraction performed in the angular dispersive mode, i.e. the angle of incidence is varied by turning the electrochemical cell with respect to the radiation source.

As already mentioned, energy dispersive measurements are possible when the spectroelectrochemical cell stays in place, whereas the wavelength of the incident radiation is varied. An additional advantage of EDXD is the high intensity of X-ray radiation in the employed energy range with an upper limit only given by the power supply of the X-ray tube. The strong X-ray absorption of standard K- and L-lines in the range of lower energies caused by electrochemical cell components is no problem, because the construction materials that are employed usually show only weak absorption at higher wavelengths [8]. The poor resolution of the solid-state detector causes a broadening of the peaks in the measured diffractogram [8]. Results reported so far deal with intercalation materials as used in batteries [8].

The application of time-resolved high energy X-ray diffraction on platinum nanoparticles in fuel electrodes has been described [28]. Results indicate severe surface reconstruction of the nanoparticle surface showing at least three types of Pt–O bonds (adsorbed OH, adsorbed atomic O and amorphous PtO_x) under oxidative conditions.

X-ray diffraction has also been used in studies of solid electrolytes as reviewed elsewhere [29]. Diffraction techniques (including X-ray and neutron diffraction), as applied in electrolyte solution studies, are described in [30].

6.1.2 Surface X-Ray Diffraction (SXD)

Fundamentals. X-ray diffraction as applied to the elucidation of the crystallographic structure of thick layers, films or other three-dimensional samples can also be applied to basically[1] two-dimensional structures on surfaces or at interfaces based on various instrumental advances, particularly the use of synchrotron radiation. This way, information on the two-dimensional periodicity of surface layers can be obtained. The required surface sensitivity can be obtained primarily by employing an angle of incidence that is less than or in the order of the critical angle. This results in a significantly reduced depth of penetration of the X-rays and scattering from the bulk of the sample is also diminished. Under these conditions, the ratio of the signal from the surface vs. the signal from the bulk (or substrate) can be enhanced by a factor of $2/\theta_c$ (with θ_c being the critical angle of total external reflection in radians). At X-ray frequencies typical for metals in contact with transparent incident phases, θ_c is about 0.5° (or 9 milliradians). These conditions are also known as total reflection Bragg diffraction [31]. If grazing angle conditions (i.e. angle of incidence θ about 3.5° or lower) are employed, the method is also called grazing incidence X-ray diffraction (GIXRD). Further enhancement of general and surface sensitivity can be achieved by applying an electrode potential modulation procedure combined with an appropriate data treatment. This way, only potential dependent features in the diffractogram will become visible, whereas features (Bragg reflexes) of the bulk that are not affected by the electrode potential will be canceled out. Because of fluctuations in the intensity of the X-ray source and other experimental components, it is insufficient to record complete diffractograms sequentially; instead, a slow potential modulation (with the frequency set to a value resulting in a time constant that is sufficiently longer than that of the occurring electrochemical process) is applied. Using computer-based data acquisition, the respective data are stored and treated; for further details and an overview, see [32].

Instrumentation. The electrochemical cells described in the preceding section can be used. A cell design with a significantly reduced radiation absorption of the electrolyte solution film as used for specular X-ray reflectivity measurements (see description below Fig. 6.10) can also be used. Electrode potentials are selected based on standard electrochemical experiments (e.g. cyclic voltammetry) with respect to well-defined changes of the electrode–solution interface (e.g. potential steps between potentials of complete desorption and maximum adsorption). Control of the potentiostat and the X-ray diffractometer as well as data acquisition, storage and manipulation are done with a suitably programmed computer.

Typical examples include studies of the underpotential deposition of various metals on metallic substrates. The structure of the upd-layer [33, 34], the position of adsorbed anions and water molecules on top of the upd-layer and the respective bond angles and lengths could be elucidated [35, 36]. Surface reconstruction caused by weakly adsorbed hydrogen [37], surface expansion effects of low-index platinum and gold surfaces correlated with adsorption/desorption of solution species [38] and

[1] The slightly gradual distinction implies possible overlap between both groups of methods.

crystallographic transformations in layers of $NiOH_2$ have been studied [39]. In a corrosion study of a $Cu_3Au(111)$ alloy single crystal that was used as a model system, initial stages of corrosion, particularly structure formation, was elucidated with surface XRD [40]. Evidence of the diffusion of gold atoms into the upd-Te layer on top of a Au(111) has been reported [41]. A study of the water layer structure on a Cu(111) electrode surface during hydrogen evolution combining SXD and IRRAS showed a stacked, close-packet structure with significantly different inter- and intra-layer specific oxygen–oxygen nearest neighbor distances [42]; the infrared spectra indicated the presence of both free and hydrogen-bonded hydroxyl ions.

6.1.3 Surface Differential X-Ray Diffraction (SDD)

Fundamentals. The scattered X-ray from crystal planes or a thin metal film can interfere coherently with scattered X-rays from an adsorbate layer. This phenomenon is called surface differential X-ray diffraction (SDD) or Bragg peak interference. The theory has been described elsewhere in detail [43, 44]. Information on the relative amount of material in the adsorbate layer (coverage) and the distance between atoms in the top layer of the substrate and in the adsorbate layer as a function of electrode potential can be derived from SDD measurements.

Instrumentation. The experimental setup uses a working electrode prepared from mica coated with the metal to be investigated (e.g. silver) by vapour deposition. Because of the mode of growth, a certain monocrystalline orientation will result (e.g. (111) in the case of silver). The working electrode is attached to an electrochemical cell equipped with counter and reference electrode and the standard ancillary equipment for electrochemical measurements. The X-ray of well-defined wavelength from a standard laboratory source or from a synchrotron (selected with e.g. a Si(220) monochromator) impinges on the working electrode through the mica substrate. Diffracted intensity is collected with a multichannel analyser, which can be done in real time during electrode potential scans. Investigated systems include predominantly metal adlayers (upd-layers) on metal surfaces [44, 45]. Data on modes of layer growth, adsorbate–atom and substrate–atom distance and relaxation of involved atom layers were obtained.

6.1.4 Neutron Diffraction

Fundamentals. Neutrons of low energy have wavelengths similar to those of X-rays (e.g. a neutron with 1 eV energy has a wavelength of $\lambda = 0.027$ nm, a neutron moderated by graphite down to a speed of about 4 $km \cdot s^{-1}$ has a wavelength of about $\lambda = 0.1$ nm). They show diffraction like X-rays as predicted in 1936 by Elsasser. Contrary to X-ray diffraction, where electromagnetic radiation (the X-rays) interacts with electrons of the involved atoms, neutrons will interact with the nuclei of these atoms. Thus the diffraction of neutrons does not depend on the number of electrons in the atom. The rapid increase of X-ray scattering factors with increasing numbers of electrons in the atoms is consequently not observed with neutrons.

Scattering factors depend only weakly on this number and do not depend at all on the direction of the scattering. Scattering is not dominated by heavy atoms. Thus neutron scattering is much better suited for detection of light atoms like hydrogen atoms. Atoms with very similar X-ray scattering factors, e.g. nickel and cobalt atoms, can be distinguished with neutron scattering. Since neutron fluxes tend to be very small (typically about 10^{-5} of X-ray radiation) and are more difficult to handle, neutron diffraction is possible only with powerful neutron sources like nuclear reactors. They are of particular use in structural studies of metals and solids wherein the position of light atoms like hydrogen has to be identified.

Instrumentation. The experimental setup is similar to that employed for *in situ* X-ray diffraction. The material under investigation is pressed into a thin sheet and mounted together with suitable counter and reference electrodes into a silica cell. In order to decrease the large incoherent scattering contributions from protons in aqueous electrolyte solutions, deuterated solutions are used. In a typical study, the reaction mechanism of $Ni(OH)_2$ (employed in nickel accumulators) was studied with neutron powder diffraction NPD [46]. A direct and continuous structural transformation of both the γ- and β-NiOOH phases into β-$Ni(OH)_2$ was observed during reduction with no direct relationship or discontinuity related to the transition from the first discharge electrode potential to the second one, which was located about 0.4 V lower.

6.2 Miscellaneous Methods

6.2.1 X-ray Standing Wave Fluorescence Analysis (XSW)

Fundamentals. Dynamical reflection of a monochromatic, longitudinally coherent planar X-ray wave at a perfect (or almost perfect) single crystal surface results in the formation of an interference field of low thickness on both sides of the interface when incident and reflected waves interfere. For every reflection, this standing wave (XSW) field has the same spatial periodicity as the respective diffraction planes. The antinodal planes of the electric field intensity lie parallel to the diffraction planes. The movement of the standing wave field, which is controlled by changing the angle of incidence, can be used to determine the position of foreign atoms on a surface by measuring their characteristic fluorescence radiation together with the shift of the antinodal planes (for further details, see [47–51]) because only those atoms inside an antinode show X-ray fluorescence. The obtained plot shows fluorescence intensity as a function of the angle of incidence for the selected wavelength of the X-rays.

Instrumentation. The necessary γ-radiation is usually supplied by a synchrotron. Selection of the proper monochromatic radiation is done with germanium and/or silicon single crystals. The angle of incidence of the radiation on the electrode surface under investigation can be varied; the reflected and scattered intensity are measured with suitable detectors. Further experimental details have been reported elsewhere [52, 53]. In the studies reported so far, underpotential deposited layers have been investigated and distances between the adlayers and the underlying metal surface were determined [54, 48]. Closely related studies of membranes have been described [56]. The use of layered synthetic microstrucutures (LSM) [57] for XSW measurements has been described [58]. Layered synthetic microstructures are prepared by sequentially depositing materials that are typically of low and high electron density (e.g. W/C, Mo/C, W/Si or Pt/C) onto a smooth surface like silicon or cleaved mica, avoiding typical difficulties encountered when handling bulk metal single crystals. The latter system (Pt/C) was employed in an investigation of iodine and subsequent competitive adsorption of copper ions on platinum.

6.2.2 Surface X-ray Scattering (SXS)

Fundamentals. X-rays scattered at a surface are reflected from a crystal like any light from a reflecting surface. The specular (mirror-like) reflection is strongest when the incident beam arrives at the glancing angle[2] (typically less than 1° at the metal/air interface). It falls off rapidly with an increase of the angle of incidence beyond this critical angle for total external reflection. When the angle of incidence passes the various Bragg angles (as given by the crystallographic data of the material under study), the intensity rises again because scattering amplitudes from every atomic layer in the crystal are in-phase and signals are added. At intermediate angles, the signals are out-of-phase and show destructive interference. A weak intensity can nevertheless be observed at these intermediate angles, which are caused by scattering from the surface layers. Atoms in the surface layer have a slightly different crystallographic environment because the crystal is truncated; the atoms in the topmost layer have dangling (unsaturated) bonds. Consequently, their scattering conditions are different from those in the bulk of the crystal. In this specular reflection, the momentum transfer of X-rays is surface normal. The measured intensity represents the square of the Fourier amplitude of the surface normal electron density. Since Fourier phases cannot be measured, a direct inversion of the measured intensities is impossible. Instead, a fit of calculated intensity based on surface structure models to the actually measured intensity provides an accurate description of the surface-normal structure.

Besides crystal planes parallel to the surface, further planes exist that might reflect impinging X-rays. The scattered X-rays are not in the plane of reflection and are called off-specular reflections. Their intensity is also high at the Bragg angles. As with the specularly reflected X-rays, the weak signal between these peak intensities contains surface-relevant information. A combined analysis of these reflections often leads to a complete picture of the surface structure.

In reciprocal space, these reflections form streaks of scattered X-ray intensity with a narrow but elongated cross section in the direction of the surface normal. With a flat surface parallel to a crystallographic axis, these streaks connect the Bragg

[2] Unfortunately there is some confusion with the use of angle of incidence. Sometimes the angle between beam and surface, sometimes the angle between beam and surface normal are meant. Here the angle between beam and surface is considered.

reflection peaks. These Bragg reflections essentially contain the crystallographic information of the bulk and the streaks (sometimes also called "rods") contain the surface-relevant information. Surface X-ray scattering measurements are sometimes called crystal truncation rod (CTR) measurements with respect to the various aspects mentioned above [59, 60]. The method is capable of providing two-dimensional and three-dimensional information at resolutions better than 0.1 Å. Further details have been reported elsewhere [61–65].

Instrumentation. Investigations require an X-ray source of sufficient intensity that provides monochromatic radiation, usually a synchrotron. The schematic setup is shown in Fig. 6.6. Typically, X-rays in the energy range 6 to 17 kV ($\lambda = 70$ to 200 pm) are used.

Two basically different cell geometries can be used. In Fig. 6.7 these designs are shown together with their analogous arrangements for single crystal studies.

In the reflection geometry (or in analogy, the Bragg geometry), the cell window is parallel to the electrolyte surface. The current distribution and *iR*-drop is

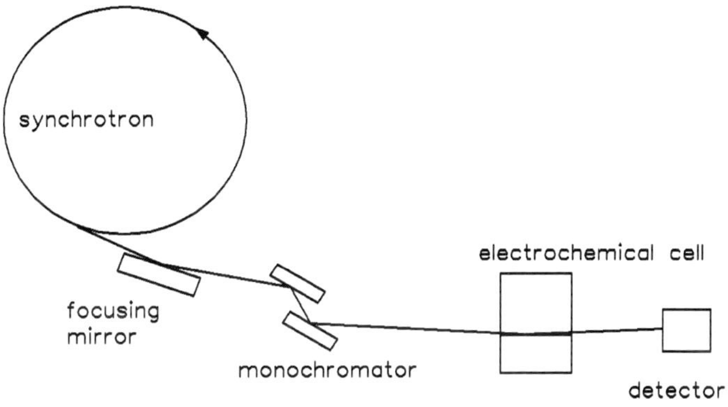

Fig. 6.6. Schematic setup for X-ray scattering experiments

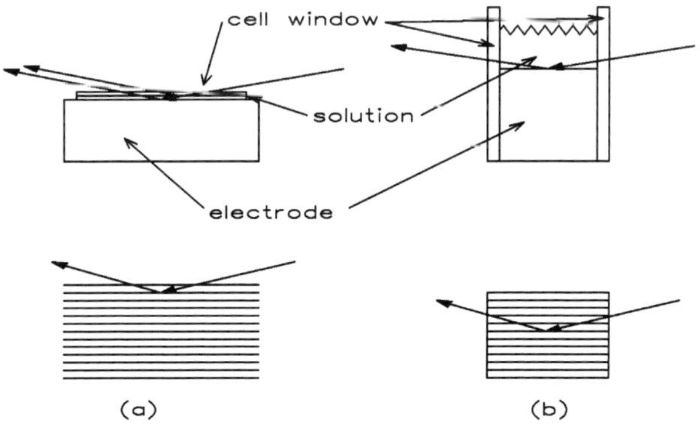

Fig. 6.7. Schematic cell designs for X-ray scattering experiments: (**a**) reflection (Bragg) geometry; (**b**) transmission (Laue) geometry

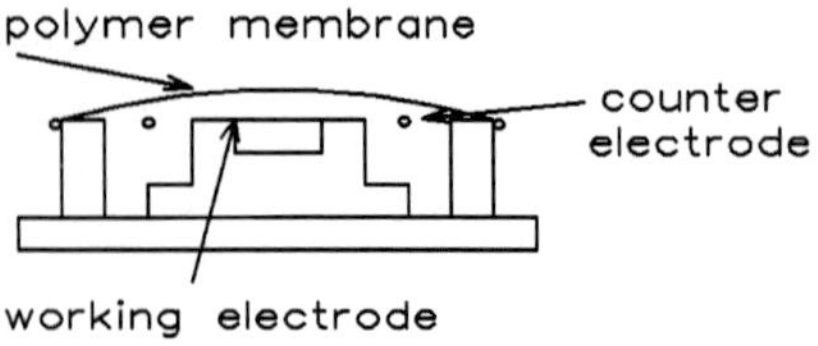

Fig. 6.8. Electrochemical cell for X-ray scattering experiments in reflection geometry (schematic cross section)

considerable because of the thin electrolyte layer between the electrode surface and the cell window, which is kept at minimum thickness by employing membranes as cell windows that were kept as close to the electrode as possible. By expanding the membrane during electrochemical manipulation, this drawback can be reduced somewhat. This is illustrated in Fig 6.8, showing a simplified cell cross section.

The electrolyte solution is pumped into the cell before electrochemical manipulation, causing the flexible membrane to rise. The solution is removed before starting X-ray measurements and the membrane collapses down on the electrode surface, leaving only a thin electrolyte solution film in place. Further typical cell designs have been described elsewhere [59, 60, 66]. In the transmission (Laue) geometry, the electrochemical properties of the cell are almost perfect and the free access of the solution/metal interface even allows for flushing the electrolyte solution in order to remove radiolysis products. Unfortunately, the beam is attenuated considerably. This situation can be improved somewhat by keeping the cell thickness as small as possible. Only X-rays reflected from the interface will reach the detector, because the X-rays impinge on the cell window at an angle excluding any reflections from the window from reaching the detector. Because the beam path through the electrolyte volume hardly changes when the angle of incidence is varied, little or no absorption correction with the reflection geometry is necessary. Typical cell designs can be found elsewhere [59, 60]. A wide-angle accessible cell [67] and an experimental setup for measurements at elevated temperatures [68] have been described.

Investigated systems include various metal deposits, like underpotentially deposited silver on Au(111) [69], thallium on a Pt(111) surface as a function of solution pH and bisulfate coadsorption [70] and other upd-systems [36, 59]. Evidence of dealloying of $Cu_3Au(111)$ has been reported [71]. Near-neighbor distances between atoms in upd-monolayers of various transition metals deposited on Ag(111) and Au(111) surfaces have been measured as a function of electrode potential [72]. Typical results of a study of thallium-upd showing the changing Tl–Tl distance as a function of electrode potential are shown in Fig. 6.9.

The incommensurate, hexagonal monolayers are compressed compared to the bulk metal and they are rotated from the substrate by several degrees. From the results, the monolayer compressibility could be calculated. The restructuring (i.e. surface reconstruction) of top layers of single crystal metal surfaces as a function of solution composition and electrode potential has been studied [73]. The induced charge density was found to be the critical parameter [74]. Structural changes during

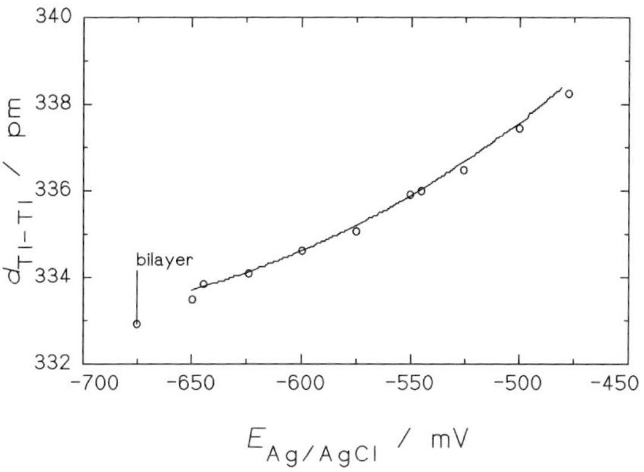

Fig. 6.9. Thallium monolayer near-neighbor distance as a function of electrode potential, aqueous electrolyte solution 0.1 **M** Na_2SO_4 + 2.5 m**M** Tl_2SO_4 (based on data in [72])

oxidation/reduction of copper surfaces and passivation have been investigated [75]. Dissolution and passivation of a Ni(111) surface have been studied with surface X-ray scattering combined with *in situ* STM measurements [76]. By combining surface X-ray diffraction and X-ray scattering, the palladium–gold bond length in a palladium monolayer on a Au(111) surface was determined to 28.2 pm, which is close to the sum of the atomic radii of the involved atoms [77]. A study of copper-upd layers on Au(111) revealed sulfate ions bound via three oxygen atoms to the copper atoms; this seems to correct previous scanning probe microscopy studies resulting in the claim, that these microscopies probed copper atoms as topmost layer constituents [78]. The coverage of monocrystalline surfaces with various adsorbates and the adsorbate structure, including the position of adsorbed ions, have been determined [79, 80]. A more detailed picture of adsorbate–substrate interactions emerges when polarization dependencies of the scattered intensity and resonant scattering are taken into account [81]. Using suitable cell designs, time-resolved measurements on a second scale have become possible; results pertaining to phase transitions of upd-layers and the kinetics of oscillating reactions have been reported [82].

6.2.3 Small Angle X-ray Scattering

Fundamentals. Particles like macromolecules in solution or particles in a heterogeneous solid scatter X-rays. Information about the particle size can be obtained from measurements of the scattered intensity as a function of scattering angle; the term "small angle" refers to the fact that the scattering is investigated at angles of only a few degrees, corresponding to particle sizes in the nanometer range. In addition, information about the shape can be derived. Scattering vectors from the atoms of interest can be selected with X-rays of tunable wavelength (e.g. from synchrotrons). The use of a microbeam of 5-μm diameter enables laterally resolved studies of layered systems [83].

Instrumentation. In studies reported so far [84], polymer nanocomposites as used in lithium batteries prepared from poly(ethylene oxide) and lithium hectorite have been investigated. Using a sample holder that could be heated, structural changes of the nanocomposites as a function of temperature could be monitored *in situ*. *Ex situ* studies of electrodeposited amorphous NiP coatings have been described [85].

6.2.4 Specular X-ray Reflection

Fundamentals. The reflectivity of a surface with respect to electromagnetic radiation in the X-ray range for a beam with equal incident and outgoing angles and both beams in one plane can be described in a kinematic approximation [73, 86, 87]. The structure factor included in this relation contains information pertaining both to the surface layer of the reflecting electrode surface and the bulk of the material, particularly to interlayer spacings and sample morphology. The parameters describing the electron density profiles and further structure-related details are optimized until the calculated reflectivity fits the experimentally observed one (for an example, see [88]).

Instrumentation. In most studies reported so far, X-ray radiation of a synchrotron collimated and monochromatized with suitable optics is used. An electrochemical cell with a moving working electrode is employed in order to minimize X-ray absorption by the electrolyte solution. A typical cell design is depicted in Fig. 6.10.

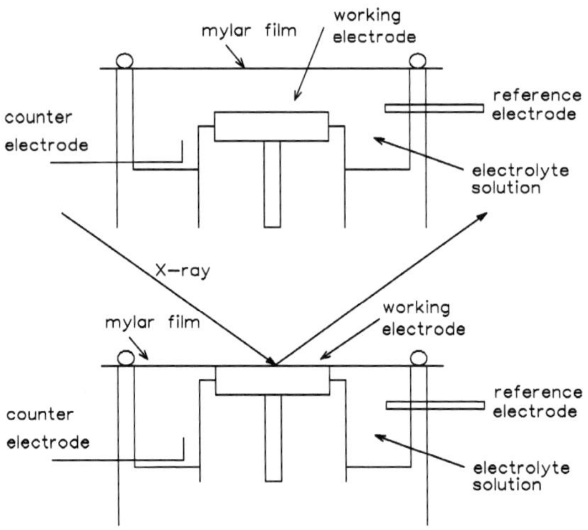

Fig. 6.10. Schematic spectroelectrochemical cell for *in situ* X-ray specular reflection measurements. *Top*: Position of central working electrode down for electrochemical measurements; *Bottom*: Working electrode up for X-ray measurements (based on [88])

For electrochemical measurements, the electrode is lowered away from the Mylar® film that serves as a cell cover and X-ray window. After moving the electrode upwards, an electrolyte solution film of about 10 μm remains with only small radiation absorption.

The surface-normal structure of epitaxially grown Te and Cd upd-layer was studied [89] and other upd-systems have been the subject of investigations [33, 88]. Kinetic studies of passive oxide formation on iron and iron alloys [90] and of Au(001) reconstruction have been reported [73].

References

1. Special issue of Electrochim. Acta **47**, 3035 ff. (2002)
2. J. Robinson, in *Spectroscopic and Diffraction Techniques in Interfacial Electrochemistry*, ed. by C. Gutiérrez, C. Melendres. NATO ASI Series C, vol. 320 (Kluwer Academic, Dordrecht, 1990), p. 313
3. J. Robinson, in *Spectroelectrochemistry*, ed. by R.J. Gale (Plenum, New York, 1988), p. 9
4. J. Zegenhagen, F. Renner, Nachr. Chem. **54**, 847 (2006)
5. W. Massa, *Kristallstrukturbestimmung* (Teubner, Stuttgart, 1996)
6. W. Clegg, *Crystal Structure Analysis* (Oxford University Press, Oxford, 2001)
7. R. Caminiti, V. Rossi Albertini, Int. Rev. Phys. Chem. **18**, 263 (1999)
8. V. Rossi Albertini, P. Perfetti, F. Ronci, B. Scrosati, Chem. Mater. **13**, 450 (2001)
9. J. Robinson, in *Spectroelectrochemistry*, ed. by R.J. Gale (Plenum, New York, 1988), p. 9
10. T. Ikeshoji, T. Iwasaki, Inorg. Chem. **27**, 1123 (1988)
11. M.S. Whittingham, Science **192**, 1126 (1976)
12. R.R. Chianelli, J.C. Scanlon, B.M.L. Rao, J. Solid State Chem. **29**, 323 (1979)
13. R.R. Chianelli, J.C. Scanlon, B.M.L. Rao, J. Electrochem. Soc. **125**, 1563 (1978)
14. R.R. Chianelli, J.C. Scanlon, B.M.L. Rao, J. Solid State Chem. **29**, 323 (1979)
15. V. Puglisi, M. Simoneau, Y. Geronov, 210. Electrochemical Society Meeting, Cancun, Mexico, 29.10–03.11.2006, Ext. Abstr. #164
16. J. Li, J.R. Dahn, 210. Electrochemical Society Meeting, Cancun, Mexico, 29.10–03.11.2006, Ext. Abstr. #261
17. J. Li, J.R. Dahn, J. Electrochem. Soc. **154**, A156 (2007)
18. M. Nishizawa, I. Uchida, Denki Kagaku **66**, 991 (1998)
19. O. Bergström, T. Gustafsson, J.O. Thomas, J. Appl. Cryst. **31**, 103 (1998)
20. Z.G. Li, R.L. Harlow, F. Gao, P. Lin, R. Miao, L. Liang, J. Electrochem. Soc. **150**, A1171 (2003)
21. Th. Koop, W. Schindler, A. Kazimirov, G. Scherb, J. Zegenhagen, Th. Schulz, R. Feidenhans'l, J. Kirschner, Rev. Sci. Instr. **69**, 1840 (1998)
22. B. Meyer, B. Ziemer, F. Scholz, J. Electroanal. Chem. **392**, 79 (1995)
23. H. Prinz, H.H. Strehblow, Electrochim. Acta **47**, 3093 (2002)
24. R. De Marco, Z.-T. Jiang, B. Pejcic, E. Poinen, J. Electrochem. Soc. **152**, B389 (2005)
25. R. de Marco, B. Pejcic, K. Prince, A. van Riessen, Analyst **128**, 742 (2003)
26. R. De Marcoa, Z.-T. Jiang, J. Martizano, A. Lowe, B. Pejcic, A. van Riessen, Electrochim. Acta **51**, 5920 (2006)

27. R. de Marco, Z.T. Jiang, B. Pejcic, A. van Riessen, Electrochim. Acta **51**, 4886 (2006)
28. H. Imai, K. Izumi, Y. Kubo, K. Kato, Y. Imai, Extended Abstracts of the 209th Meeting of the Electrochemical Society Spring Denver, Colorado, USA, 07.05–11.05.2006, Ext. Abstr. #302
29. J.B. Wagner Jr., in *Techniques for Characterization of Electrodes and Electrochemical Processes*, ed. by R. Varma, J.R. Selman (Wiley, New York, 1991), p. 3
30. J.E. Enderby, in *Techniques for Characterization of Electrodes and Electrochemical Processes*, ed. by R. Varma, J.R. Selman (Wiley, New York, 1991), p. 327
31. W.C. Marra, P. Eisenberger, A.Y. Cho, J. Appl. Phys. **50**, 6927 (1979)
32. J. Zegenhagen, F. Renner, Nachr. Chem. **54**, 847 (2006)
33. H. Kawamura, M. Takahasi, N. Hojo, M. Miyake, K. Murase, K. Tamura, K. Uosaki, Y. Awakura, J. Mizuki, J. Electrochem. Soc. **149**, C83 (2002)
34. M. Fleischmann, B.W. Mao, J. Electroanal. Chem. **247**, 297 (1988)
35. M. Nakamura, O. Endo, T. Ohta, M. Ito, Y. Yoda, Surf. Sci. **514**, 227 (2002)
36. E. Herrero, S. Glazier, H.D. Abruna, J. Phys. Chem. B **102**, 9825 (1998)
37. M. Fleischmann, B.W. Mao, J. Electroanal. Chem. **247**, 311 (1988)
38. C.A. Lucas, Electrochim. Acta **47**, 3065 (2002)
39. M. Fleischmann, A. Oliver, J. Robinson, Electrochim. Acta **31**, 899 (1986)
40. F.U. Renner, A. Stierle, H. Dosch, D.M. Kolb, T. L. Lee, J. Zegenhagen, Nature **439**, 707 (2006)
41. H. Kawamura, M. Takahasi, N. Hojo, M. Miyake, K. Murase, K. Tamura, K. Uosaki, Y. Awakura, J. Mizuki, J. Electrochem. Soc. **149**, C83 (2002)
42. M. Ito, M. Yamazaki, Phys. Chem. Chem. Phys. **8**, 3623 (2006)
43. T. Rayment, R.K. Thomas, G. Bomchil, J.W. White, Mol. Phys. **43**, 601 (1981)
44. E.D. Chabala, T. Rayment, Langmuir **10**, 4324 (1994)
45. A.R. Ramadan, E.D. Chabala, T. Rayment, Phys. Chem. Chem. Phys. **1**, 1591 (1999)
46. F. Bardé, M.R. Palacin, Y. Chabre, O. Isnard, J.-M. Tarascon, Chem. Mater. **16**, 3936 (2004)
47. M.J. Bedzyk, D. Bilderback, J. White, H.D. Abruna, M.G. Bommarito, J. Phys. Chem. **90**, 4926 (1986)
48. G. Materlik, M. Schmäh, J. Zegenhagen, W. Uelhoff, Ber. Bunsenges. Phys. Chem. **91**, 292 (1987)
49. H.D. Abruna, G.M. Bommarito, H.S. Yee, Acc. Chem. Res. **28**, 273 (1995)
50. M.J. Bedzyk, G.M. Bommarito, M. Caffrey, T.L. Penner, Science **248**, 52 (1990)
51. H.D. Abruna, G.M. Bommarito, D. Acevedo, Science **250**, 69 (1990)
52. A. Krolzig, G. Materlik, J. Zegenhagen, Nucl. Instrum. Methods **208**, 613 (1983)
53. A. Krolzig, G. Materlik, M. Swars, J. Zegenhagen, Nucl. Instrum. Methods **219**, 430 (1984)
54. G. Materlik, J. Zegenhagen, W. Uelhoff, Phys. Rev. B **32**, 5502 (1985)
55. H.D. Abruna, J.H. White, M.J. Albarelli, G.M. Bommarito, M.J. Bedzyk, M. McMillan, J. Phys. Chem. **92**, 7045 (1988)
56. M. Caffrey, J. Wang, Faraday Discuss. **94**, 283 (1992)
57. for an introduction to LSM see: T.W. Barbee, in *Low Energy X-ray Diagnostics*, ed. by D.T. Atwood, B.L. Henke (AIP, New York, 1981)
58. H.D. Abruna, J.H. White, M.J. Albarelli, G.M. Bommarito, M.J. Bedzyk, M. McMillan, J. Phys. Chem. **92**, 7045 (1988)
59. H. You, Z. Nagy, Curr. Top. Electrochem. **2**, 21 (1993)
60. Z. Nagy, H. You, Electrochim. Acta **47**, 3037 (2002)

61. B.M. Ocko, J. Wang, in *Synchrotron Techniques in Interfacial Electrochemistry*, ed. by C.A. Melendres, A. Tadjeddine (Kluwer Academic, Dordrecht, 1994), p. 127
62. B.M. Ocko, in *Spectroscopic and Diffraction Techniques in Interfacial Electrochemistry*, ed. by C. Gutiérrez, C. Melendres. NATO ASI Series C, vol. 320 (Kluwer Academic, Dordrecht, 1990), p. 343
63. M. Takahashi, Y. Hayashi, J. Mizuki, K. Tamura, T. Kondo, H. Naohara, K. Uosaki, Surf. Sci. **461**, 213 (2000)
64. J.X. Wang, R.R. Adzic, B.M. Ocko, in *Interfacial Electrochemistry*, ed. by A. Wieckowski (Dekker, New York, 1999), p. 175
65. D.M. Kolb, Angew. Chem. Int. Ed. [Angew. Chem.] **40**[**113**], 1162[1199] (2001[2001])
66. K.M. Robinson, W.E. O'Grady, Rev. Sci. Instr. **64**, 1061 (1993)
67. Z. Nagy, H. You, R.M. Yonco, Rev. Sci. Instr. **65**, 2199 (1994)
68. V. Stamenkowvic, C. Lucas, D. Tripkovic, D. Strmenik, N.M. Markovic, 210. Electrochemical Society Meeting, Cancun, Mexico, 29.10.–03.11.2006, Extended Abstract #1913
69. T. Kondo, J. Morita, M. Okamura, T. Saito, K. Uosaki, J. Electroanal. Chem. **532**, 201 (2002)
70. R.R. Adzic, J.X. Wang, O.M. Magnussen, B.M. Ocko, J. Phys. Chem. **100**, 14721 (1996)
71. F.U. Renner, T.L. Lee, S. Warren, B.C.C. Cowie, D.M. Kolb, A. Stierle, I. Costina, H. Dosch, R. Felici, Extended Abstracts of the 203rd Meeting of the Electrochemical Society, Paris, France, 27.04–02.05.2003, Ext. Abstr. #287
72. M.F. Toney, J.G. Gordon, M.G. Samant, G.L. Borges, O.R. Melroy, D. Yee, L.B. Sorensen, J. Phys. Chem. **99**, 4733 (1995)
73. B.M. Ocko, J. Wang, A. Davenport, H. Isaacs, Phys. Rev. Lett. **65**, 1466 (1990)
74. J. Wang, B.M. Ocko, A.J. Davenport, H.S. Isaacs, Phys. Rev. B **46**, 10321 (1992)
75. H. You, C.A. Melendres, Z. Nagy, V.A. Maroni, W. Yun, R.M. Yonco, Phys. Rev. B **45**, 11288 (1992)
76. J. Scherer, B.M. Ocko, O.M. Magnussen, Electrochim. Acta **48**, 1169 (2003)
77. H. Furukawa, M. Takahashi, M. Ito, J. Electroanal. Chem. **256**, 213 (1988)
78. M.F. Toney, J.N. Howard, J. Richer, G.L. Borges, J.G. Gordon, O.R. Melroy, D. Yee, L.B. Sorensen, Phys. Rev. Lett. **75**, 4472 (1995)
79. T. Wandlowski, J.X. Wang, B.M. Ocko, J. Electroanal. Chem. **500**, 418 (2001)
80. B.M. Ocko, G.M. Watson, J. Wang, J. Phys. Chem. **98**, 897 (1994)
81. Y.V. Tolmachev, A. Menzel, A. Tkachuk, Y.S. Chu, H. You, Extended Abstracts of the 203rd Meeting of the Electrochemical Society, Paris, France, 27.04–02.05.2003, Ext. Abstr. #2840
82. J.X. Wang, K. Tamura, B.M. Ocko, R.R. Adzic, Extended Abstracts of the 203rd Meeting of the Electrochemical Society, Paris, France, 27.04–02.05.2003, Ext. Abstr. 2839
83. S.V. Roth, P. Müller-Buschbaum, M. Burghammer, H. Walter, P. Panagiotou, A. Diethert, C. Riekel, Spectrochim. Acta B **59**, 1765 (2004)
84. G. Sandi, H. Joachin, R. Kizilel, S. Seifert, K.A. Carrado, Chem. Mater. **15**, 838 (2003)
85. D. Tatchev, R. Kranold, S. Armyanov, J. Electrochem. Soc. **150**, C502 (2003)
86. D. Gibbs, B.M. Ocko, D.M. Zehner, S.G.J. Mochrie, Phys. Rev. B **38**, 7303 (1988)
87. B.M. Ocko, in *Spectroscopic and Diffraction Techniques in Interfacial Electrochemistry*, ed. by C. Gutiérrez, C. Melendres. NATO ASI Series C, vol. 320 (Kluwer Academic, Dordrecht, 1990), p. 343
88. H. Kawamura, M. Takahasi, J. Mizuki, J. Electrochem. Soc. **149**, C586 (2002)

89. H. Kawamura, M. Takahasi, J. Mizuki, Extended Abstracts of the 203rd Meeting of the Electrochemical Society, Paris, France, 27.04–02.05. 2003, Ext. Abstr. #2837
90. D.H. Kim, S.S. Kim, H.W. Jang, D.Y. Noh, Extended Abstracts of the 203rd Meeting of the Electrochemical Society, Paris, France, 27.04–02.05. 2003, Ext. Abstr. #270

7 Surface Analytical Methods

In previous chapters a broad variety of methods has been treated which could be easily grouped into spectroscopic methods or techniques employing some sort of interaction between electromagnetic radiation and the electrochemical interface. Various techniques remain which are hard to assign to any of the families of methods presented so far. This is partly caused by the fact, that some of these methods combine optical, i.e. spectroscopic, details with experimental features providing spatial resolution or to the fact, that surface properties like e.g. electrical conductivity are measured which have no direct or obvious relationship to spectroscopic methods.

The advent of ever smaller electrochemically cells (microcells, capillary cells) which can be placed on selected areas of an electrode surface allows spatially resolved measurements of local properties. Spectroscopic methods modified in such a way like e.g. locally resolved electrochemical mass spectrometry have been treated in previous sections. Optical methods incorporating scanning probes will be treated below. Classical electrochemical methods like e.g. impedance measurements employing these miniaturized cells [1] thus providing localized information will not be treated in this book. The same applies to scanning electrodes employed in localized electrochemical impedance measurements (LEIS).

7.1 Topographic Methods

A topographic picture of a surface can be obtained in various ways. Mostly optical methods yielding pictures based on secondary electrons that are emitted from the investigated surface or other probes and signals have been employed so far. Most of these methods are applicable only under vacuum and the obtained pictures are indirect, i.e. they are based upon the interaction between the surface and a probe (e.g. an electron beam). Directly scanning the surface on a microscopical scale using an appropriately small probe (tip) has become possible with micromanipulators of sufficient mechanical resolution and reproducibility (see [2, 3]). A first approach that closely resembles the well-known technology employed in phonographic pickup cartridges has been described [4, 5]. The device employs piezoelectric actuators. Its operation is based on the principle of a field emission probe. A field emitter with a typical radius of 100–10,000 Å that is close to a conducting surface shows an electric field strength given by the Fowler–Nordheim equation when a constant current is passed between tip and surface. The amplified voltage is applied to the Z-axis piezo actuator. Thus a constant distance between tip and surface are maintained. The applied voltage is a measure of the altitude of the surface. By scanning the position of the tip with X- and Y-servos, this altitude can be measured across the surface; a topographic picture of the surface results. The device operates only under vacuum conditions. A vertical resolution of 30 Å and a horizontal resolution of 4000 Å have been achieved. Experimental results have been reviewed [6].

7.2 Scanning Probe Methods

The advent of micropositioners capable of moving a probe (e.g. a tiny metal tip) in closest proximity to the solid surface to be investigated (i.e. in nanometer distance) with high spatial resolution based predominantly on piezocrystal-driven actuators has made a variety of scanning probe methods, or scanning probe microscopies (SPM), possible. Methods are named depending on the principle of measurement and the type of probe. Several methods can be used in different modes of operation (e.g. the scanning tunneling microscope can be run in the constant distance and the constant current mode). In addition, some methods have been developed into further variations. The following overview of established methods starts with a general description of the most often used variant of a method; variations are included wherever it seemed appropriate. A classification of scanning probe microscopies has been provided [7, 8].

The use of light (photons) instead of electrons results in the photon scanning tunneling microscope (PSTM) [9]. A fiber tip with an apex shape similar to the one used in other SPMs with electrons is positioned close to the investigated surface. As outlined in Fig. 7.1, the surface is illuminated from the backside with a laser beam or even with a beam of white light at an angle larger than the critical angle.

On top of the hemispherical cylinder, the sample with the surface to be investigated is mounted with an index matching gel. Thus the upper surface becomes the interface where internal reflection occurs. Light tunneling through the surface apex gap is guided via the optical fiber to the photomultiplier. Its intensity is directly related to the distance in an exponentially decaying way. Thus the intensity can be used as a height indicator. In advanced setups, Raman spectroscopy or optical ab-

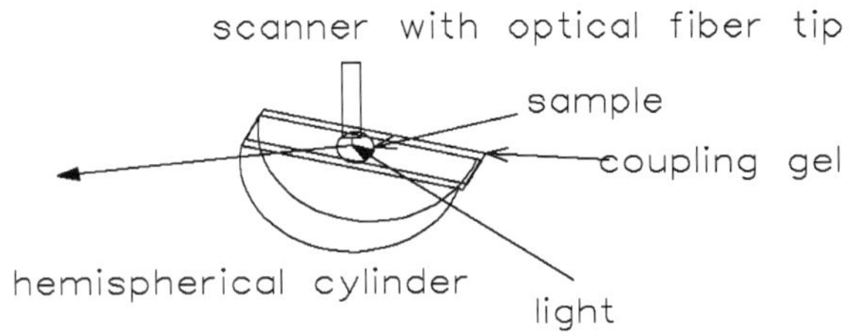

Fig. 7.1. Schematic setup for photon scanning tunneling microscopy

sorption spectroscopy of species at the interface have been proposed. Uses in an electrochemical environment have not been reported so far.

Scanning the surface with a laser beam in a confocal arrangement can result in topographic information about the investigated surface; since the confocal arrangement and further optical features dominate this experimental approach, whereas the scanning feature is of lesser importance, this microscopy is treated in the Sect. 7.3.

7.2.1 Scanning Tunneling Microscopy (STM)

Fundamentals. An STM uses an atomically sharp probe tip of an electronically conducting material in close proximity ($\cong$ 1 nm) to the surface under investigation. The tip is rastered (scanned) relative to the surface using piezoelectric devices. Thus an STM can be used to directly monitor the local density of electronic surface states with atomic resolution. The current flowing between the tip and the surface when a small voltage is applied depends on the exponential dependency on the tip-surface distance that is characteristic of an electronic tunneling process. This results in the remarkable vertical resolution of the apparatus. In a simple approximation [10], the tunneling current I can be given in terms of local density of states (LDOS) $\rho_s(E_F)$ at the Fermi level, distance d, applied voltage U and decay constant κ:

$$I \approx U\rho_s(E_F)\exp(-2\kappa d).$$

Using average values for a semi-classical square potential barrier and effective barrier height of 4 eV yields a value of $\kappa \cong 1$ Å^{-1}. Further details are provided elsewhere [11]. This causes the tunneling current to drop by about an order of magnitude per Å of increased distance, which finally results in the indicated vertical resolution. The very pronounced distance–current relationship contributes also in a very specific way to the high resolution: Although tips may be very sharp, they nevertheless show even in the perfect case a curvature on an atomic scale. Instead, in the real world the tip is generally composed of several atoms because it is prepared by cutting a wire at an oblique angle or by etching a thin wire [12]. Nevertheless, there will most likely be one atom protruding somewhat beyond the neighbors. As the tunneling currents drops by about an order of magnitude when the distance increases by 1 Å (100 pm), the dominant fraction of the tunneling current will flow across this particular atom. In the case of extremely blunt tips, the image of the tip will be recorded instead of the investigated surface.

Probing the LDOS can be done in two fundamentally different ways:

1. The tip can be scanned at constant height (distance) over the surface. This is possible only when the surface is smooth on an atomic scale, otherwise the tip might crash into surface features.
2. Alternatively the tip can be scanned in a constant tunneling current mode. In this case the actually measured current is fed into an electronic regulation circuit which adjusts the actual tip-surface distance to a value resulting in a constant value of the tunneling current at all probed places. The signal fed into the z-axis piezodrive provides information about the local elevation. This mode works

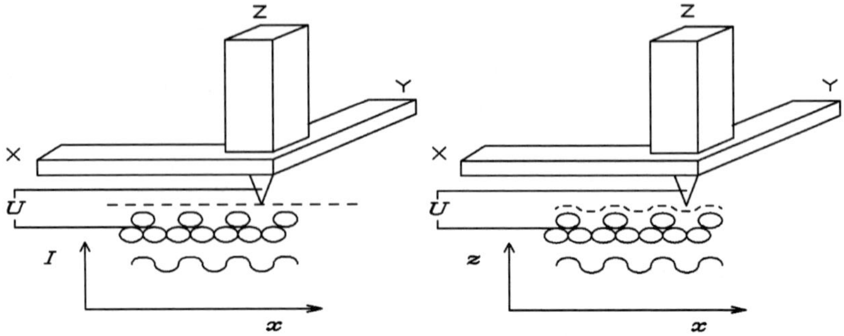

Fig. 7.2. Scheme of scanning tunneling microscopy in the constant height mode (*left*) and the constant current mode (*right*); X, Y and Z designate the respective piezodrives, *dashed line* indicates tip position

well even with strongly structured surfaces; it is inherently slower because of the employed feedback circuitry.

At first glance, both methods seem to be equivalent, i.e. the results should be the same. Upon closer inspection, differences appear. The first method, which operates at constant height, yields results wherein the observed tunneling current taken as a measure of the tip-surface distance is influenced by the distance–current relationship mentioned above. Interpreting the observed pictures requires a sophisticated understanding of this relationship for the system under investigation.[1] This problem is absent in the second mode. In order to get a correct topographic picture, a precise knowledge of the relationship between LDOS and actual atomic surface structure on an atomic level is needed. Both modes of operation are shown schematically in Fig. 7.2.

Both designs were initially developed and applied under vacuum conditions, yielding microscope pictures with atomic resolution [13, 14]. Very soon it was found that this design was also suitable for measurements at ambient temperature or even in the presence of an electrolyte solution [15]. The need to maintain the tip at a certain potential (bias) with respect to the surface (i.e. the electrode) under investigation and to keep this electrode itself at a selected potential adds to the complexity of the experiment. The tip acts as a probe for tunneling microscopy and as an ultramicroelectrode in the electrolyte solution. Attention has to be paid to conceivable Faradaic processes occurring at the tip. This current is usually minimized by coating the tip with an insulating material with only its apex exposed to the solution. A typical current of the STM of about 1 to 10 nA corresponds to 10^6 A cm^{-2} flowing between the tip and the probed section of the surface under investigation (typically 10^{-14} cm^2). Any Faradaic current at the tip that is caused by an electrochemical process would flow across the whole exposed surface area of the tip (about 10^{-8} to 10^{-10} cm^2),[2] thus 10 nA would correspond to about 10 to 100 A cm^{-2}. This Faradaic current

[1] This plays a major role in scanning tunneling spectroscopy (see p. 277).

[2] In addition, the tip is covered with an insulating material (wax, resin) that leaves only the apex of the tip exposed to the electrolyte solution.

density is much smaller than the tunneling current and no distortion of the tunneling current should be expected.

Chemical modification of the microscope tip most frequently prepared from metals like tungsten or gold results in surface properties that are useful for transferring "chemical sensitivity" to the tip. Chemical modification of the tip (by coating with polypyrrole or with self-assembled monolayers (SAM) [16]) resulting in enhanced hydrogen bond or coordination bond interaction with species on the scanned surface results in enhanced tunneling electron transfer and increased brightness of the observed surface location or in higher contrast [17].

Instrumentation. In order to operate a STM under *in situ* conditions, i.e. in the presence of an electrolyte solution, some conditions have to be fulfilled. The design of the STM must allow investigation of a horizontal surface at the bottom of the microscope. The tip has to be coated as completely as possible in order to minimize the Faradaic current. Since the potential of the electrode surface under investigation has to be maintained at a fixed, controlled potential with respect to a reference electrode, a four-electrode arrangement requiring a corresponding bipotentiostat is necessary. The schematic drawing of the electrochemical cell as depicted in Fig. 7.3 shows the major components.

The peripheral components necessary for this experiment are indicated in Fig. 7.4.

An experimental setup specifically dedicated to electrochemical *in situ* investigations has been described in detail [18]. The preparation of suitable tips preferably prepared from metals like Pt, Pt–Ir, Ir, and W has been reviewed elsewhere [10]. The importance of operation under a controlled gas atmosphere, in particular the

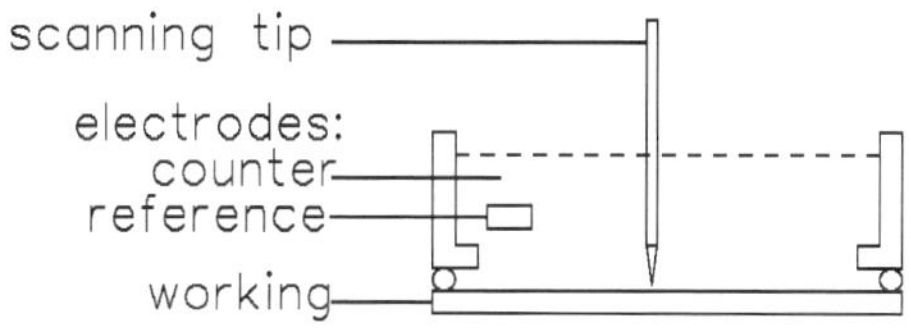

Fig. 7.3. Scheme of an electrochemical cell for *in situ* investigations with an STM

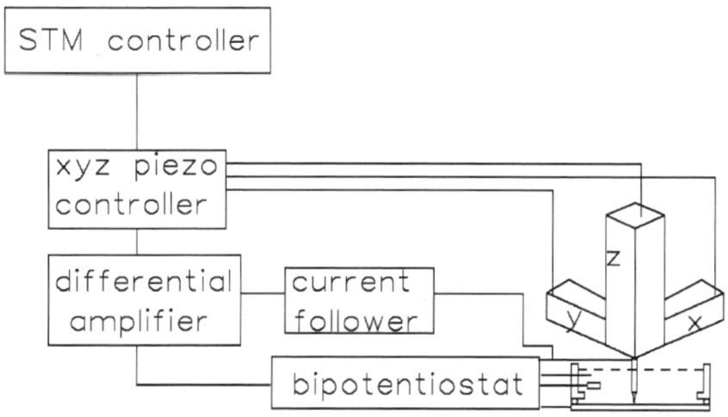

Fig. 7.4. Scheme of a setup for electrochemical *in situ* investigations with an STM

Fig. 7.5. STM pictures of a Au(111) electrode surface in contact with an aqueous solution of 0.1 **M** H_2SO_4 at various electrode potentials as indicated (picture provided by D. Kolb, University of Ulm)

influence of dioxygen, has been examined extensively [19]. An experimental setup for measurements at elevated temperatures [20] has been described.

Fast scan measurements, i.e. for investigations of the dynamics of surface diffusion or reconstruction are done preferably in constant height instead of constant current mode because no electronic feedback circuit, limiting response time and scan speed, is involved in this mode. Obviously this works only with very smooth electrode surfaces. An electronic setup (bipotentiostat) that allows fast transient methods combined with scanning probe microscopies has been reported [21].

The formation of a well-ordered adsorbate layer could be demonstrated using an STM. The dependence of the formation of the adsorbate layer on the electrode potential, i.e. in the case of an anion at a sufficiently positive electrode potential, is obvious in Fig. 7.5.

Scanning tunneling microscope pictures (Fig. 7.6) obtained during the deposition of copper on Au(111) electrode surfaces provide evidence that copper deposition proceeds almost exclusively at the steps [22].

The hexagonal arrangement of surface atoms after reconstruction of a Au(111) surface in contact with an aqueous solution of 0.1 **M** H_2SO_4 can be shown with STM (Fig. 7.7).

General overviews of STM studies of metal electrodes have been provided [10, 23–25]; in addition, an extensive review that focused on ordered anion monolayers on metal electrode surfaces has been published [26]. Charge-induced surface phase transitions on ordered Au(111) caused by increasing iodide adsorption from an aqueous electrolyte solutions have been observed [27]. The formation of copper sulfide nanostripe patterns on a Au(111) electrode surface formed by exposure of a single copper monolayer on this electrode and exposed to bisulfide ions in the electrolyte solution has been studied with *in situ* STM [28]. Correlations between

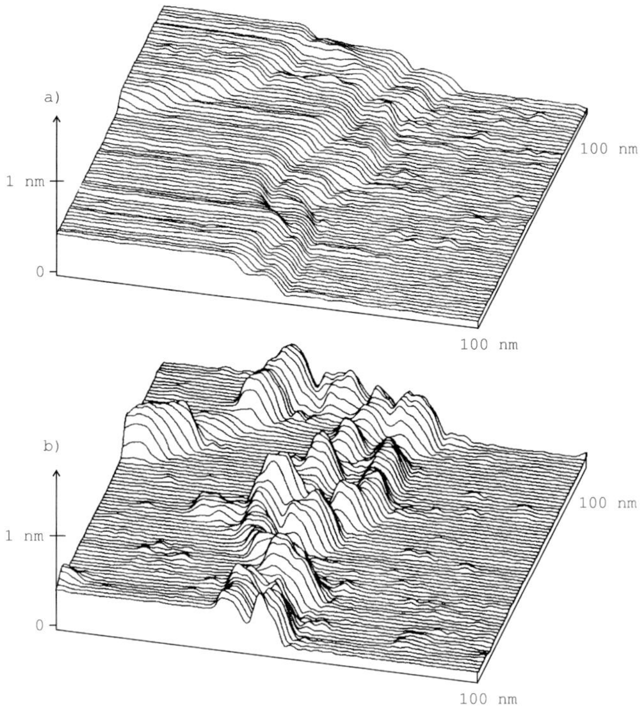

Fig. 7.6. STM pictures of a Au(111) electrode surface before (*top*) and during (*bottom*) copper deposition from an aqueous solution of 5 m**M** $H_2SO_4 + 0.05$ m**M** $CuSO_4$ (pictures provided by D. Kolb, University Ulm; for further details see [22])

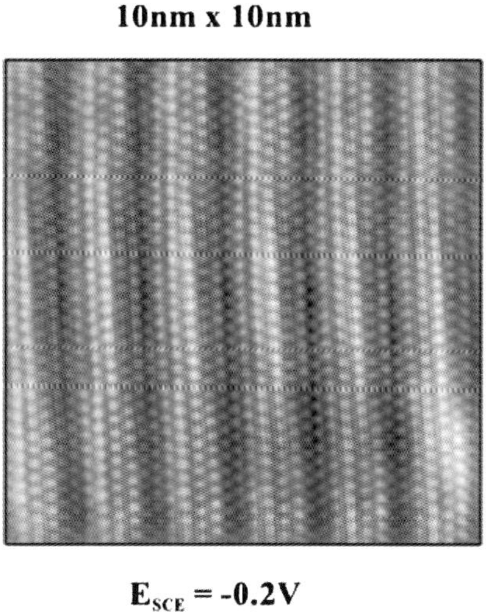

Fig. 7.7. STM picture of a reconstructed Au(100) surface (picture provided by D. Kolb, University of Ulm)

STM and SEM pictures have been discussed [29] and implications for reversible oxidative roughening have been pointed out [30]. Relationships between observed topography, electrochemical roughening parameters and Raman spectroscopic features have been discussed [31]. Metal deposition, including underpotential deposition processes on metal substrates [32–35] and on highly ordered pyrolytic graphite [36], has been frequently studied. The influence of solution additives (both organic molecules and inorganic anions) has been investigated [37–39] and organic adsorbate layers have been reviewed elsewhere [40]. Applications of STM to disordered (polycrystalline) materials, including metallic glasses, have been described [41]. Topography and local barrier height of metallic glass surfaces were measured. Step and island dynamics at liquid/solution interfaces have been studied and comparisons with the solid/gas interface were drawn [42]. Step and kink energies on Au(100) electrode surfaces could be derived from island studies with an STM [43]. Real-time observations of surface reactivity and mobility that were initially made *ex situ* only [44] have been extended to *in situ* observations[3] recently. Reported examples included CO adsorption on Au(111) electrodes [45], adsorption of alkanethiols on Au(111) [46] and reductive desorption of self-assembled monolayers of hexanethiol from Au(111) surfaces [47, 48]. In the latter study it was observed that desorption initiates at defects in the SAM, at missing rows and edges of vacancy islands. Both formation and final structure of self-assembling osmium-bipyridine complexes were monitored [49]. Adsorption and subsequent monolayer film formation of various protoporphyrins on a HOPG surface have been studied with both STM and AFM [50]. Studies of active metal dissolution [51] have been reported. In an investigation of electrodeposition of bismuth on a graphite surface, formation of small particles of about 10 nm diameter was observed initially [52]. Upon making contact with neighbors, these particles coalesce. The coarsening of platinum island deposits in the electrochemical double layer potential region has been studied with STM [53]. Underpotential deposition of lead on a Cu(100) electrode surface initially revealed a high surface mobility [54]. Subsequent deposition of lead caused numerous changes of both structure and dynamics at the interface. Copper deposition and stripping at a gold electrode has been investigated [55]. Dissolution of highly polished copper surfaces showed roughening and formation of facets [56]. During selective dissolution of copper from Cu–Au alloys, pit formation and finally porosity were observed. AuCN-adlayers were found during adsorption from aqueous solutions of $KAu(CN)_2$ on Au(111) surfaces [57]. Adsorption of NO from a KNO_2-containing electrolyte solution on a Rh(111) surface and the subsequent reduction were monitored in real time [58]. The reaction proceeds preferably at atomically flat terraces, not at surface defects. Initial reaction fronts were spatially concentrated, not randomly distributed.

[3] There seems to be some debate over the limit between static and dynamic measurements related to the question of the frequency of image registration (frames per second) at which a true real time observation is done. Obviously this judgement—if it is necessary at all— has to be made with reference to the rate of change at the surface. No attempt is made in this text to enter this discussion. Consequently, no attempt is made in this text to enter into this debate: "video STM" will not be separated from other "real time STM".

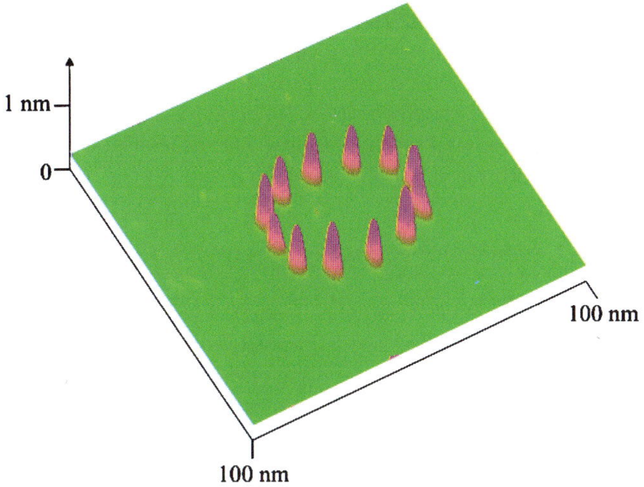

Fig. 7.8. STM pictures of a Au(111) in contact with an aqueous solution of 0.1 **M** H_2SO_4 at various electrode potentials as indicated (picture provided by D. Kolb, University of Ulm)

A computational tool for analysing observations obtained with STM based on a self-consistent semiempirical molecular orbital model has been described [59]. The electrodeposition of aluminum and titanium–aluminum alloys that are potentially useful as corrosion protection layers from room temperature molten salt electrolytes has been studied with STM [60]. Underpotential deposition of cadmium on Ag(111) surfaces from ionic liquids has been monitored *in situ* with STM [61], spinodal decomposition and surface alloying were observed. The use of STM beyond simple surface imaging, including molecular identification, investigation of molecular reactivity, electron transfer kinetics and nanofabrication have been reviewed elsewhere [62, 63]. Investigations of the semiconductor/solution interface beyond topographic ones [64] with varied tunnel gap distances and tip potentials allowed the separation of the effects of the tunneling barrier and the Schottky barrier at this interface [65].

An STM can be used to study the effects of illumination of a semiconductor surface by measuring local photovoltages resp. photocurrents [66]; surprisingly, this approach has been employed so far only in *ex situ* studies. Gold/*n*-Si(111) nanocontacts (interface area about 10^{-12} cm^2) that were prepared by electrodeposition of gold onto *n*-Si(111):H substrates have been studied with an STM [67] in the presence of an aqueous solution of 0.02 **M** $HClO_4$. By varying the tip voltage and the electrode potential of the silicon substrate, a Schottky diode behavior of the nanocontact was verified.

An emerging application of the STM is the structuring of an electrode surface on a nanometer scale (nanostructuring). In a representative example (see Fig. 7.8), selective copper deposition resulted in a regular arrangement of metallic tips [68].

The use of an STM to deposit silver inside a polymer electrolyte film (Nafion®) has been reported [69]. A combination of an STM with an SECM (see also below for this method) has been described [70]. The specifically adapted tip can be operated both in tunneling mode and as a probe electrode for scanning the surface over a large distance in the feedback mode, measuring diffusion-controlled oxidation of a mediator. This way, the topography and local reactivity can studied with a single instrument and with high spatial reproducibility. The influence of the geometry of a nanometer-sized electrode has been discussed [71].

7.2.2 Differential Conductance Tunneling Spectroscopy (DCTS)

Fundamentals. Modulation of the tip bias voltage of an STM at a frequency much larger than the time constant of the STM's tip-positioning constant-current feedback circuit results in a modulation of the tunneling current. At a small amplitude modulation, this signal corresponds to the density of tunneling states at the bias voltage [72]. This signal forms the basis of DCTS. The DCTS image of a surface obtained in this way can be understood as the variation of the density of tunneling states on the surface.

Instrumentation. An STM is equipped with suitable additional electronics to generate the desired bias modulation and to detect the modulation of the tunneling current [72]. Differential Conductance Tunneling Spectroscopy data that was obtained for a platinum film electrode have been interpreted in terms of step density and surface disorder [72].

7.2.3 Atomic Force Microscopy (AFM)

Fundamentals. Beyond the tunneling current flowing between the tip and the surface, further interactions are effective between the tip and the surface. Spence et al. [73] have observed strain fringes on a graphite surface interacting with an STM tip that extends 200 nm from the tunnel junction. This observation led to the development of the atomic force microscopy (AFM)[4] by Binnig et al. [74]. Depending on the design (including surface coating) of the tip van der Waals forces, electrostatic or magnetic forces can be monitored [75]. Generally, forces between 10^{-9} and 10^{-6} N are measured; there have been reports describing measurements down to $3\ 10^{-13}$ N [76]. They can be attractive or repulsive. When considering interatomic interactions, the force reaches a minimum at the mechanical point contact and, at smaller distances, the repulsive interactions measured in the contact mode dominate; at greater distances, the attractive interactions observed in the non-contact mode dominate. In the contact mode, the tip actually touches the surface. Obviously the electronic conductivity of the surface and the tip play no role in the operation, thus non-conducting surfaces that are not suitable for the use of STM can be studied. The presence of a liquid between tip and surface provides no fundamental problem;

[4] The terms SFM (scanning force microscopy) and AFM (atomic force microscopy) are used synonymously; the former term is used less frequently. In the initial stages of development, the latter term was exclusively used for setups providing atomic (or better) resolution.

of course, the mechanical properties of the whole setup are changed. Nevertheless electrochemical measurements *in situ* are possible.

Chemical sensitivity can be conferred to AFM by coating the tip with covalently linked monolayers which affect the tip–surface interaction; the method is called chemical force microscopy [77]. Additional modulation of the piezo actuator operating in z-direction and evaluation of the force signal can be used to measure the adhesion force between a surface and a chemically modified AFM tip [78]. Metal coated AFM tips can be used in a scanning electrochemical microscopy (SECM, see p. 264) mode [79] in studies of crystal dissolution or growth where surface processes are associated with considerable fluxes of species.

Instrumentation. A cantilever with a sharp tip interacting with the surface under investigation is used. The actual bending of the cantilever is measured with a laser beam deflected from a mirror-like surface spot on the back of the cantilever towards a position-sensitive photodetector. The measured signal is used to control the piezo actuators. A constant force mode in which the cantilever–surface distance is kept at a preset interaction force and a constant height mode of scanning operation are possible. The principle of operation is schematically outlined in Fig. 7.9.

The mechanical properties of the tip-cantilever assembly are of central importance. Caused by the forces that are effective between surface and tip, the cantilever is deformed. This deformation controls the overall performance of the microscope. The spring constant k and the resonance frequency ω_0 of the cantilever are particularly important. In order to be insensitive to mechanical noise from the environment, a high resonance frequency is desirable. A small spring constant in turn is required to detect weak forces. To obtain high resonance frequencies, stiff materials (silicon, silicon oxide or silicon nitride) are used for the cantilever. A small spring constant can be maintained by limiting the mass of the device to a minimum by microfabrication techniques. A typical cantilever has a length of 0.1 mm, a thickness of 1 μm, a spring constant around 0.1 to 1 nm^{-1} and a resonance frequency around 10 to 100 kHz. The development of single-wall carbon nanotubes (SWNT) as tips for AFM to also be used in electrochemical investigations has been described [80]. An alternative mode of operation without optical detection as described above employs a tip attached to a vibrating fork-like assembly. This approach has resulted in very high resolution; unfortunately it cannot be employed in an electrochemical environment because of the dampening effect of the electrolyte solution. The integration of an ultramicroelectrode into a tip for atomic force microscopy has been accomplished

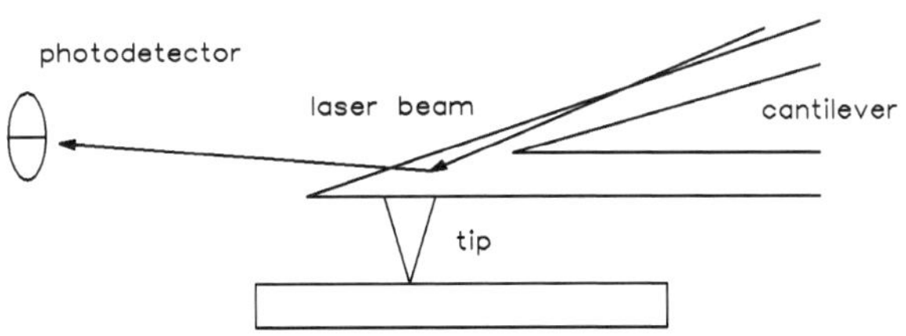

Fig. 7.9. Scheme of an atomic force microscope

[81]. The electrochemically active area is located as a ring around the tip. It has been used in SECM measurements; an AFM picture was simultaneously obtained.

In actual operation in the contact mode, the tip touches the surface like the stylus of a record player. In the non-contact mode, the cantilever is oscillated at a frequency close to the resonance frequency with a large amplitude. In this mode, vertical long-range forces are probed, whereas lateral forces (friction-like forces in the plane of the sample surface) are almost non-effective. These forces have been employed in lateral force microscopy (LFM).

Investigations published so far include metal dissolution studies (relevant to corrosion and corrosion inhibition) [82], underpotential metal deposition (upd) [83] and overpotential deposition (opd) [84]. Structural features of deposits, the influence of electrolyte composition, electrode potential, etc. were reported. In a study employing both AFM and LFM, specific adsorption and phase changes at the polycrystalline silver/halide-containing electrolyte solution were investigated [85]. Whereas AFM provided topological imaging, LFM enabled detailed studies of adsorption and chemical reactions in adsorbate layers. Specifically, adsorbed halide anions with their hydration shells stripped provided higher friction values probed with LFM and hydrated anions in the outer Helmholtz layer that were not adsorbed to specific sites and maintained intact hydration shells caused lower friction values. Using a colloidal probe (a silica particle attached to the cantilever of an AFM), the diffuse layer properties of a thiol-modified gold electrode has been investigated [86].

With chemically modified AFM tips, adhesion forces between the tip and a two-component self-assembled monolayer on a gold electrode have been studied [87]. Utilizing the different strengths of interaction between the modified tip (methyl and carboxyl terminating group functionalized), SAM areas with methyl and carboxyl end groups could be distinguished.

Several reviews dealing with the fundamentals, experimental aspects and applications have been published [88–90]. Operated in the constant force mode, the AFM can monitor changes in the thickness of a film (e.g. a metal hydroxide, which shows swelling/shrinking during redox processes) [91]. Dimensional changes of highly oriented pyrolytic graphite (HOPG) during lithium ion intercalation/deintercalation have been studied with an AFM [92]. During the first intercalation cycle, an irreversible increase of layer spacing was found. In the following cycles, a reversible change of 17% of the layer spacing was measured. Roughness effects caused by the formation of a solid electrolyte interface were taken into account by statistical analysis of the data. Electrochemical deposition and dissolution of molecular crystals of organic conductors have been studied [93]. Morphological changes occurring during electropolishing of stainless steel in an ionic liquid have been identified with AFM [94].

Atomic force microscopy has been combined with nano-indentation and nano-scratching studies [95]. The hardness (and, to a similar extent, the friction coefficient) of passivated titanium was three to four times higher under *in situ* conditions, this was assigned to a much faster repassivation process in the presence of the passivating electrolyte solution. Nanotribology, particularly surface friction forces measurements of electrode surfaces modified with submonolayer foreign metal (upd deposits with AFM have been reported [96]). An AFM operated in the contact mode was used to scratch a surface of the aluminum alloy AA2024-T3 in contact with electrolyte solutions of different compositions (with/without chloride, dichromate) and under varying experimental conditions (stagnant/flowing solution) to gain insights into corrosion, protection and breakdown [97].

7.2.4 Scanning Kelvin Probe Force Microscopy (SKPFM)

Fundamentals. An AFM can also be used to probe the local Volta potential. Using a metal-coated silicon tip (e.g. Co–Cr40), first the topography of the surface under investigation is mapped using the tapping mode. In a second scan, the tip is moved along and kept at a constant distance of 50 nm above every point on the surface. An AC voltage is applied to the tip, generating an oscillating dipole. In the presence of an external field this will in turn create a mechanical oscillation of the cantilever, which can be detected using the standard features of the AFM. At every point of the scan a DC ramp is added to the AC modulation. At the DC voltage, where the oscillation of the cantilever vanishes, the potential on the tip and on the surface are the same. Thus a map of the surface Volta potentials with respect to the tip is created. Because the potential of the tip might be unstable and could vary from experiment to experiment, calibration is necessary. A particularly reproducible reference is a nickel surface exposed to deionized water before the measurement [98]. In the absence of further calibration this is the point of reference. The method cannot be applied in the presence of an electrolyte solution because of the large voltages applied to the tip, which would cause Faradaic reactions. Data from measurements of Volta potentials at corroding surfaces could be related to corrosion potentials of the same surface in contact with a solution because the linear correlation has been established before [98]. Nevertheless studies at air or in the presence of ultrathin electrolyte films (i.e. under conditions frequently encountered in atmospheric corrosion) are possible. The general advantages of SKPFM in comparison with SKP, in particular the greatly enhanced spatial resolution, have been discussed in detail [99]. A critical review of the applications of SKPFM that focuses on corrosion science with particular attention to possible artifacts and a comparison with SKP has been provided [100].

Instrumentation. The experimental procedure for an AFM equipped with a suitably coated tip has been outlined above. In a study of an aluminum alloy AA2024-T3, intermetallic particles and the matrix phase could be separated clearly [98]. The different surface films on these phases could be associated with their corrosion behavior. Inclusions and their corrosive behavior have been studied with a combination of SKPFM and AFM [101]. The effect of chloride-containing solution on corrosion at the matrix and the intermetallic particles was studied with SKPFM, in addition, light scratching with the AFM in the contact mode was applied to study the effect of the mechanical destabilization [102]. The intermetallic particles dissolved immediately after the film on their surface had been destabilized by mechanical abrasion.

A general study of the influence of experimental parameters applied during emersion of the electrode, distance between tip and surface, influence of oxide coverages, etc. on the observed Volta potentials has been reported [103]; relationships to previous studies at emersed electrodes (see [104–107]) and the topic of the adherence of the electrochemical double layer on an emersed electrode have been discussed. The influence of aluminum in magnesium alloys on atmospheric corrosion (in the absence/presence of CO_2) was studied with SKPFM [108] and a corrosion mechanism was suggested. Applications and limitations of SKPFM in studies of the surface of aluminum alloys have been reviewed thoroughly [109]. The surface of cast AlSi(Cu) alloys has been characterized with SKPFM [110]. Numerous particles of different composition were detected and they showed a positive Volta potential difference relative to the matrix with the actual value depending on the matrix composition.

Filiform corrosion on epoxy-coated 1045 carbon steel was investigated with SKPFM [111]. Under coatings of 150 and 300 nm thickness at 93% relative humidity, samples were studied under air. Separation of active anode and cathode locations in the head of the filament could be identified. Microscopic and even submicroscopic aspects of electrochemical delamination have been studied with SKPFM [99].

7.2.5 Scanning Electrochemical Microscopy (SECM)

Fundamentals. A microelectrode[5] with a small diameter (e.g. 10–20 μm, such an electrode is sometimes also called ultramicroelectrode (UME) [112–116]) is exposed to an electrolyte solution containing an electrochemically active substance. The electrode potential is adjusted to a value sufficiently negative to drive the electrochemical reaction $O + ne^- \rightarrow R$ under diffusion control. Diffusion of reactive species to the electrode surface is hemispherical instead of planar, as in the case of large electrodes. The current I flowing across the solid/electrolyte solution interface of the microelectrode tip quickly reaches a steady state value $I_{T\infty} = 4nFDcr$ with n as the number of electrons transferred in the electrochemical reaction step, F the Faraday constant, D the diffusion coefficient of the reacting species, c its concentration and r the tip radius. The experimental setup is pictured schematically in Fig. 7.10.

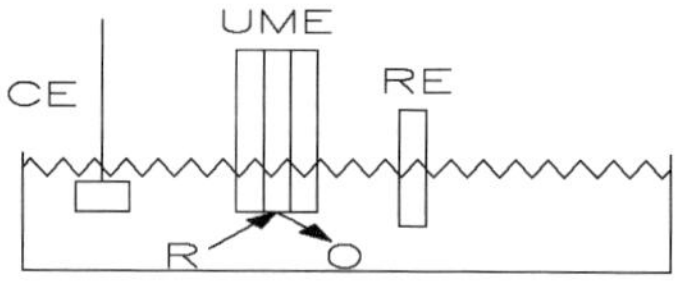

Fig. 7.10. Schematic setup for voltammetry with an ultramicroelectrode (UME); CE: counter electrode; RE: reference electrode

[5] A microelectrode is a conductive material with an active surface area of a few μm^2 that is embedded in an insulating material. Microelectrodes are commonly fabricated by coating a metal or carbon fiber with a polymer or glass sheath.

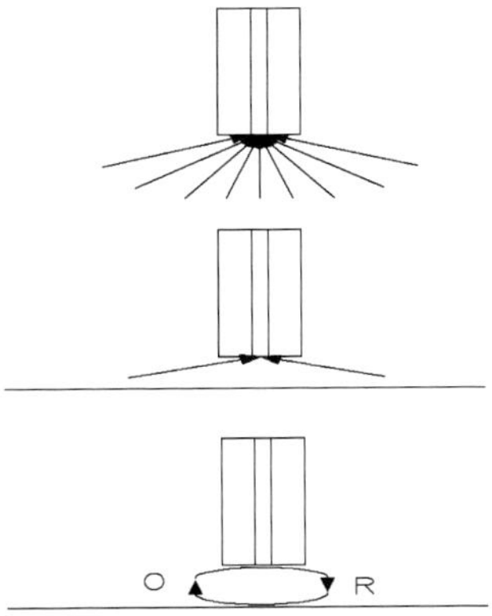

Fig. 7.11. Principle of SECM with the UME far away from a surface (*top*), approaching an insulating surface (*middle*) and approaching a conductive surface kept at a suitable electrode potential (*bottom*)

When the UME is moved close to an insulating surface, the current drops to a lower value I_T because the surface and the insulating sheath of the UME block transport of active species O. This effect is sometimes called negative feedback and is further enhanced by the fact that no reoxidation of R can occur at insulating parts of the surface. Approaching a conductive surface kept at an electrode potential where reoxidation of R is possible causes an opposite effect (positive feedback) and I_T is enhanced with a closer distance. Both possibilities are schematically depicted in Fig. 7.11. A similar effect may be observed with an unbiased (not kept at any specific potential, but instead at open circuit) surface. Because the large surface area is in contact with the solution containing a supply of O, the surface electrode potential is essentially controlled by the Nernst equation. At the potential established by the concentration of O, the reduced species R created at the UME will be reoxidized, whereas further O is reduced elsewhere on the surface.

In an operational mode where species R are generated at the surface and collected and reduced at the UME, an SG/TC experiment is done (substrate generation/tip collection). This mode is particularly interesting when localized phenomena or properties of an inhomogeneous surface are studied. The change of the current I normalized with respect to the current at infinite distance as a function of the distance d of the microelectrode from the surface normalized with respect to the radius of the microelectrode is depicted in Fig. 7.12 for a conductive and an insulating substrate (for a brief overview of the associated mathematics, see also [117]). An introductory overview has been published [118]. This rather general description of the principles of an SECM will become more complicated when further constraints (e.g. electron transfer at the tip or surface is charge transfer controlled instead of diffusion controlled) are considered [119–121]. The described mode of operation is based on current measurement and is called amperometric mode. Localized catalytic acitivity of the particles of a catalyst deposited on an inert current collector support can be monitored with SECM in the redox competition mode (RC-SECM)

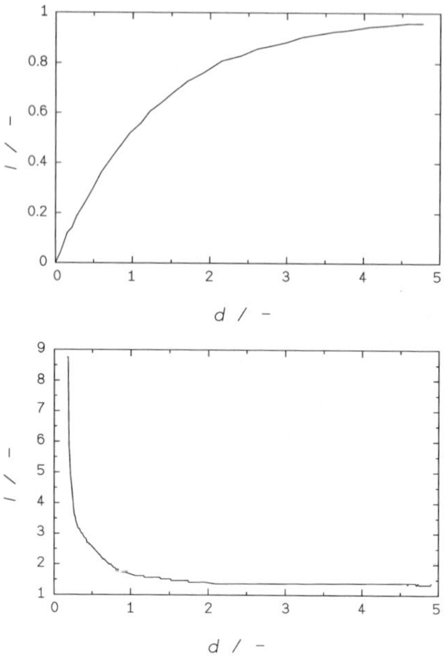

Fig. 7.12. Current I measured at a microelectrode normalized with respect to the current at infinite distance between tip and surface as a function of distance d between microelectrode and surface (normalized with respect to the radius of the microelectrode) for a conductive substrate (*top*) and an insulating substrate (*bottom*)

[122]. With an SECM positioned above a surface with deposited nanoparticles of dioxygen reduction electrocatalysts in a solution saturated with dioxygen, significant decreases of tip current are observed. Variations in the nature of catalyst spots, including inhomogeneities, could be localised. Further modes include the potentiometric mode with an ion selective UME that is used to probe the local composition of the solution. This method is basically equivalent to the scanning ion-sensitive electrode technique SIET (see p. 270, particularly pH microscopy).

An AC voltage can be applied to the UME and a counter electrode (AC-SECM). The AC current response can be evaluated and it can provide information about local surface conductivity of the surface under investigation [123–125]. This setup has been applied to interrogate living cells [126]. Enhanced spatial resolution may be obtained by using a shear force-based distance control to operate the UME at submicrometer distance.

Instrumentation. A suitable microelectrode [119] or nanoelectrode [127] is attached to a piezo-driven micropositioner. It is connected as the working electrode with a potentiostat. A counter electrode and a reference electrode are wired in a three-electrode arrangement. Investigations with conducting substrates require the use of a bipotentiostat. The surface to be investigated is immersed into the electrochemical cell together with the other electrodes. The position of the microelectrode and the flowing current are controlled and monitored by a computer equipped with

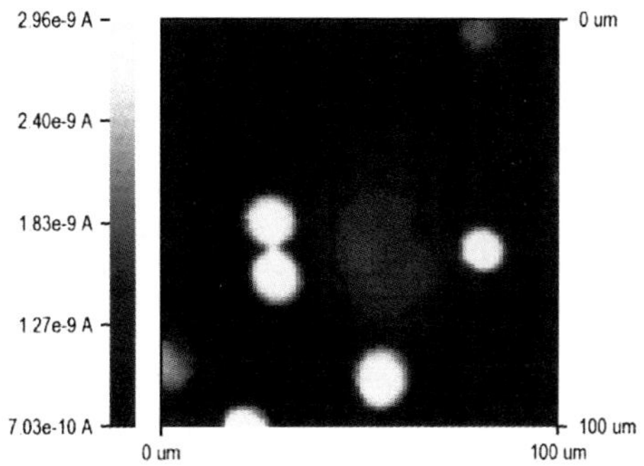

Fig. 7.13. SECM picture of a polycarbonate membrane

necessary interface cards and software [128]. As an example, the SECM image of a filtration membrane prepared from polycarbonate with an average hole size of 10 μm in a solution containing $[Fe(CN)_6]^{4-}$-ions scanned with a 2 μm UME is shown in Fig. 7.13.

Lateral charge propagation in a monolayer of polyaniline has been monitored with an SECM [129]; kinetic data could be extracted by modeling. The charge transfer between a dissolved redox mediator and polyalkylterthiophene films has been studied [130]. In the oxidized (*p*-doped) state of the film, redox reactions proceeded at the film/solution interface, not inside the film. In the reduced state the film behaved like a completely passivating film and penetration of redox mediator ions into the film was obviously completely inhibited.

The combination of SECM and a quartz microbalance has been reported [119]. The amount of information obtained at any given point of the electrode surface can be greatly increased by recording a cyclic voltammogram at every spot [131, 132]. At a high scan rate (about 100 V s^{-1}), the actual SECM picture acquisition rate is not impeded significantly. A microelectrode array that is useful for parallel imaging has been described [133]. A broad variety of systems has been investigated with SECM; for examples and reviews, see [119–139]. These studies cover electron transfer processes [140], mapping of local reactivities [141], local conductivities of intrinsically conducting polymer film layers [142] and efficiency of corrosion inhibitors [117], including formation of inhibitive benzotriazole films on copper [143] and coupled measurements of electrolyte solution resistance [144]. The SECM has also been used for surface modification and microstructuring of carbon surfaces [145]. Improvements in the preparation and application of small microelectrodes, i.e. nanoelectrodes or nanodes, have enabled nanostructuring of surfaces with an SECM [146]. A combination of AFM and SECM as described [147] has been used to study the dissolution of calcite in an aqueous solution; the dissolution of the (100) cleavage plane of potassium bromide has also been investigated [148]. The platinum-coated tip of the AFM serves both as topographical sensor and as an electrode for a SECM. Alternatively, a partially paint-coated platinum tip was used for this purpose [149]. The integration of an ultramicroelectrode into a tip for atomic

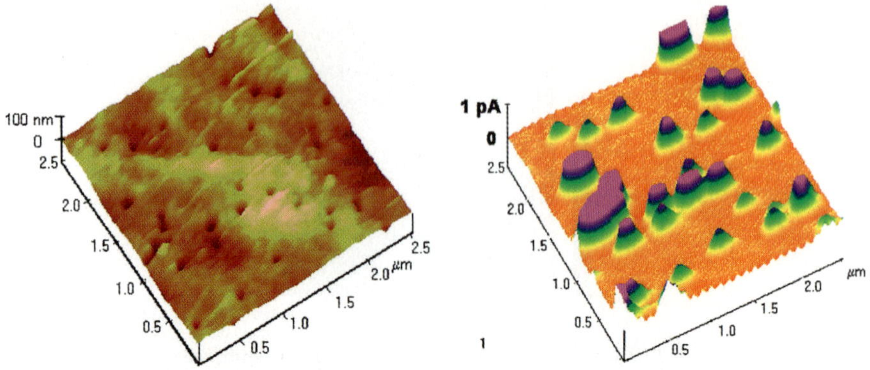

Fig. 7.14. Current (*right*) and topographic (AFM, *left*) pictures of a track-etched polycarbonate membrane in contact with an aqueous solution of $[IrCl_6]^{3-}$, AFM tip of platinum coated with Si_3N_4 (pictures kindly provided by J.V. Macpherson, University of Warwick, UK)

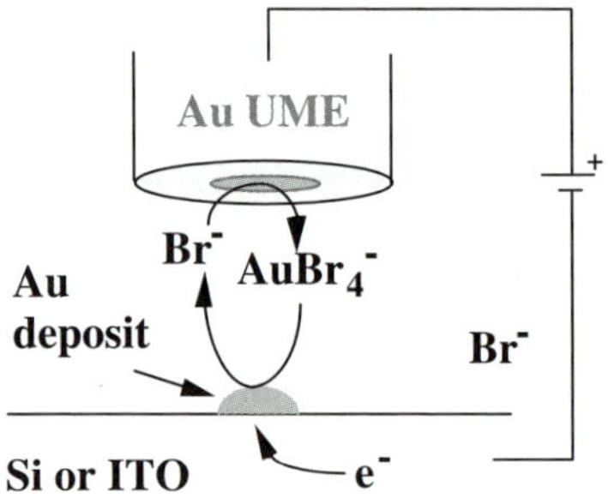

Fig. 7.15. Scheme of gold deposition with an SECM (picture kindly provided by D. Mandler, Hebrew University, Jerusalem, Israel)

force microscopy has been accomplished in a third way [81]: The electrochemically active area is located as a ring around the tip. This method has been used in SECM measurements; simultaneously, an AFM picture was obtained (for an example, see Fig. 7.14). An introductory overview of the AFM/SECM combination has been provided [150].

A combination of an STM with an SECM (see also below for this method) has been described [70]; for details, see above. The SECM can also be used for surface structuring. In order to deposit gold on a surface that is spatially resolved, the experimental setup schematically depicted in Fig. 7.15 was used. The current flowing between the ultramicroelectrode and the surface is displayed in Fig. 7.16. Its distance dependence resembles exactly the behavior observed with a conductive surface, as discussed above. The deposited gold microdots are shown in Fig. 7.17.

The generation of palladium clusters on a surface of Au(111) with an SECM has been reported [152]. More stable and larger clusters were found at closer tip–surface distances. Associated computer simulations suggest that larger clusters are composed of a gold–palladium mixture. The dissolution of clusters proceeds from

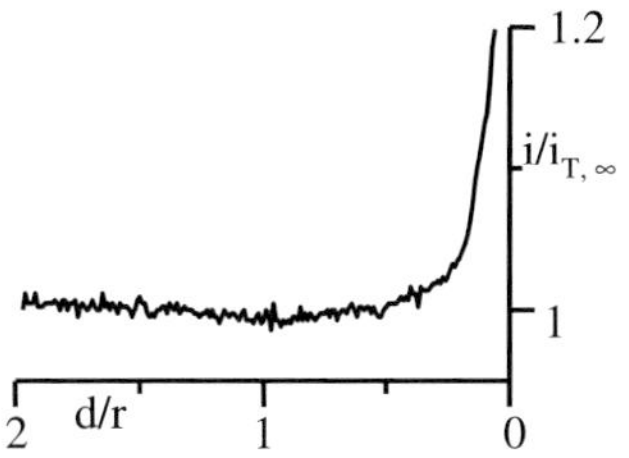

Fig. 7.16. Normalized current vs. normalized distance plot for the SECM setup depicted in Fig. 7.15 (figure kindly provided by D. Mandler, Hebrew University, Jerusalem, Israel)

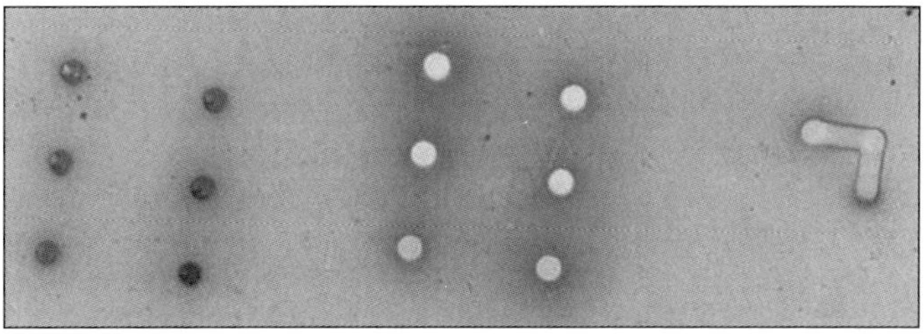

Fig. 7.17. Gold dots deposited from a solution of 0.01 **M** HCl and 0.01 **M** HBr with an SECM (Picture kindly provided by D. Mandler, Hebrew University, Jerusalem, Israel) [151]

the edges, not layer-by-layer. The formation of polypyrrole towers of about 150 μm diameter and 120 μm height using an SECM has been described [153]; without an SECM, localized electropolymerization at a considerably lower spatial resolution can be obtained with short voltage pulses [154]. An overview of the use of SECM for the modification of surfaces has been provided [155].

In a study of the corrosion of stainless steel with MnS inclusions, dissolution products of the sulfide could be localized and identified using the redox couple iodide/triiodide as mediator [156]. Pitting corrosion starting in the vicinity of sulfide inclusions on Ni200 was also studied with SECM [157]. Other applications of the SECM include the characterization of thin films and membranes, liquid/liquid interfaces, the fabrication of nanostructured devices [158], the characterization of microelectrode arrays [159] and the imaging of biological systems (like photosystem I) on surfaces patterned with discrete regions of methyl- and hydroxyl-terminated self-assembled monolayers [160]. The use of an SECM in a study of superoxide generation during the electroreduction of dioxygen in aprotic media has been reported [161].

The use of an SECM to probe the surface conductivity of ultrathin films has been proposed [162]. Lateral conductivity of poly(3-hexylthiophene) films containing gold nanoparticles in addition was measured with an SECM in the feedback mode [163].

A microelectrode scanned in the vicinity of a macroelectrode surface has been employed in an investigation of concentration profiles and the associated electrochemical processes of organic species being created and/or consumed at the interface has been described [164]. In an early study, the concentration profiles within the diffusion layer adjacent to an electrode were mapped with a spatial resolution of 2 μm [165].

7.2.6 pH-Microscopy[6]

Fundamentals. Based on the functional principles of the scanning electrochemical microscope, other scanning probe methods used to determine localized surface properties of the electrode under investigation or of the solution phase adjacent to this surface have been developed utilizing suitable microelectrodes. A pH-sensitive microelectrode based on a glass capillary filled with a pH-constant buffer solution and containing an internal reference electrode that has a tip filled with a proton-selective ionophor cocktail is scanned across the surface. The potential of the internal reference electrode with respect to an external reference electrode is directly correlated to the local pH value. A schematic cross section of this microelectrode is shown in Fig. 7.18.

Instrumentation. A setup employing the described microelectrode positioned at a distance of about 1 μm [166] above the investigated steel surface has been used to study pH-gradients developing in front of a corroding surface as a function of nitrite concentration [167]. The microelectrode showed Nernst factors ranging from −58 mV at room temperature to −69 mV at 60°C. pH shifts of 0.8 pH units at a concentration $c = 11.6$ m**M** of $NaNO_3$ were found. Localized acidification in a neutral aqueous solution of 0.5 **M** Na_2SO_4 over reinforced 6092 aluminum composites indicating anodic regions were identified with SIET [168]. The corrosion of zinc and iron at cut edges of galvanized steel have been studied with a pH-microelectrode

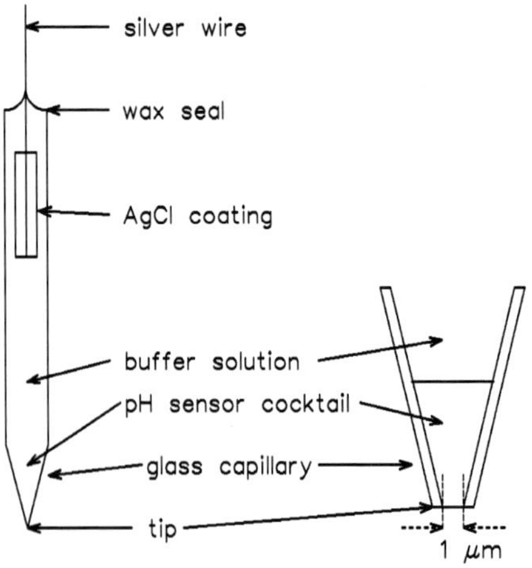

Fig. 7.18. Schematic cross section of a pH-sensitive microelectrode (*left*); tip details (*right*)

[6] This device is also called a scanning ion-sensitive electrode technique (SIET).

[169]. Large pH variations between $7 < pH < 11$ were observed and assigned to hydroxyl ions formed in the cathodic reaction occurring on the steel surface. Localization of corrosion product deposition and pH-change could be identified in the presence of a solution of NaCl [170]; in the case of a solution of $(NH_4)_2SO_4$, no such direct correlation was seen.

A somewhat similar approach with an open "pipette"-like tip has been suggested for imaging and controlled release of species on a nanometer scale [171].

7.2.7 Scanning Ion-Conductance Microscopy

Fundamentals. A micropipette having an aperture of about 0.25 μm diameter is placed above the sample under investigation, which is immersed in an electrolyte solution. On the bottom of the cell vessel, two electrodes (one besides the sample, one underneath the sample) are mounted (see Fig. 7.19). The electrical current flowing between the electrode inside the micropipette and the two electrodes on the cell bottom is measured. It is used as a feedback signal for the standard scanning probe microscope electronics operating the piezoactuators that are moving the micropipette across the sample surface. The vertical movement of the z-actuator follows the topography of the sample, thus providing its image [172, 173].

Instrumentation. A setup has been reviewed [172]. A modified setup using a vibrating micropipette and an AC electronic circuitry that allows better (more precise) position control of the tip and its aperture has been described [174]. Applications reported so far deal with living cells [175, 176] and the internal and external pore structure of membranes [177].

7.2.8 Scanning Reference Electrode Technique

Fundamentals. Localized very small variations of the electrode potential that are caused by current flow across the metal/solution interface over the surface of an electrochemically active material (e.g. a corroding metal) can be measured with a scanning reference electrode [178]. The local variations are picked up by a pair of very fine tips about 10 μm above the surface. The response of a twin platinum electrode has been modelled and results could be matched satisfactorily with real

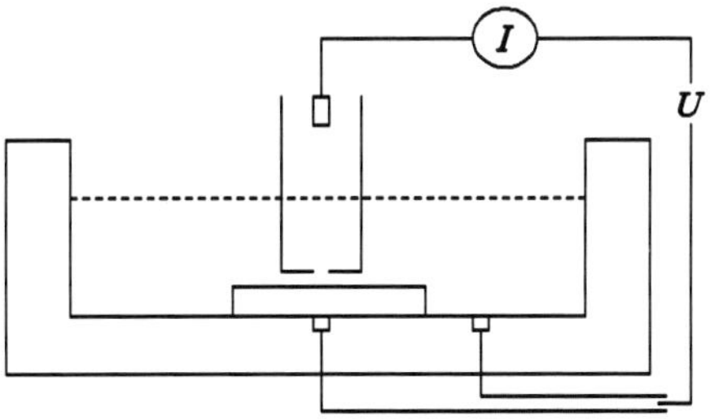

Fig. 7.19. Schematic of a scanning ion conductance microscope

scans across localized events [179]. Instead of real reference electrodes, pseudo-reference electrodes like platinum or iridium tips or wires may be used. The tips pick up potential gradients normal to the current flux lines, which are caused by the current flowing across the interface and subsequently amplified and displayed. By scanning the tips across the surface, a map of local potential variations emerges. In more recent versions, a single tip is used. It vibrates at a frequency of 80 Hz using a piezocrystal. The potential change is picked up and amplified using a lock-in amplifier. A spatial resolution of 0.5 μm is possible [180]. At high electrolyte concentrations, the effect of ion flux caused by localized corrosion is swamped out and the sensitivity of the method is diminished.

Instrumentation. A commercial setup has been reported [181]. Operation with a fixed reference electrode and a rotating sample or a flat, fixed sample and a moving electrode has been described and the particular advantages and limitations have been reviewed [182]. In most applications the rotating electrode systems appeared to be superior. Reported studies include investigations of corrosion protection coatings [180, 183], weld metal corrosion [184], pit initiation [185] (including hydrogen-promoted pitting [186]) and localized corrosion [187–189]. The method has been identified as being highly sensitive, compared to other methods in short time results can be obtained [190]. For inherent limitations and a comparison with SVET see below. Measurements of electrode potentials at higher resolution with a scanning probe setup as used for STM and AFM have been reported [191].

7.2.9 Scanning Vibrating Electrode Technique[7] (SVET)

Fundamentals. A small electrode (typically a microelectrode of about 20 μm diameter) is scanned across the surface under investigation in a distance of about 100 μm. Any current flow across the sample/solution interface causes a potential drop in the solution that is probed by the microelectrode. Using previously established calibration data (with known current densities from a point source electrode), the measured voltage vectors are converted into current vectors [192]. Magnitude and direction of currents above the interface can thus be mapped. Cathodic and anodic processes can be localized; overlaying the obtained maps onto optical micrographs allows detection of visible surface features related to localized corrosion phenomena [193, 194]. The technique is similar to the scanning reference electrode technique (SRET) (see above). In comparison to the initially used two microreference electrodes with SRET, which were later replaced by a single vibrating microelectrode, the actual way of vibrating the scanning electrode is different in SVET [195]. This results in a limitation to measurement of DC currents only, which in turn have restricted the use of SRET to bare metals or coated metals with defects [196, 197].

Instrumentation. Various possibilities for manufacturing the required microelectrodes and their piezoelectric drivers as well as the associated electronics have been described [194, 198]. Reported studies deal mainly with corrosion processes. With

[7] This method is also called current density probe (CDP).

chromate-containing epoxy coatings on both steel and aluminium surfaces, a significant delay in the onset of anodic corrosion currents at a defect site was observed, whereas chromate-free epoxy coatings did not show this delay [194]. With the steel sample, the cathodic current was observed at the defect site only with the chromate-free coating; on a chromate-containing epoxy, a cathodic current was also observed on this coating. With coatings of an intrinsically conducting polymer (ICP), e.g. poly(3-octyl pyrrole), a further onset of any detectable current was observed both with coated steel and aluminum alloy [199]. Current density maps with the coated steel sample showed reduction currents on the polymer surface and oxidation was confined to the defect site. With the coated aluminum alloy sample, no significant oxidation was observed at the defect. Instead, reduction was observed both on the polymer coating and the defect site and concomitant oxidation was observed locally under the coating. The localization of this process seemed to be associated with specific interactions between the polymer and locally enriched copper (an alloy constituent). With pure aluminum, the oxidation current was more distributed over the coated surface, but still as far away from the defect as possible. The metal surface showed no pitting after removal of the coating. The influence of the application of the ICP coating (direct deposition by anodic electropolymerization onto the sample or casting from a polymer solution) was apparent in a study of polypyrrole-coated aluminum and aluminum alloys [200]. The former method yielded better corrosion protection because of stronger electronic (conductive) coupling. The action of dissolved cerium ions as corrosion inhibitors on steel has been investigated [201]. Contrary to previously published assumptions, these ions act as anodic inhibitors. Corrosion studies at metal matrix composites (MMC, reinforced 6092 aluminium composites) with SVET revealed corrosion initiation at localized anodic regions [168]. In a study of the corrosion at cut edges of galvanized steel zinc oxide, zinc carbonate and zinc hydroxide were suggested as reaction products of the anodic current observed with SVET over the zinc surface [169]. The cathodic current observed over the steel surface showed a behavior typical of a diffusion limited oxygen reduction current; consequently, localized pH-shifts were observed with a pH-microelectrode. In the presence of $SrCrO_4$ in the electrolyte solution, an increase of the Tafel slope of the anodic current was found, which is indicative of the passivating effect of this inhibitor. In a solution of NaCl the spatial patterns of deposition of corrosion products and anodic/cathodic currents could be matched with the SVET [170]. No such localized behavior was found with an aqueous solution of $(NH_4)_2SO_4$.

Localized impedance measurements with an SVE have been described [202] and limitations and artifacts observed when using an SVE under specific corrosion conditions were discussed in detail [203].

7.2.10 Scanning Kelvin Probe (SKP)

Fundamentals. The surface (or contact) potential of a solid or a liquid film covered solid can be measured with a Kelvin probe [204, 205]. Essentially, the Volta potential difference $\Delta\Psi$ between the two employed surfaces, as described below, is measured. In common abbreviation, this is also called measurement of a Volta

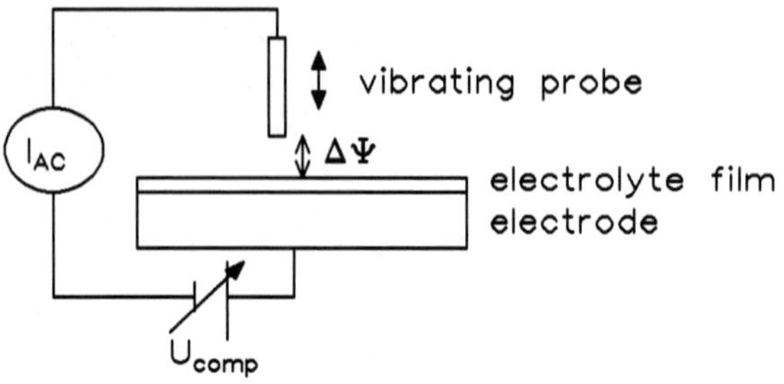

Fig. 7.20. Scheme of Kelvin probe measurement

potential Ψ (e.g. in [206]). As depicted in Fig. 7.20, a probe tip is brought close to the surface under investigation.

The tip and the adjacent surface form a condenser. When the distance between the tip and the surface is changed by vibrating the probe, an AC current flows; its magnitude depends on the existing potential difference. By adjusting the external bias voltage U_{comp}, this potential difference can be compensated. As a consequence the AC current vanishes. In most cases, relative changes of the local surface potential are of interest. In order to remove any unwanted influence of the probe surface potential, a material with constant surface potential is used (typically an etched Ni/Cr wire tip or a cylindrical probe of this material), thus the measured local Volta potential depends only on the surface potential of the sample. A calibration of the probe is accomplished by measuring the corrosion potential with a conventional reference electrode that touches the electrolyte film-covered surface under investigation with a Luggin capillary [207]. Simultaneous measurements with an SKP yield the desired Volta potential difference, which differs from the measured corrosion potential only by a constant difference that is typical of the experimental setup. The Volta potential is thus closely and directly related to the local corrosion potential [207–210]. Thus spatially resolved measurements are useful in studies of localized processes like corrosion on heterogeneous surfaces. The required resolution can be obtained with piezodrives. The required compensation voltages are high enough to cause Faradaic processes in aqueous solution in the gap between tip and surface; obviously this method will work only in the absence of bulk solution. It works well with thin electrolyte solution films coating the corroding surface under investigation as frequently encountered in atmospheric corrosion [211, 212]. In its described setup, the distance between the tip (needle) and the surface is kept constant on a macroscopic level. In the case of very rough, bent or otherwise deformed surfaces, this might prove insufficient. A mode of operation with a height-regulated probe has been proposed (HR-SKP) [213].

Instrumentation. Reported examples are mostly related to corrosion studies; the particular problems in relating Volta potential differences as measured with an SKP and local corrosion potentials have been treated in detail [209]. The effect of barrier layers and metal surface pretreatment has been investigated [214, 215]. In a study of the effect of a corrosion protection primer that contains the intrinsically conductive

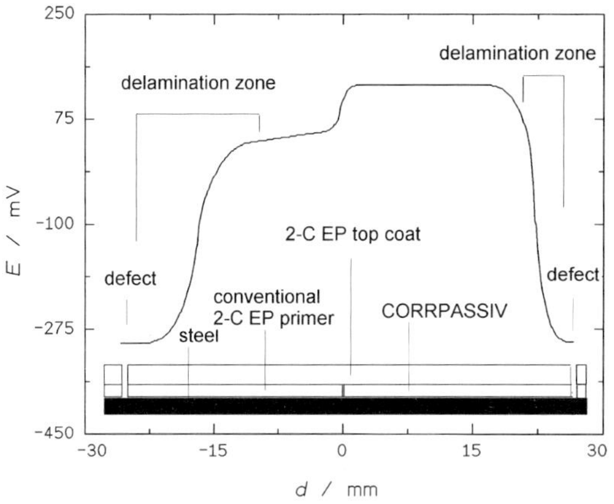

Fig. 7.21. Potential profile measured with a SKP at a steel surface coated alternatively with standard primer and a new primer (see text for details) and top coat (based on data in [206])

polyaniline (CORRPASSIV™) [206], the positive shift of the potential of the steel surface coated with this primer in comparison to the surface coated with a standard primer is evident and the positive effect of the primer on the extent of the (much narrower) delamination zone is obvious (see Fig. 7.21). The delamination zone is larger by a factor of two with the standard primer.

In a study of zinc-coated steel covered with a polymer topcoat, the mechanism of topcoat delamination was elucidated with high spatial resolution [216]. Depending on the details of the defect and the composition of the corroding atmosphere, the rate and type of delamination could be described. A similar study with a coated iron surface has been reported [217]. A comparison of results obtained with SKP, electrochemical impedance measurements and cyclic voltammetry with respect to validity as a corrosion prediction tool has been reported [218].

Differences in detected Volta potentials between pristine and corroded Al–Mg alloy surfaces could be related to the factors influencing thickness and conductivity of the corrosion product layers [219]. Corrosion layers developed in the presence of ion-containing solutions yielded lower Volta potentials and showed higher conductivity. Cathodic delamination of polyaniline-based organic coatings on iron have been studied with SKP [220]. The role of dioxygen reduction and of the polyaniline fraction in the coating were included in a proposed corrosion mechanism.

The surface topography (i.e. the distance between the actual surface and the needle when the latter is kept at a constant distance from the sample itself) can be mapped simultaneously with the local potentials with a Kelvin probe equipped with an additional modulation setup [213]. A typical example of a zinc-coated iron surface with a thin polymer top coat shows the change of height (about 8 μm based on the deposition data) at the edge of the zinc coating (see Fig. 7.22). The expected change of the Volta potential upon changing from zinc to iron is also observed. The roughness of the metal surface is visible in the plot.

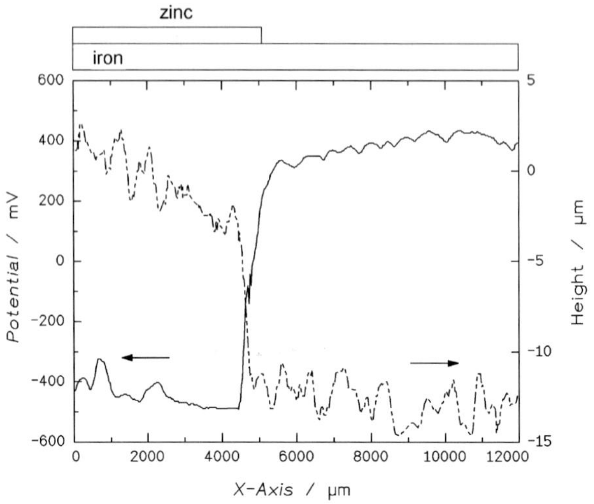

Fig. 7.22. Height and potential profile measured with a HR-SKP at an iron surface coated partially with zinc (based on data in [213])

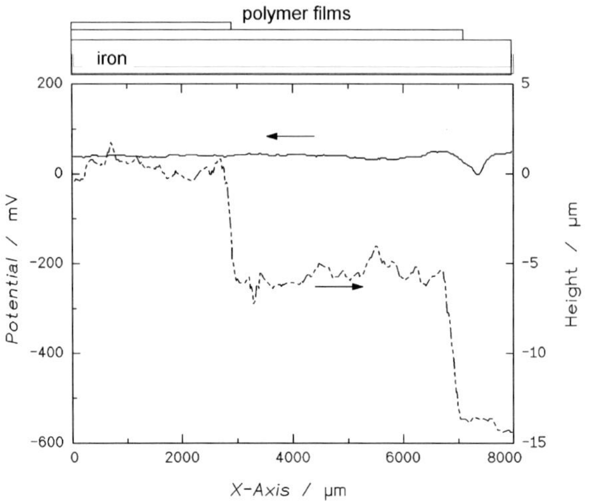

Fig. 7.23. Height and potential profile measured with an HR-SKP at an iron surface coated with layers of latex (based on data in [213])

An iron surface coated with layers of latex of varying thickness yields a considerably different topographic picture (Fig. 7.23). The change of height is registered when an edge in the coating is passed. The changes of height (5 µm for every step) is well defined; the potential remains unchanged because the polymer coating has no influence.

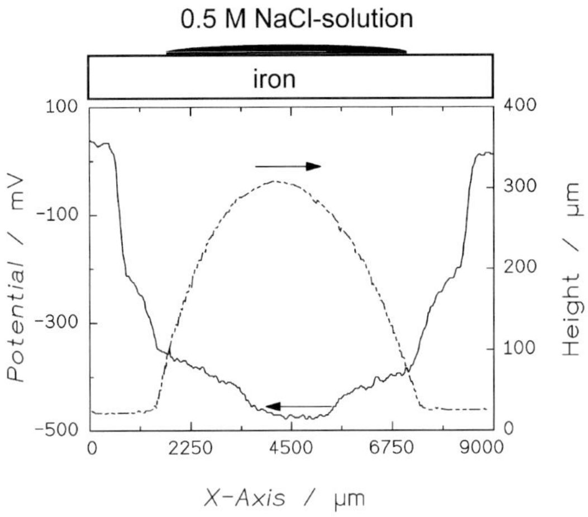

Fig. 7.24. Height and potential profile measured with a HR-SKP at an iron surface with a drop of an aqueous solution of 0.5 **M** NaCl (based on data in [213])

In a setup similar to the Evans drop experiment, the height and the potential of an iron surface with a drop of an aqueous solution of 0.5 **M** NaCl were scanned; results are displayed in Fig. 7.24.

The lowest potential is measured in the center, where corrosion (i.e. anodic dissolution of iron) attacks most aggressively. At the edges, the potential increases somewhat; in this zone oxygen reduction proceeds. The potential changes around the drop imply the presence of an ultrathin electrolyte film because the potential reaches values of the bare iron surface only at a considerable distance from the edges of the macroscopically observed drop [213]. Filiform corrosion of automotive aluminium alloy AA6016 has been studied with SKP [221].

In earlier studies using a fixed Kelvin probe, corrosion kinetics and mechanisms were studied without the spatial resolution possible with the SKP [222, 223]. The use of a Kelvin probe as a reference electrode in corrosion studies with very thin electrolyte films (2 µm) has been described [224]. The use of Kelvin probes to control and to monitor the potential has been reviewed [225].

7.2.11 Scanning Tunneling Spectroscopy[8] and Related Methods

Fundamentals. The tunneling current depends at fixed voltage across the gap between tip and probed surface on the distance and the local probability of electron transfer. This in turn is a function of local electron density, work function or related localized electronic properties (for an overview of tunneling spectroscopies,

[8] Although this method is a spectroscopy, the obtained vibrational spectra and other data contain localized information pertaining to the probed point of the surface. Thus it can also be considered surface analytical and spatially resolved.

see [226]). Depending on the polarity of the applied voltage (also called bias) the tunneling electrons probe different surface atomic states (see a study of the surface of n-doped TiO_2 [227]). With positive sample bias, the net tunneling current is caused by electrons tunneling from occupied states of the negatively biased tip into empty states of the surface (this may include LUMOs of molecules, too). At negative sample bias, electrons tunnel from occupied (HOMO) states of the sample into empty tip states. Because the states with the highest energy have the longest decay lengths into vacuum, in both cases electrons close to the Fermi level of the respective emitter are most likely to contribute to the tunneling current. Consequently STM pictures obtained at the two possible bias polarities can be obtained and must be interpreted accordingly. With continuous modulation of voltage (in addition to switching polarity), densities of state (DOS) can be probed quantitatively [voltage tunneling spectroscopy (VTS)]. Any desired chemical contrasts beyond DOS require far greater energetic resolutions, as are generally required in inelastic tunneling spectroscopy (IETS, which is also known as ITS) [226]. A few meV are generally assumed to be necessary (see also HREELS [228]). This would require operation at liquid helium temperature or lower. Examples of feasibility seem to be limited so far. Obviously, since measurements at modulated bias voltage may result only in data of DOS or possibly also other localized information, the term "scanning tunneling spectroscopy" is used in ambiguous ways, including both possibilities. An STM can definitely be operated in a way that yields data of the local surface density of states (so-called I–V spectroscopy with modulation of the bias between tip and sample) and the effective barrier height (I–S characteristics) can be obtained [11, 229]. An overview of STM and related spectroscopies as applied to electrochemical systems has been provided [230].

Instrumentation. For electrochemical measurements of DOS and effective barrier height, a standard setup as used for STM is employed. In a study of the Au(111) surface in contact with a solution containing copper ions and chloride ions with distance tunneling spectroscopy (i.e. measurements of I–S characteristics, abbreviated DTS) the double layer outside the inner Helmholtz layer was probed [59]. Examples of double layer studies employing DTS have been briefly reviewed [231]. The electrical conductance of n-alkanethiol and α,ω-alkanedithiols has been studied with DTS [232]. Molecules of the latter type assume two distinctly different orientations in the gap between the STM tip and the surface (break junction), resulting in significantly different electrical conductivities. On the contrary, n-alkanethiols show only one conductivity value for a given alcohol, which is almost identical with the lower value of the respective α,ω-alkanedithiol. In a study of decanethiolate adsorbed on a Au(111) electrode, an electronic model of the thiol layer and possible conduction mechanism are proposed [233]. Investigations of distance and voltage dependencies of tunneling currents have suggested indirect tunneling via intermediate states [234]. These states are related to dipole resonances of well-ordered water dipoles located in the electrochemical double layer at the tip-solution and the solution–substrate interface. The chemical nature and the crystallographic orientation of the substrate and (when present) of 2-D upd adlayers influence the effective barrier height. Energetic inhomogeneities on single-crystal and textured platinum surfaces were detected with a combination of DTS and VTS measurements [235]. For a brief overview of methods for the generation and investigation of nanostructures at electrochemical interfaces, with particular attention paid to DTS and VTS, see [231]. In an *ex situ* study of electropolymerized polypyrrole, the band structure was studied as a function of polymer doping [236]. Currently, results of true scanning tunneling spectroscopy under electrochemical conditions have not been reported; the described fundamental requirements and conditions make this actually rather unlikely.

7.3 Near Field and Confocal Optical Methods

The optical resolution of a microscope is traditionally described based on the Abbé equation [237] as being equal to half the wavelength of the employed electromagnetic radiation, i.e. in the visible part of the spectrum, the resolution amounts to 250–300 nm. Actual values might be better, but generally experimental conditions might conspire towards worse results. Methods to overcome this limitation include the use of near field optics. Figure 7.25 shows a schematic comparison of the situation with an optical device, which is limited by the Abbe equation and in addition to possible near field approaches.

The use of confocal microscopy is predominantly driven by the investigation of localized phenomena beyond the simple investigation of structures and topographies. The resolution in this case is not necessarily better than the optical resolution discussed initially above. A serious problem, and even a limitation, is the need to prepare a tip with an aperture of the required diameter right at the tip apex. The light passing through the aperture stays confined for a length approximately equal to the aperture diameter. As has been reported elsewhere, the resolution is limited to $\lambda/10$ or 50 nm [238]. The same arguments apply to the use of a scanning near field optical microscope (SNOM), where the tip is scanned over the surface and in scanning near field infrared (spectroscopy) microscopy (SNIM). A practical resolution limit in the middle infrared is about 1 μm. A promising alternative, not yet explored in electrochemical studies, is the use of apertureless tips. A sharp, metallic needle that acts like an optical antenna supplying a concentrated electric field at its apex is used

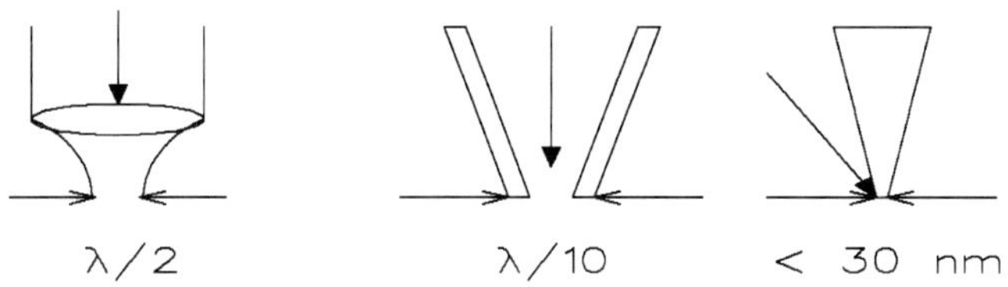

Fig. 7.25. Focusing principles in microscopy: *left*: classical objective; *middle*: near field with aperture, *right*: near field scattering tip; resolution as indicated, arrow indicates path of light

[239]. Infrared radiation is focused on the tip–surface gap and the scattered light is detected using standard techniques [240].

7.3.1 Near Field Methods

Fundamentals. Based on a suggestion by Synge [241], a light beam created by a hole in an opaque material very close to the specimen to be studied, with a diameter much smaller than the optical wavelength, is used to image the sample surface. Using single mode optical fibers acting as tips in a scanning atomic force microscope (AFM), this illumination has become possible. The light passing the sample or created at the interface by fluorescence is collected by suitable optics of an optical microscope. Further details of the method, instrumentation, etc. have been reported elsewhere [242, 243]. Scanning near field optical microscopy (SNOM) has been employed in material sciences using fluorescence images in studies of polymer blends [244].

7.3.2 Confocal Optical Methods

Fundamentals. In a confocal optical arrangement, light coming from the surface/interface under investigation is collected and guided by suitable (e.g. microscopy) objectives towards an optical detector (a photomultiplier, a photodiode, etc.). In front of this detector a pinhole is placed. If properly located, this pinhole will only allow light to pass from the focal point on the surface and all other light that is "out of focus" will be rejected. By scanning the surface in x- and y-directions and storing the optical information (e.g. scattered light intensity), an image of the surface can be generated [245–247]. Illumination of the investigated surface can be done in various ways; this of course includes monochrome laser light. In this case, laser light is focused with suitable lenses to a diffraction limited spot [confocal laser scanning microscopy (CLSM)]. Optical properties of this spot are investigated with an optical microscope or other means. These applications include microscope-assisted fluorescence photon correlation measurements [related to and based on fluorescence correlation spectroscopy (FCS)]. This technique is a method used to study intermolecular interaction. It is based on the statistical evaluation of the fluctuation of concentration based on measurements of fluorescent light [248]. Although it has no direct and obvious relevance to electrochemists, this method has been used starting from locally resolved surface enhanced Raman spectroscopy (see Sect. 5.2.12). The closely related confocal Raman microscopy is treated in Sect. 5.2.15.

Instrumentation. In a typical CLSM experiment, a narrow laser beam scans a surface horizontally, i.e. at constant height, in x- and y-directions using piezo-driven mirrors. A small pinhole is located in front of an optical detector at a position conjugate to the focal point in the sample plane. This way, the detector measures the intensity of light reflected from the surface at every scanned position. Light scattered from out-of-focus positions is focused outside the pinhole and thus does not reach the detector. An image of scattered and/or reflected laser light intensity is created

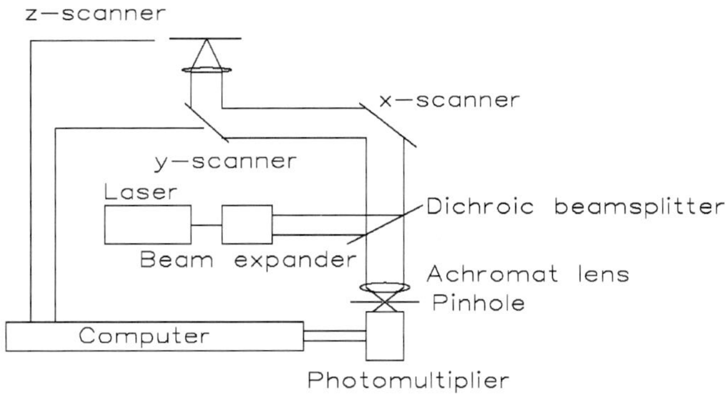

Fig. 7.26. Schematic setup of a confocal scanning laser microscope; for details, see text

for one height z for every point. Repetition of this procedure at different values of z results in a three-dimensional image of the sample. In electrochemical investigations this can be helpful in corrosion studies of rough surfaces [249]. The distance between the surface and the front end of the optical arrangement is about 1 mm. A schematic setup is shown in Fig. 7.26.

In a study of the effect of thiourea (an additive to electroplating solutions) on copper deposition it was observed that thiourea acted as a mediator in the electrochemical reactions; cathodic consumption was found to be negligible. The additive acted as a brightening agent and did not show any leveling capability [250]. The adsorption of the organic dye 3,3′-dihexyloxacarbocyanine iodide, $DiOC_6(3)$, on a polycrystalline gold surface was tracked with CLSM [251]. The dye inhibited cathodic processes (copper deposition) at sites where it was adsorbed. At sufficiently negative electrode potentials, a distinctive influence of the dye on copper deposition was observed. Besides CLSM, with the scanning of the illuminating laser light beam as an important feature, confocal laser-assisted microscopy has been used to investigate two- and three-dimensional samples as reviewed elsewhere [252]. Underfilm corrosion of epoxy-coated steel AA2024-T3 has been investigated with CLSM [253]. Corrosion metrology could be described in detail.

7.4 Surface Conductivity Measurements

Fundamentals. According to the Drude equation, the electrical resistance of a conducting material depends on the concentration, the specific charge and the mobility of charge carriers in the conducting medium. A more thorough examination of the mobility takes into account the way the charged species (electron, hole) are moving in the conductor. Any distortion of the material, particularly in the solid phase (defects in crystals, etc.), will affect the conductivity. In the case of a very thin conducting film on an insulating substrate, the influence of particles sitting on the surface (e.g. adsorbed molecules) of the thin film will be particularly strong because their effect will extend into almost the complete conducting layer. Consequently the electrical resistance is called surface resistance[9]. Species adsorbed on the surface will introduce additional "defect-like" changes in the film, resulting in changes of the film conductivity. More recently, the change of the DC resistance was explained by Schumacher et al. [254] by considering the adsorbate-induced density of states at the Fermi level and the half-width of the Newns–Anderson resonance corresponding to the lowest unoccupied orbital of the adsorbate molecule. Because the adsorption of neutral species depends on the electrode potential and reaches a maximum around the potential of maximum adsorption, measurements of the thin film resistance (or conductance) as a function of adsorbate concentration in the solution phase and electrode potential will provide the necessary information to construct an adsorption isotherm (see also [255–259]). As the mode and intensity of adsorbate–surface interaction may be different for different surface places (differing in local energy, topography, atomic arrangement), results may differ from those of other methods not sensitive to local differences in surface properties. In addition to studies of very thin metal films, conductivity measurements of layers of intrinsically conducting polymers have been made frequently [260, 261]. Changes of polymer conductivity result from a multitude of causes: degree of oxidation (doping), type and composition of electrolyte solution and pH-value are only a selection. The use of a SECM as a probe for the conductivity of ultrathin films has been proposed [162].

Instrumentation. Thin films of the material to be studied (mostly metals) are deposited onto insulating substrates (glass, mica); for a typical setup, see Fig. 7.27. Films of intrinsically conducting polymers are deposited by casting from their solution or by electropolymerization [262, 263].

The electronic resistance of polyaniline changes reversibly from the poorly conducting reduced (undoped) through the medium oxidized (doped) state to the strongly oxidized state and back without any significant difference between the initial and

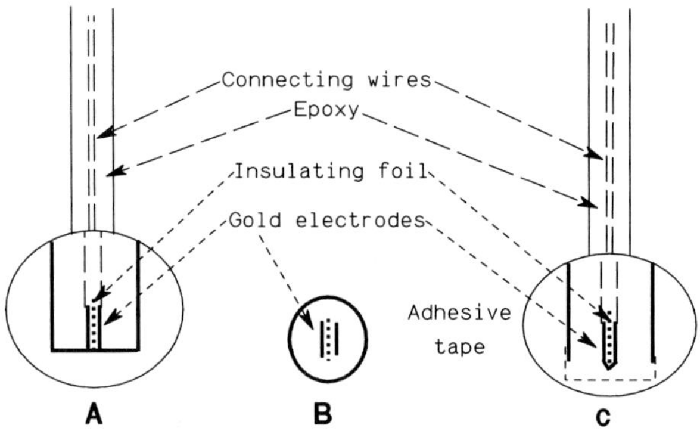

Fig. 7.27. Two band electrode as used for *in situ* DC conductivity measurements with polymer films deposited on the gap between the two electrodes; **A**: cross section after finished preparation; **B**: top view after preparation; **C**: after embedding, before grinding [262]

[9] The terms "surface conductivity" and "surface resistance" are used equally.

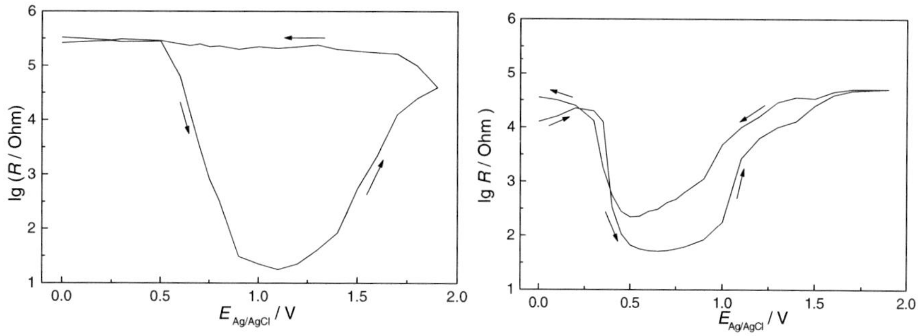

Fig. 7.28. Resistance of a film of polyaniline (*left*) and polythiophene (*right*) measured *in situ* as a function of electrode potential in a nonaqueous electrolyte solution [264]

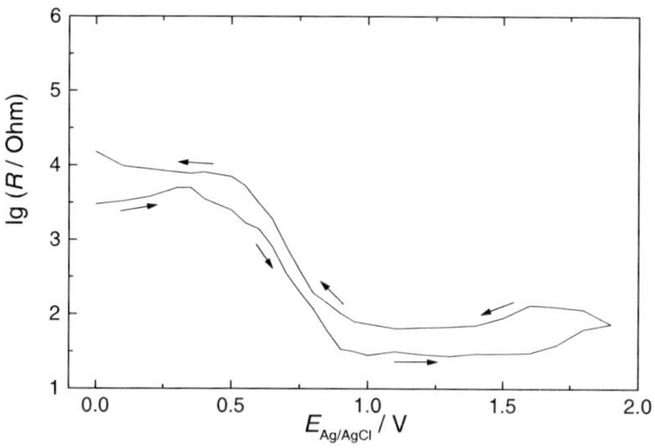

Fig. 7.29. Resistance of a film of a 1:1 copolymer of aniline and thiophene measured *in situ* as a function of electrode potential in a nonaqueous electrolyte solution [264]

the final state (Fig. 7.28, left). Polythiophene shows an irreversible change when exposed to electrode potentials as positive as established before with polyaniline (Fig. 7.28, right). A different situation emerges when an electrochemically formed copolymer of both thiophene and aniline is used (Fig. 7.29).

A lower resistance in the oxidized state and a broader electrode potential range wherein no irreversible change occurs are observed.

Adsorbates of molecules or ions from solution onto the electrode as a function of electrode potential or concentration in solution can be investigated. The electrode to be investigated is prepared as a thin film by vapor deposition onto a suitable insulating (monocrystalline, if desired) substrate. For simultaneous measurements of film resistance and electrode potential an electrode design as depicted in Fig. 7.30 provides an optimum compromise between current distribution and a minimum of interference between both measurements [258].

The highly conductive silver cement contact between the strips providing connection to the electrochemical circuitry (i.e. potentiostat) does not interfere with the conductivity measurements and any conceivable contribution of the electrochemical

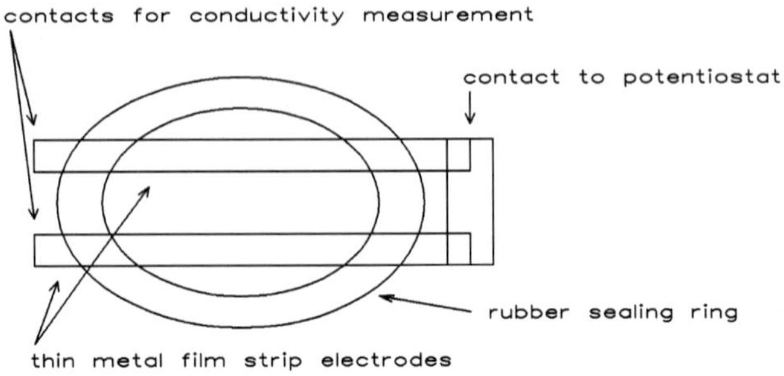

Fig. 7.30. Electrode array for *in situ* thin film conductivity measurements (according to [258])

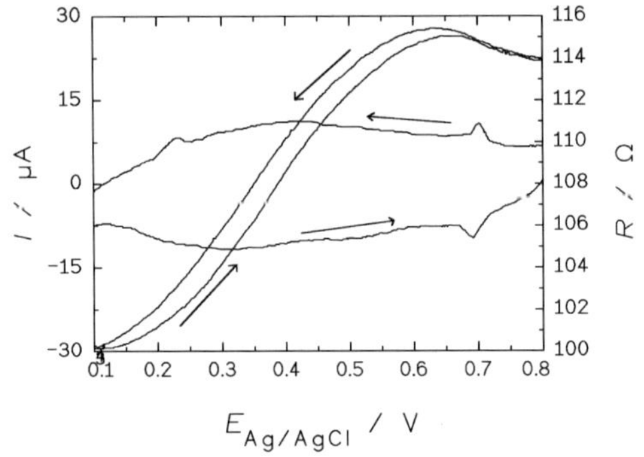

Fig. 7.31. Cyclic voltammogram and thin film resistance of a Au(111) electrode in contact with an aqueous solution of 0.1 **M** H_2SO_4 (according to [258])

current flowing across the metal film/solution interface will compensate because of opposite signs at both strips. The electrode array is pressed towards an opening of the electrochemical cell vessel with the rubber ring providing an adequate seal.

Besides adsorption of ions and organic molecules on metal surfaces [265], studied systems include underpotential metal deposition [266, 267], intercalation processes [268] and surface reconstruction phenomena. Interactions between various metal adsorbates (e.g. upd-Zn on polycrystalline gold with Cu adatoms already present [269]) have been studied. The application of surface resistance measurements in heavy metal analysis in aqueous solution has been reported[10] [270]. A rule (surface Linde rule) correlating the surface concentration of a species with the relative resistance change has been found to be effective in many cases [269]. In the case of specifically adsorbed halide ions, the resistance seems to depend on the number of ions directly adsorbed on the surface, as determined with surface X-ray scattering (see Chap. 6) and not with the coverage determined by classical thermodynamic electrochemical methods like chronocoulometry [271]. Figure 7.31 shows results with features in the CV, including the current peak around $E_{Ag/AgCl} = 230$ mV, which is indicative of the lifting of surface reconstruction; a broad current wave

[10] The method has been called voltohmmetry by the author.

around $E_{Ag/AgCl} = 400$ mV, which is assigned to sulfate adsorption; and a peak pair around $E_{Ag/AgCl} = 695$ mV, which is associated with the ordering/disordering of the sulfate adsorbate layer. The simultaneous increase of the film resistance is caused by sulfate ion adsorption [258].

7.5 Interfacial Conductivity Measurements[11]

Fundamentals. The electric (i.e. generally the Ohmic) resistance observed between the connection to the electrochemical measuring device (e.g. a potentiostat) and the Faradaic impedance assigned to the entirety of the electrochemical process and the double layer (or interfacial) capacitance coupled in parallel to the latter impedance is generally assumed to be constant. As depicted in Fig. 7.32, it is connected in an equivalent circuit in series with both latter elements.

The resistance is composed of the electrical resistance of the wiring and the bulk of the electrode, both of which are assumed to be constant and independent of time and electrode potential. The solution phase resistance encountered between the outer electrode surface (i.e. the actual interface between the electronically conducting and the ionically conducting phase) and the reference electrode is also an entity assumed to be constant in most experiments. (It should be kept in mind that gas evolution or rapid creation or depletion of ionic species by electrochemical reactions might cause totally different observations.) A third component is the Ohmic resistance of films or deposits present, formed or removed during electrochemical reactions and thus causing changes of the resistance as a function of time and/or electrode potential. A method to determine this change, called cyclic resistometry, has been reported [272].

Instrumentation. Using a fast electronic switching circuit short (45 μs) galvanostatic (100 mA) pulses with fast risetime (<1 μs) are applied to the electrochemical cell; the potentiostat is switched to a dummy cell for this time [272]. Under these conditions, the observed potential drop is essentially caused only by R_{comp}. From the acquired voltage response, the desired value R_{comp} is calculated. This procedure is repeated frequently enough to yield continuous R_{comp} vs. electrode potential plots. After the measurement pulse an inverted galvanostatic pulse was applied to avoid unwanted leakage of charge into the double layer.

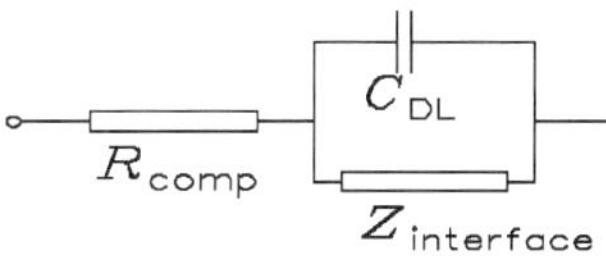

Fig. 7.32. Simplified equivalent circuit of the electrochemical interphase containing the Faradaic impedance $Z_{interface}$, the double layer capacitance C_{DL} and the composed Ohmic resistance R_{comp}

[11] This method is sometimes also called the contact electroresistance (CER) method.

Investigated examples include film formation on lead electrodes, various metal dissolution processes, redox electrochemistry of electrochromic films of IrO_3 [272] and various intrinsically conducting polymers [273–276]. A review covering experimental aspects and results pertaining to ion adsorption, hydride and oxide film formation and hydrophilicity of metals has been provided elsewhere as well as further reports [277–284].

7.6 Microradiology

Fundamentals. Using suitably selected monochromatic synchrotron radiation, the local absorption caused by a selected species (e.g. an ion) can be monitored in real time with high lateral resolution. This method is closely related to X-ray microscopy and tomography with X-rays (see Chap. 6). Further resolution and mapping, using suitably focused synchrotron radiation, has been described [285, 286]; applications seem to be limited so far to non-electrochemical environments.

Instrumentation. A miniaturized cell used for studies of metal-deposition dynamics has been described [287]. Results pertaining mostly to metal-deposition processes have revealed that zinc can deposited on hydrogen bubbles developed at sufficiently negative electrode potentials, thus explaining common defects in zinc plating.

References

1. M. Pilaski, T. Hamelmann, A. Moehring, M.M. Lohrengel, Electrochim. Acta **47**, 2127 (2002)
2. G. Binnig, H. Rohrer, Angew. Chem. **99**, 622 (1987)
3. R. Wiesendanger, *Scanning Probe Microscopy* (Springer, Heidelberg, 1998)
4. R. Young, J. Ward, F. Scire, Rev. Sci. Instr. **43**, 999 (1972)
5. R. Young, J. Ward, F. Scire, Phys. Rev. Lett. **27**, 922 (1971)
6. R.D. Young, Physics Today **1971** (11) 42
7. G. Friedbacher, H. Fuchs, Pure Appl. Chem. **71**, 1337 (1999)
8. G. Friedbacher, H. Fuchs, Angew. Chem. **115**, 5804 (2003)
9. S.L. Sharp, R.J. Warmack, J.P. Goudonnet, I. Lee, T.L. Ferrell, Acc. Chem. Res. **26**, 377 (1993)
10. T.P. Moffat, *Electroanalytical Chemistry*, vol. 21, ed. by A.J. Bard, I. Rubinstein (Dekker, New York, 1999), p. 211
11. D.A. Bonnel (ed.), *Tunneling Microscopy and Spectroscopy* (VCH, New York, 1993)
12. G. Legett, *Surface Analysis – The Principal Techniques*, ed. by J.C. Vickerman (Wiley, Chichester, 1997), p. 393
13. G. Binnig, H. Rohrer, Ch. Gerber, E. Weibel, Phys. Rev. Lett. **49**, 57 (1982)
14. G. Binnig, H. Rohrer, Phys. Bl. **39**, 176 (1983)
15. P. Lustenberger, H. Rohrer, R. Christoph, H. Siegenthaler, J. Electroanal. Chem. **243**, 225 (1988)
16. T. Ito, P. Bühlmann, Y. Umezawa, Anal. Chem. **71**, 1699 (1999)

17. T. Nishino, T. Ito, Y. Umezawa, J. Electroanal. Chem. **550–551**, 125 (2003)
18. M. Wilms, M. Kruft, G. Bermes, K. Wandelt, Rev. Sci. Instrum. **70**, 3641 (1999)
19. J. Zhang, J. Ulstrup, J. Electroanal. Chem. **599**, 213 (2007)
20. V. Stamenkovic, C. Lucas, D. Tripkovic, D. Strmenik, N.M. Markovic, 210th Electrochemical Society Meeting, Cancun, Mexico, 29.10–03.11.2006, Ext. Abstr. #1913
21. W. Schindler, E. Bucharsky, C. Behrend, Extended Abstracts of the 203rd Meeting of the Electrochemical Society, Paris, France, 27.04–02.05. 2003, Ext. Abstr. #2314
22. D.M. Kolb, Angew. Chem. **113**, 1199 (2001)
23. A.A. Gewirth, K.J. Hanson, Interface **2**, 37 (1993)
24. R. Sonnenfeld, J. Schneir, P.K. Hansma, Modern Aspects of Electrochemistry, vol. 21, ed. by R.E. White, J.O'M. Bockris, B.E. Conway (Plenum, New York, 1990), p. 1
25. A.J. Arvia, in *Spectroscopic and Diffraction Techniques in Interfacial Electrochemistry*, ed. by C. Gutiérrez, C. Melendres. NATO ASI Series C, vol. 320 (Kluwer Academic, Dordrecht, 1990), p. 449
26. O.M. Magnussen, Chem. Rev. **102**, 679 (2002)
27. X.P. Gao, G.J. Edens, M.J. Weaver, J. Phys. Chem. **98**, 8074 (1994)
28. D. Friebel, C. Schlaup, P. Broekmann, K. Wandelt, Phys. Chem. Chem. Phys. **9**, 2142 (2007)
29. J. Gómez, L. Vázquez, A.M. Baró, C. Alonso, E. González, J. Gónzalez-Velasco, A.J. Arvía, J. Electroanal. Chem. **240**, 77 (1988)
30. E. Holland-Moritz, J. Gordon II, G. Borges, R. Sonnenfeld, Langmuir **7**, 301 (1991)
31. L.M. Siperko, J. Electrochem. Soc. **137**, 2791 (1990)
32. D. Carnal, U. Müller, H. Siegenthaler, J. Phys. IV **4**, 297 (1994)
33. W. Obretenov, U. Schmidt, W.J. Lorenz, G. Staikov, E. Budevski, D. Carnal, U. Müller, H. Siegenthaler, E. Schmidt, J. Electrochem. Soc. **140**, 692 (1993)
34. W. Obretenov, U. Schmidt, W.J. Lorenz, G. Staikov, E. Budevski, D. Carnal, U. Müller, H. Siegenthaler, E. Schmidt, J. Electrochem. Soc. **140**, 692 (1993)
35. R.J. Nichols, D. Schröer, H. Meyer, Scanning **15**, 266 (1993)
36. Z.F. Chen, J. Li, E.K. Wang, J. Electroanal. Chem. **373**, 83 (1994)
37. W. Haiss, D. Lackey, J.K. Sass, H. Meyer, R.J. Nichols, Chem. Phys. Lett. **200**, 343 (1992)
38. R.J. Nichols, C.E. Bach, H. Meyer, Ber. Bunsenges. Phys. Chem. **97**, 1012 (1993)
39. Z.L. Wu, Z.H. Zang, S.L. Yau, Langmuir **16**, 3522 (2000)
40. K. Itaya, Progr. Surf. Sci. **58**, 121 (1998)
41. R. Wiesendanger, M. Ringger, L. Rosenthaler, H.R. Hidber, P. Oelhafen, H. Rudin, H.J. Güntherodt, Surf. Sci. **181**, 46 (1987)
42. M. Giesen, Prog. Surf. Sci. **68**, 1 (2001)
43. S. Dieluweit, M. Giesen, J. Electroanal. Chem. **524**, 194 (2002)
44. X.C. Guo, R.J. Madix, Acc. Chem. Res. **36**, 471 (2003)
45. C.-H. Shue, L.-Y.O. Yang, S.-L. Yau, K. Itaya, Langmuir **21**, 1942 (2005)
46. R. Yamada, K. Uosaki, Langmuir **14**, 855 (1998)
47. H. Wano, K. Uosaki, Langmuir **17**, 8224 (2001)
48. H. Wano, K. Uosaki, Langmuir **21**, 4024 (2005)
49. J.E. Hudson, H.D. Abruna, J. Phys. Chem. **100**, 1036 (1996)
50. N.J. Tao, G. Cardenas, F. Cunha, Z. Shi, Langmuir **11**, 4445 (1995)
51. R.S. Robinson, J. Electrochem. Soc. **135**, 143C (1988)
52. R. Nishitani, A. Kasuya, Y. Nishina, Z. Phys. D **26**, S42 (1993)
53. Q. Xu, T. He, D.O. Wipf, Langmuir **23**, 9098 (2007)
54. H.C. Wu, S.L. Yau, J. Phys. Chem. B **105**, 6965 (2001)

55. J.E.T. Andersen, P. Møller, J. Electrochem. Soc. **142**, 2225 (1995)
56. Y.C. Wu, H.W. Pickering, D.S. Gregory, S. Geh, T. Sakurai, Surf. Sci. **246**, 468 (1991)
57. T. Sawaguchi, T. Yamada, Y. Okinaka, K. Itaya, J. Phys. Chem. **99**, 14149 (1995)
58. Z.H. Zang, Z.L. Wu, S.L. Yau, Langmuir **15**, 8750 (1999)
59. T. Kawamoto, Phys. Chem. Chem. Phys. **6**, 4913 (2004)
60. C.L. Aravinda, I. Mukhopadhyay, W. Freyland, Phys. Chem. Chem. Phys. **6**, 5225 (2004)
61. G.-B. Pan, W. Freyland, Phys. Chem. Chem. Phys. **9**, 3286 (2007)
62. N.J. Tao, C.Z. Li, H.X. He, J. Electroanal. Chem. **492**, 81 (2000)
63. A. Gonzalez-Martin, R.C. Bhardwaj, J.O'M. Bockris, J. Appl. Electrochem. **23**, 531 (1993)
64. K. Uosaki, M. Koinuma, Faraday Discuss. **94**, 361 (1992)
65. R. Hiesgen, M. Krause, D. Meissner, Electrochim. Acta **45**, 3213 (2000)
66. R. Hiesgen, D. Meissner, Adv. Mater. **10**, 619 (1998)
67. M. Hugelmann, W. Schindler, Appl. Phys. Lett. **85**, 3608 (2004)
68. for further details see e.g.: D.M. Kolb, R. Ullmann, T. Will, Science **275**, 1097 (1997)
69. D.H. Craston, C.W. Lin, A.J. Bard, J. Electrochem. Soc. **135**, 785 (1988)
70. T.H. Treutler, G. Wittstock, Elcctrochim. Acta **48**, 2923 (2003)
71. O. Sklyar, T.H. Treutler, N. Vlachopoulos, G. Wittstock, Surf. Sci. **597**, 181 (2005)
72. for an overview see: R.S. Robinson, C.A. Widrig, Langmuir **8**, 2311 (1992)
73. J.C.H. Spence, W. Lo, M. Kuwabara, Ultramicroscopy **33**, 69 (1990)
74. G. Binnig, C.F. Quate, C. Gerber, Phys. Rev. Lett. **63**, 2669 (1989)
75. N.A. Burnham, R.J. Colton, in *Scanning Tunneling Microscopy and Spectroscopy*, ed. by D.A. Bonnell (VCH, New York, 1993), p. 191
76. Y. Martin, C.C. Williams, Y. Wickramsinghe, J. Appl. Phys. **61**, 4723 (1987)
77. for a review see: A. Noy, D.V. Vezenov, C.M. Lieber, Annu. Rev. Mater. Sci. **27**, 381 (1997)
78. A. Rosazeiser, E. Weilandt, S. Hild, O. Marti, Meas. Sci. Technol. **8**, 1333 (1997)
79. M.N. Holder, C.E. Gardner, J.V. Macpherson, P.R. Unwin, J. Electroanal. Chem. **585**, 8 (2005)
80. N.R. Wilson, D.H. Cobden, J.V. Macpherson, J. Phys. Chem. B **106**, 13102 (2002)
81. C. Kranz, G. Friedbacher, B. Mizaikoff, A. Lugstein, J. Smoliner, E. Bertagnolli, Anal. Chem. **73**, 2491 (2001)
82. B.J. Cruickshank, A.A. Gewirth, R.M. Rynders, R.C. Alkire, J. Electrochem. Soc. **139**, 2829 (1992)
83. C.H. Chen, S.M. Vesecky, A.A. Gewirth, J. Am. Chem. Soc. **114**, 451 (1992)
84. S. Manne, P.K. Hansma, J. Massie, V.B. Elings, A.A. Gewirth, Science **251**, 183 (1991)
85. W. Kautek, S. Dieluweit, M. Sahre, J. Phys. Chem. B **101**, 2709 (1997)
86. S. Rentsch, H. Siegenthaler, G. Papastavrou, Langmuir **23**, 9083 (2007)
87. K.M. Balss, G.A. Fried, P.W. Bohn, J. Electrochem. Soc. **149**, C450 (2002)
88. R. Sonnenfeld, J. Schneir, P.K. Hansma, in *Modern Aspects of Electrochemistry*, vol. 21, ed. by R.E. White, J.O'M. Bockris, B.E. Conway (Plenum, New York, 1990), p. 1
89. T.R.I. Cataldi, I.G. Blackham, G.A.D. Briggs, J.B. Pethica, H.A.O. Hill, J. Electroanal. Chem. **290**, 1 (1990)
90. H. Takano, J.R. Kenseth, S.-S. Wong, J.C. O'Brien, M.D. Porter, Chem. Rev. **99**, 2845 (1999)
91. P. Häring, R. Kötz, J. Electroanal. Chem. **385**, 273 (1995)
92. F.P. Campana, R. Kötz, J. Vetter, P. Novak, H. Siegenthaler, Electrochem. Commun. **7**, 107 (2005)

93. A.C. Hillier, M.D. Ward, Science **263**, 1261 (1994)
94. A.P. Abbott, G. Capper, K.J. McKenzie, A. Glidle, K.S. Ryder, Phys. Chem. Chem. Phys. **8**, 4214 (2006)
95. M. Seo, Y. Kurata, Electrochim. Acta **48**, 3221 (2003)
96. M. Nielinger, H. Baltruschat, Phys. Chem. Chem. Phys. **9**, 3965 (2007)
97. P. Schmutz, G.S. Frankel, J. Electrochem. Soc. **146**, 4461 (1999)
98. P. Schmutz, G.S. Frankel, J. Electrochem. Soc. **145**, 2285 (1998)
99. M. Rohwerder, E. Hornung, M. Stratmann, Electrochim. Acta **48**, 1235 (2003)
100. M. Rohwerder, F. Turcu, Electrochim. Acta **53**, 290 (2007)
101. P. Leblanc, G.S. Frankel, J. Electrochem. Soc. **149**, B239 (2002)
102. P. Schmutz, G.S. Frankel, J. Electrochem. Soc. **145**, 2295 (1998)
103. V. Guillaumin, P. Schmutz, G.S. Frankel, J. Electrochem. Soc. **148**, B163 (2001)
104. D.M. Kolb, D.L. Rath, R. Wille, W.N. Hansen, Ber. Bunsenges, Phys. Chem. **87**, 1108 (1983)
105. W.N. Hansen, D.M. Kolb, J. Electroanal. Chem. **100**, 493 (1979)
106. E.M. Stuve, A. Krasnopoler, D.E. Sauer, Surf. Sci. **335**, 177 (1995)
107. D.L. Rath, D.M. Kolb, Surf. Sci. **109**, 641 (1981)
108. D. Bengtsson Blücher, J.-E. Svensson, L.-G. Johansson, M. Rohwerder, M. Stratmann, J. Electrochem. Soc. **151**, B621 (2004)
109. T.H. Muster, A.E. Hughes, J. Electrochem. Soc. **153**, B474 (2006)
110. L.E. Fratila-Apachitei, I. Apachitei, J. Duszczyk, Electrochim. Acta **51**, 5892 (2006)
111. P.P. Leblanc, Frankel, J. Electrochem. Soc. **151**, B105 (2004)
112. A.M. Bond, Analyst **119**, R1 (1994)
113. F. Magno, I. Lavagnini, Anal. Chim. Acta **305**, 96 (1995)
114. R.J. Forster, Chem. Soc. Rev. **23**, 289 (1994)
115. K. Aoki, Electroanalysis **5**, 627 (1993)
116. J. Heinze, Angew. Chem. **105**, 1327 (1993)
117. K. Mansikkamäki, P. Ahonen, G. Fabricius, L. Murtomäli, K. Kontturi, J. Electrochem. Soc. **152**, B12 (2005)
118. P. Sun, F.O. Laforge, M.V. Mirkin, Phys. Chem. Chem. Phys. **9**, 802 (2007)
119. A.J. Bard, F.F. Fan, M.V. Mirkin, in *Electroanalytical Chemistry*, vol. 15, ed. by A.J. Bard (Dekker, New York, 1994), p. 243
120. J.L. Amphlett, G. Denuault, J. Phys. Chem. B **102**, 9946 (1998)
121. A.J. Bard, F.-R.F. Fan, M.V. Mirkin, in *Physical Electrochemistry*, ed. by I. Rubinstein (Dekker, New York, 1995), Chap. 5
122. K. Eckhard, X. Chen, F. Turcu, W. Schuhmann, Phys. Chem. Chem. Phys. **8**, 5359 (2006)
123. M. Etienne, A. Schulte, W. Schuhmann, Electrochem. Commun. **6**, 288 (2004)
124. see also: M.A. Alpuche-Aviles, D.O. Wipf, Anal. Chem. **73**, 4873 (2001)
125. P.M. Diakowski, A.S. Baranski, Electrochim. Acta **52**, 854 (2006)
126. P.M. Diakowski, Z. Ding, Phys. Chem. Chem. Phys. **9**, 5966 (2007)
127. B.B. Katemann, W. Schuhmann, Electroanalysis **14**, 22 (2002)
128. A.J. Bard, F.-R.F. Fan, D.T. Pierce, P.R. Unwin, D.O. Wipf, F. Zhou, Science **68**, 254 (1991)
129. D. Mandler, P.R. Unwin, J. Phys. Chem. B **107**, 407 (2003)
130. M. Tsionsky, A.J. Bard, D. Dini, F. Decker, Chem. Mater. **10**, 2120 (1998)
131. D.O. Wipf, M. Alpuche-Aviles, L. Diaz-Ballote, 205th Electrochemical Society Meeting, San Antonio, USA, 09–13.05.2004, Ext. Abstr. #837
132. L. Diaz-Ballote, M. Alpuche-Aviles, D.O. Wipf, Electrochim. Acta **52**, 17 (2007)

133. A.L. Barker, P.R. Unwin, J.W. Gardner, H. Rieley, Electrochem. Commun. **6**, 91 (2004)
134. A.J. Bard, F.R. Fan, Faraday Discuss. **1992**, 1
135. A.J. Bard, D.E. Cliffel, C. Demaille, F.R.F. Fan, M. Tsionsky, Ann. Chim. Rome **87**, 15 (1997)
136. K. Borgwarth, C. Ricken, D.G. Ebling, J. Heinze, Ber. Bunsenges. Phys. Chem. **99**, 1421 (1995)
137. G. Nagy, L. Nagy, Fresenius J. Anal. Chem. **366**, 735 (2000)
138. C. Ricken, K. Borgwarth, J. Heinze, Nachr. Chem. Tech. Lab. **44**, 502 (1996)
139. A.L. Barker, M. Gonsalves, J.V. Macpherson, C.J. Slevin, P.R. Unwin, Anal. Chim. Acta **385**, 223 (1999)
140. I. Turyan, B. Orel, R. Reisfeld, D. Mandler, Phys. Chem. Chem. Phys. **5**, 3212 (2003)
141. G. Wittstock, in *Appl. Phys.*, vol. 85, ed. by K. Wandelt, S. Thurgate (Springer, Berlin, 2003), p. 335
142. J. Kwak, C. Lee, A.J. Bard, J. Electrochem. Soc. **137**, 1481 (1990)
143. K. Mansikkamäki, U. Haapanen, C. Johans, K. Konturri, M. Valden, J. Electrochem. Soc. **153**, B311 (2006)
144. C. Gabrielli, F. Huet, M. Keddam, P. Rousseau, V. Vivier, 205th Electrochemical Society Meeting, San Antonio, USA, 09 13.05.2004, Ext. Abstr. #848
145. R.C. Tenent, D.O. Wipf, J. Electrochem. Soc. **150**, E131 (2003)
146. J. Ufheil, C. Heß, K. Borgwarth, J. Heinze, Phys. Chem. Chem. Phys. **7**, 3185 (2005)
147. C.E. Jones, P.R. Unwin, J.V. Macpherson, Chem. Phys. Chem. **4**, 139 (2003)
148. J.V. Macpherson, P.R. Unwin, A.C. Hillier, A.J. Bard, J. Am. Chem. Soc. **118**, 6445 (1996)
149. J.V. Macpherson, P.R. Unwin, Anal. Chem. **72**, 276 (2000)
150. C.E. Gardner, J.V. Macpherson, Anal. Chem. **74**, 576A (2002)
151. E. Ammann, D. Mandler, J. Electrochem. Soc. **148**, C533 (2001)
152. M. Del Popolo, E. Leiva, H. Kleine, J. Meier, M. Mariscal, W. Schmickler, Appl. Phys. Lett. **81**, 2635 (2002)
153. C. Kranz, H.E. Gaub, W. Schuhmann, Adv. Mater. **8**, 634 (1996)
154. M.T. Giacomini, R. Schuster, Phys. Chem. Chem. Phys. **7**, 518 (2005)
155. V. Radtke, J. Heinze, Z. Phys. Chem. **218**, 103 (2004)
156. C.H. Paik, H.S. White, R.C. Alkire, J. Electrochem. Soc. **147**, 4120 (2000)
157. C.H. Paik, R.C. Alkire, J. Electrochem. Soc. **148**, B276 (2001)
158. J.F. Zhou, D.O. Wipf, J. Electrochem. Soc. **144**, 1202 (1997)
159. C.G. Zoski, N. Simjee, O. Guenat, M. Koudelka-Hep, Anal. Chem. **76**, 62 (2004)
160. M. Ciobanu, H.A. Kincaid, G.K. Jennings, D.E. Cliffel, Langmuir **21**, 692 (2005)
161. S. Bollo, P. Jara-Ulloa, S. Finger, L.J. Nunez-Vergara, J.A. Squella, J. Electroanal. Chem. **577**, 235 (2005)
162. A.L. Whitworth, D. Mandler, P.R. Unwin, Phys. Chem. Chem. Phys. **7**, 356 (2005)
163. P.G. Nicholson, V. Ruiz, J.V. Macpherson, P.R. Unwin, Phys. Chem. Chem. Phys. **8**, 5096 (2006)
164. R.C. Engstrom, T. Meaney, R. Tople, R.M. Wightman, Anal. Chem. **59**, 2005 (1987)
165. R.C. Engstrom, M. Weber, D.J. Wunder, R. Burgess, S. Winquist, Anal. Chem. **58**, 844 (1986)
166. E. Klusmann, J.W. Schultze, Electrochim. Acta **42**, 3123 (1997)
167. E. Klusmann, J.W. Schultze, Electrochim. Acta **48**, 3325 (2003)
168. H. Ding, L.H. Hihara, J. Electrochem. Soc. **152**, B161 (2005)
169. K. Ogle, V. Baudu, L. Garrigues, X. Philippe, J. Electrochem. Soc. **147**, 3654 (2000)
170. K. Ogle, S. Morel, D. Jacquet, J. Electrochem. Soc. **153**, B1 (2006)

171. L. Ying, A. Bruckbauer, D. Zhou, J. Gorelik, A. Shevchuk, M. Lab, Y. Korchev, D. Klenerman, Phys. Chem. Chem. Phys. **7**, 2895 (2005)
172. P.K. Hansma, B. Drake, O. Martio, S.A.C. Gould, C.B. Prater, Science **243**, 641 (1989)
173. C.B. Prater, P.K. Hansma, M. Tortonese, C.F. Quate, Rev. Sci. Instrum. **62**, 2634 (1991)
174. D. Pastre, H. Iwamoto, J. Liu, G. Szabo, Z.F. Shao, Ultramicroscopy **90**, 13 (2001)
175. Y.E. Korchev, C.L. Bashford, M. Milanovic, I. Vodyanov, M.J. Lab, Biophys. J. **73**, 653 (1997)
176. Y.E. Korchev, M. Milovanovic, C.L. Bashford, D.C. Bennett, E.V. Sviderskaya, I. Vodyanov, M.J. Lab, J. Microsc. **188**, 17 (1997)
177. R. Proksch, R. Lal, P.K. Hansma, D. Morse, G. Stucky, Biophys. J. **71**, 2155 (1996)
178. K.R. Trethewey, D.A. Sargeant, Metal Mater. **8**, 378 (1992)
179. S.J. Badger, S.B. Lyon, S. Turgoose, J. Electrochem. Soc. **145**, 4074 (1998)
180. P.J. Kinlen, V. Menon, Y.W. Ding, J. Electrochem. Soc. **146**, 3690 (1999)
181. EG&G Instruments, Princeton Applied Research, P.O. Box 2565, Princeton, NJ 08543, USA
182. D.A. Sargeant, Corros. Prevent. Control **44**, 91 (1997)
183. R. Gasparac, C.R. Martin, J. Electrochem. Soc. **148**, B138 (2001)
184. V.S. Voruganti, H.B. Luft, D. DeGeer, S.A. Bradford, Corrosion **47**, 343 (1991)
185. D.A. Sargeant, C. Ford, J. Corderoy, Corros. Prev. Control **38**, 12 (1991)
186. J.G. Yu, J.L. Luo, P.R. Norton, Langmuir **18**, 6637 (2002)
187. K.R. Trethewey, D.A. Sargeant, D.J. Marsh, A.A. Tamimi, Corros. Sci. **35**, 127 (1993)
188. G. Bellanger, J.J. Rameau, Electrochim. Acta **40**, 2519 (1995)
189. D.A. Sargeant, J.G.C. Hainse, S. Bates, Mater. Sci. Technol. **5**, 487 (1989)
190. V.S. Voruganti, H.B. Luft, D. DeGeer, S.A. Bradford, Corrosion **47**, 343 (1991)
191. VEECO data sheet DS57, Rev. A0, 2003
192. H.S. Isaacs, J. Electrochem. Soc. **138**, 722 (1991)
193. H.S. Isaacs, A.J. Davenport, A. Shipley, J. Electrochem. Soc. **138**, 390 (1991)
194. J. He, V.J. Gelling, D.E. Tallman, G.P. Bierwagen, J. Electrochem. Soc. **147**, 3661 (2000)
195. D.A. Sargeant, Corros. Prevent. Control **44**, 91 (1997)
196. I. Sekine, M. Yuasa, K. Tanaka, T. 1 Tsutsumi, F. Koizumi, N. Oda, H. Tanabe, M. Nagai, Shikizai **67**, 424 (1994)
197. I. Sekine, Prog. Org. Coat. **31**, 73 (1997)
198. C. Scheffey, Rev. Sci. Instrum. **59**, 787 (1988)
199. J. He, V.J. Gelling, D.E. Tallman, G.P. Bierwagen, G.G. Wallace, J. Electrochem. Soc. **147**, 3667 (2000)
200. J. He, D.E. Tallmann, G.P. Bierwagen, J. Electrochem. Soc. **151**, B644 (2000)
201. H.S. Isaacs, A.J. Davenport, A. Shipley, J. Electrochem. Soc. **138**, 390 (1991)
202. E. Bayet, F. Huet, M. Keddam, K. Ogle, H. Takenouti, J. Electrochem. Soc. **144**, L87 (1997)
203. E. Bayet, F. Huet, M. Keddam, K. Ogle, H. Takenouti, Electrochim. Acta **44**, 4117 (1999)
204. L. Kelvin, Philos. Mag. **46**, 82 (1898)
205. A.W. Adamson, *Physical Chemistry of Surfaces* (Wiley, New York, 1990)
206. B. Wessling, J. Posdorfer, Electrochim. Acta **44**, 2139 (1999)
207. M. Stratmann, H. Streckel, Corros. Sci. **30**, 681 (1990)
208. M. Stratmann, M. Wolpers, H. Streckel, R. Feser, Ber. Bunsenges. Phys. Chem. **95**, 1365 (1991)
209. S. Yee, R.A. Oriani, M. Stratmann, J. Electrochem. Soc. **138**, 55 (1991)

210. M. Stratmann, M. Wolpers, H. Streckel, R. Feser, Ber. Bunsenges. Phys. Chem. **95**, 1365 (1991)
211. G. Grundmeier, W. Schmidt, M. Stratmann, Electrochim. Acta **45**, 2515 (2000)
212. M. Stratmann, H. Streckel, R. Feser, Farbe Lack **97**, 9 (1991)
213. K. Wapner, B. Schoenberger, M. Stratmann, G. Grundmeier, J. Electrochem. Soc. **152**, E114 (2005)
214. G. Grundmeier, M. Stratmann, Appl. Surf. Sci. **141**, 43 (1999)
215. V. Barranco, P. Thiemann, H.K. Yasuda, M. Stratmann, G. Grundmeier, Appl. Surf. Sci. **229**, 87 (2004)
216. W. Fürbeth, M. Stratmann, Fresenius J. Anal. Chem. **353**, 337 (1995)
217. M. Stratmann, R. Feser, A. Leng, Electrochim. Acta **39**, 1207 (1994)
218. J. Posdorfera, B. Wessling, Fresenius J. Anal. Chem. **367**, 343 (2000)
219. E. Juzeliunas, A. Sudavicius, K. Jüttner, W. Fürbeth, Electrochem. Commun. **5**, 154 (2003)
220. G. Williams, R.J. Holness, D.A. Worsley, H.N. McMurray, Electrochem. Commun. **6**, 549 (2004)
221. H.N. McMurray, A.J. Coleman, G. Williams, A. Afseth, G.M. Scamans, J. Electrochem. Soc. **154**, C339 (2007)
222. M. Stratmann, Corros. Sci. **27**, 869 (1987)
223. A. Leng, M. Stratmann, Corros. Sci. **34**, 1657 (1993)
224. M. Stratmann, H. Streckel, K.T. Kim, S. Crockett, Corros. Sci. **30**, 715 (1990)
225. G.S. Frankel, M. Stratmann, M. Rohwerder, A. Michalik, B. Maier, J. Dora, M. Wicinski, Corr. Sci. **49**, 2021 (2007)
226. R.J. Hamers, in *Scanning Tunneling Microscopy and Spectroscopy*, ed. by D.A. Bonnell (VCH, New York, 1993), p. 51
227. F.-R. Fan, A.J. Bard, J. Phys. Chem. **94**, 3761 (1990)
228. H. Ibach, *Electron Energy Loss Spectroscopy and Surface Vibrations* (Academic Press, New York, 1982)
229. G. Nagy, T. Wandlowski, Langmuir **19**, 10271 (2003)
230. A.J. Bard, F.F. Fan, in *Scanning Tunneling Microscopy and Spectroscopy*, ed. by D.A. Bonnell (VCH, New York, 1993), p. 287
231. M. Hugelmann, P. Hugelmann, W.J. Lorenz, W. Schindler, Surf. Sci. **597**, 156 (2005)
232. E. Wierzbinski, K. Slowinski, Langmuir **22**, 5205 (2006)
233. C. Vericat, I. Diez-Pérez, P. Gorostiza, F. Sanz, Electrochemical Society Spring Meeting 209th, Denver, Colorado, USA, May 7–11, 2006, Ext. Abstr. #1213
234. J. Halbritter, G. Repphun, S. Vinzelberg, G. Staikov, W.J. Lorenz, Electrochim. Acta **40**, 1385 (1995)
235. E.V. Kasatkin, E.B. Neburchilova, Russ. J. Electrochem. **34**, 1039 (1998)
236. R. Yang, W.H. Smyrl, D.F. Evans, W.A. Hendrickson, J. Phys. Chem. **96**, 1428 (1992)
237. E. Abbe, Archiv Mikroskop. **9**, 413 (1873)
238. B. Hecht, B. Sick, U.P. Wild, J. Chem. Phys. **112**, 7761 (2000)
239. B. Knoll, F. Keilmann, Opt. Commun. **162**, 177 (1999)
240. F. Keilmann, Vib. Spectrosc. **29**, 109 (2002)
241. E.H. Synge, Phil. Mag. **6**, 356 (1928)
242. R.C. Dunn, Chem. Rev. **99**, 2891 (1999)
243. A. Bek, R. Vogelgesang, K. Kern, Rev. Sci. Instr. **77**, 43703 (2006)
244. R. Stevenson, R. Riehn, R.G. Milner, D. Richards, E. Moons, D.J. Kang, M. Blamire, J. Morgado, F. Cacialli, Appl. Phys. Lett. **79**, 833 (2001)
245. M. Minsky, Scanning **10**, 128 (1988)

246. T. Wilson, C.J.R. Sheppard, *Scanning Optical Microscopy* (Academic Press, London, 1984)
247. T.G. Corle, G.S. Kino, *Confocal Scanning Optical Microscopy and Related Techniques* (Academic Press, New York, 1996)
248. M. Völcker, A. Schnetz, R. Grub, Tech. Messen **63**, 128 (1996)
249. O. Schneider, G.O. Ilevbare, J.R. Scully, R.G. Kelly, Electrochem. Solid State Lett. **4**, B35 (2001)
250. M.A. Alodan, W.H. Smyrl, J. Electrochem. Soc. **145**, 957 (1998)
251. D.S. Chung, R.C. Alkire, J. Electrochem. Soc. **144**, 1529 (1997)
252. B. Faltermeier, Tech. Messen **63**, 4 (1996)
253. O. Schneider, G.O. Ilevbare, R.G. Kelly, J.R. Scully, J. Electrochem. Soc. **154**, C397 (2007)
254. D. Schumacher, A. Otto, B.N.J. Persson, Chem. Phys. Lett. **178**, 204 (1991)
255. J.O'M. Bockris, B.D. Cahan, G. Stoner, Chem. Inst. **1**, 273 (1969)
256. R.I. Tucceri, D. Posadas, Electrochim. Acta **32**, 27 (1987)
257. D. Körwer, D. Schumacher, A. Otto, Ber. Bunsenges. Phys. Chem. **95**, 1484 (1991)
258. P. Lilie, *Cyclovoltammetrische und transiente Messung des Gleichstromwiderstandes während elektrochemischer Prozesse an Gold- und Silberschichten* (Shaker Verlag, Aachen, 2002)
259. R. Tucceri, Surf. Sci. Rep. **56**, 85 (2004)
260. R. Holze, in *Handbook of Advanced Electronic and Photonic Materials*, vol. 2, ed. by H.S. Nalwa (Gordon and Breach & OPA N.V., Singapoore, 2001), p. 171
261. R. Holze, in *Handbook of Electronic and Photonic Materials and Devices*, vol. 8, ed. by H.S. Nalwa (Academic Press, San Diego, 2000), p. 209
262. R. Holze, J. Lippe, Synth. Met. **38**, 99 (1990)
263. R.I. Tucceri, J. Electroanal. Chem. **505**, 72 (2001)
264. S. Vogel, R. Holze, unpublished results
265. R.I. Tucceri, D. Posadas, Electrochim. Acta **32**, 27 (1987)
266. F.M. Romeo, D. Posadas, R.I. Tucceri, J. Electrochem. Soc. **134**, 145C (1987)
267. C. Hanewinkel, D. Schumacher, A. Otto, J. Electroanal. Chem. **554–555**, 325 (2003)
268. M.D. Levi, E. Levi, Y. Gofer, D. Aurbach, E. Vieil, J. Serose, J. Phys. Chem. B **103**, 1499 (1999)
269. M.H. Fonticelli, D. Posadas, R.I. Tucceri, J. Electroanal. Chem. **565**, 359 (2004)
270. M.J. Schöning, in *Ultrathin Electrochemical Chemo- and Biosensors* in: ed. by O.S. Wolfbeis, V.M. Mirsky. Springer Series on Chemical Sensors and Biosensors (Springer, Berlin, 2004), p. 117
271. C. Hanewinkel, A. Otto, T. Wandlowski, Surf. Sci. **429**, 255 (1999)
272. R.L. Deutscher, S. Fletcher, J.A. Hamilton, Electrochim. Acta **31**, 585 (1986)
273. R. John, G.G. Wallace, J. Electroanal. Chem. **354**, 145 (1993)
274. A. Talaie, G.G. Wallace, Solid State Ion. **70**, 692 (1994)
275. A. Talaie, Solid State Ionics **74**, 219 (1994)
276. A. Talaie, J.Y. Lee, K. Adachi, T. Taguchi, J. Romagnoli, J. Electroanal. Chem. **468**, 19 (1999)
277. V.A. Marichev, Surf. Sci. Rep. **44**, 53 (2001)
278. V.A. Marichev, Russ. J. Electrochem. **36**, 240 (2000)
279. V.A. Marichev, Electrochim. Acta **41**, 2551 (1996)
280. V.A. Marichev, Russ. J. Electrochem. **33**, 990 (1997)
281. V.A. Marichev, Electrochim. Acta **43**, 2203 (1998)
282. V.A. Marichev, Russ. J. Electrochem. **35**, 434 (1999)

283. N.D. Sverdlova, W. Schäfer, G.N. Mansurov, O.A. Petrii, Russ. J. Electrochem. **31**, 227 (1995); Elektrokhimiya **31**, 250 (1995)
284. V.A. Marichev, Russ. J. Electrochem. **35**, 417 (1999)
285. J.-D. Grunwaldt, S. Hannemann, C.G. Schroer, A. Baiker, J. Phys. Chem. B **110**, 8674 (2006)
286. S. Hannemann, J.-D. Grunwaldt, N. van Vegten, A. Baikeroer, P. Boye, C.G. Schroer, Catal. Today **126**, 54 (2007)
287. W.L. Tsai, P.C. Hsu, Y. Hwu, C.H. Chen, L.W. Chang, S.K. Seol, J.H. Je, G. Margaritondo, Extended Abstracts of the 203rd Meeting of the Electrochemical Society, Paris, France, 27.04–02.05.2003, Ext. Abstr. #2845

Index

Printing: Krips bv, Meppel, The Netherlands
Binding: Stürtz, Würzburg, Germany